T0331047

World Scientific Series on Emerging Technologies:
Avram Bar-Cohen Memorial Series – Volume 5

POWER BEAMING

HISTORY, THEORY, AND PRACTICE

World Scientific Series on Emerging Technologies: Avram Bar-Cohen Memorial Series

Print ISSN: 2737-5862
Online ISSN: 2737-5870

Series Editors: Jens Rieger *(Independent Consultant, and former Senior Vice President of BASF SE, Advanced Materials & Systems Research, Technology Scouting & Incubation, Ludwigshafen, Germany)*
Eran Sher *(Technion – Israel Institute of Technology, Israel)*

This compendium provides a comprehensive collection of the emergent applications of big data, machine learning, and artificial intelligence technologies to present day physical sciences ranging from materials theory and imaging to predictive synthesis and automated research. This area of research is among the most rapidly developing in the last several years in areas spanning materials science, chemistry, and condensed matter physics.

Written by world renowned researchers, the compilation of two authoritative volumes provides a distinct summary of the modern advances in instrument — driven data generation and analytics, establishing the links between the big data and predictive theories, and outlining the emerging field of data and physics-driven predictive and autonomous systems.

Published

More information on this series can also be found at https://www.worldscientific.com/series/wset

World Scientific Series on Emerging Technologies:
Avram Bar-Cohen Memorial Series – Volume 5

POWER BEAMING

HISTORY, THEORY, AND PRACTICE

Paul Jaffe
US Naval Research Laboratory, USA

Tom Nugent
PowerLight Technologies, USA

Bernd Strassner II
Massive Light LLC, USA

Mitchel Szazynski
Virtus Solis Technologies, USA

NEW JERSEY · LONDON · SINGAPORE · BEIJING · SHANGHAI · HONG KONG · TAIPEI · CHENNAI · TOKYO

Published by

World Scientific Publishing Co. Pte. Ltd.

5 Toh Tuck Link, Singapore 596224

USA office: 27 Warren Street, Suite 401-402, Hackensack, NJ 07601

UK office: 57 Shelton Street, Covent Garden, London WC2H 9HE

Library of Congress Control Number: 2023057797

British Library Cataloguing-in-Publication Data
A catalogue record for this book is available from the British Library.

World Scientific Series on Emerging Technologies:
Avram Bar-Cohen Memorial Series — Vol. 5
POWER BEAMING
History, Theory, and Practice

Copyright © 2024 by World Scientific Publishing Co. Pte. Ltd.

ISBN 978-981-12-4310-3 (hardcover)
ISBN 978-981-12-4311-0 (ebook for institutions)
ISBN 978-981-12-4312-7 (ebook for individuals)

For any available supplementary material, please visit
https://www.worldscientific.com/worldscibooks/10.1142/12438#t=suppl

Desk Editors: Balasubramanian Shanmugam/Steven Patt

Typeset by Stallion Press
Email: enquiries@stallionpress.com

Printed in Singapore

This book is dedicated to the memory of Jordin Kare (1956–2017), Phillip Jenkins (1956–2019), and Avi Bar-Cohen (1946–2020). They are some of the giants on whose shoulders we stand.

Foreword

This book will further the fortunes of Power Beaming (PB) as a very useful text for teaching and research and development. By providing clear definitions of PB and beginning standards developments, this work will lead to longer-range, more efficient, and safer systems. The authors have also done a great service in documenting the history and recent accomplishments of Beamed Power. The references and contact information will prove useful to students and developers.

Knowing well most of the authors of this important book and interacting with them as technology has grown, I was honored when I was asked by Paul Jaffe to write this Foreword to this book.

The authors have had real field and space experience developing PB technology theory, analytic modeling, hardware fabrication, performing demonstrations, and contributing to safety and operational equipment advances. Now with this book is the time to set new records and engender safe, practical, and useful applications of PB soon.

I was lucky to be in the right place at the right time as the Supervisor of the High-Power Transmitter Group of Caltech/JPL/NASA Deep Space Network site at Goldstone, CA. The 500-kW continuous wave S-Band Klystron transmitter and 26-m diameter Cassegrain-fed parabolic antenna used not only for emergency transmission to spacecraft but also for Planetary Radar, was temporarily repurposed with the site's collimation tower a mile away, to perform a demonstration of wireless power transmission, as it was then called. (*Search*: 1975 NASA Goldstone Demo of Wireless Power Transmission.)

Hopefully, my nearly 50-year record of the most usefully recovered microwave-beamed power (34 kW DC power at 1.55 km at 2.388 GHz in June 1975) will soon be broken. I must comment on the use of light bulbs to show visually the effect of microwave beam movement in the demonstration. As Project Manager, I was originally opposed to the design of sharing some of the DC power from a well-calibrated pure resistor load with the bank of nonlinear lamps (possibly compromising the accuracy of measuring total power) arranged in the geometry of the rectenna arrays in 1975. However, lighting light bulbs has been seen to be an objective of some subsequent demonstrations.

These authors deal well with definitions of power transfer efficiency which, as they point out, need to be standardized. Accuracy of power measurements was of the utmost importance in the March 1975 end-to-end efficiency measurement at Raytheon Waltham, MA. Principally because my JPL lab Director Dr. Pickering had said to me upon reviewing what was proposed using the newly developed device termed a rectenna. "Dick, any good physicist knows that a good absorber is also a good radiator and you will never get greater than 50% power transfer efficiency". In Dr. Pickering's defense, other persons commented that the use of $1/2$ wave rectifier diodes would further reduce the efficiency to 25%. And even others when informed of our proposed use of $1/2$ wave dipole antennas remarked that 12.5% efficiency would be the best NASA could hope for. Thanks to Bill Brown's tireless research and other laws of physics (narrow RF spectrum, uniform polarization, sinusoidal RF input riding on DC output, etc.) 54% was measured. The results were certified by Jet Propulsion Laboratory Quality Control personnel with instrument calibrations traceable to the National Bureau of Standards.

Another doubter of PB efficiency was the VP of Technology of Hughes, who said SPS was impossible because of the known huge space loss beaming from GEO to Earth. Perhaps he didn't reflect enough on the huge, proposed transmitter diameter of 1 km and huge diameter of the 10 km rectenna at 2.45 GHz.

Sam Fordyce of NASA HQ pushed for funding to do the larger power, longer range demonstration at Goldstone to quiet the doubters of the technology scaling up from the laboratory's mere 495 W output at 170 cm range to something much larger. There were those who said the lab tests were possibly affected by near-field and/or resonant effects between the Potter horn and the rectenna array or within the rectennas. The low VSWR measurements showed otherwise. Because funds were limited, we could not build a rectenna large enough to capture all of the beam from the 26-m diameter parabolic antenna at Deep Space Station 13. In fact, the rectenna array originally planned to consist of 18 panels had to have production curtailed at 17 — thus the gap-tooth opening at the top. End-to-end efficiency was not this demonstration's goal. More Power was.

The authors provide a useful collection of industry contacts related to PB activities which will be valuable for teachers, students, and entrepreneurs. The template for documenting demonstrations is highly recommended. I wish I had it in 1975.

Their electromagnetic beam propagation and laser and microwave safety discussions are invaluable for system design and operation. Especially for Project and Program managers and Safety Officers involved with PB activities. Interlocks and fail-safe systems are imperative, and encrypted-coded pilot beams or retrodirective systems are recommended at high-power levels. Also pairing beam safety with energy storage due to beam interruptions will be required for effective PB Systems as the authors point out. In PB system designs, care must also be taken with the resulting voltage transients in the DC circuits accompanying beam interruptions.

I am eagerly awaiting the first bi-directional PB system utilizing essentially the same equipment at either end of the link to compete with the well-established wired power systems that can send power in either direction. The laser and microwave diodes can be designed to operate either as oscillators (Impatt for microwaves) or rectifiers (rectennas). Due care must be exercised with the resonators for transmitter coherence. The current flows similarly in the devices for either mode of operation. The polarity of the DC voltage changes.

My time spent with others in this and allied fields has been most intellectually rewarding. Especially John Mankins (Solar Power Satellites), James McSpadden (harvesting rectennas from the Goldstone array for his thesis), Berndie Strassner (circular polarization rectenna impetus), Paul Jaffe (responding to his requests for key data), and many others around the World particularly in Japan, France, Germany, Austria, England, Canada, South Korea. Russia, and China. With a large tip of the hat to Kevin Parkin, I would like to dedicate this Foreword to Bob Forward whose writing has shown us the interstellar way to use PB.

Richard M. Dickinson
La Crescenta, CA, USA
September 10, 2023

Preface

The sun represents by many orders of magnitude the largest source of energy in our solar system. The means by which the sun has transmitted its energy to Earth for eons, via electromagnetic waves traveling through free space, offers compelling advantages and enables novel applications for energy delivery for many modern purposes.

Despite the long-standing precedent of power sent wirelessly to the earth from the sun, it often comes as a surprise to laypeople that meaningful amounts of energy can be sent without the need for a medium. In part, this stems from the relative obscurity of the research and technology development that has occurred in this field. Recently, this has been changing as the technology has matured and potential applications have begun to resolve themselves. Developments in the consumer market that have increased the prevalence of wireless power using inductive and capacitive means have also raised broader awareness that power can be sent wirelessly.

This book arose from the recognition of the largely unfilled niche for a modality-agnostic reference on the emerging technology of power beaming. Power beaming is a subset of wireless power transmission that focuses on the delivery of energy over relatively long distances. Functionally, power beaming can involve a range of modalities that have historically been confined to their respective specialized technical communities. Though wireless power transmission, energy harvesting, and directed energy have received ample treatment in both technical books and peer-reviewed literature, power beaming has largely been an afterthought.

In attempting to fill this void, the authors have approached power beaming as a capability rather than as an outgrowth of a specific engineering subdiscipline. While treated generally irrespective of the specific technology or modality employed, the principal focus in this text is on those means of power beaming that employ electromagnetic waves.

The intended audience for this text includes technologists, engineers, technicians, and students. In general, mathematics has been deliberately kept at a level accessible to undergraduates in technical fields, with a greater focus on conveying qualitative understanding. References to sources with well-established and rigorous underpinnings of optics and electromagnetics are included for those wishing to delve more deeply into theory.

As it is anticipated that developments in power beaming and its component technologies will accelerate in the coming years, readers are strongly advised to consult contemporary literature for matters concerning demonstrated device and link efficiencies. Without a doubt, even within the relatively short time between the completion of this book and its publication, there will likely be many new advances.

As with nearly any first edition, we anticipate readers will identify improvement opportunities in the text. We welcome all feedback, which can be sent to powerbeaming@ieee.org.

Contents

Chapter 1

Introduction

Advanced civilization is based on advances in energy.

— Bill Gates [1]

1.1 The Imperative of Energy

Energy fundamentally underpins every human activity. Despite this, from the dawn of recorded history, energy's role in enabling the means for meeting humanity's most basic needs has often been overlooked, underappreciated, or simply taken for granted. Whether it comes in the form of the sunlight that allows plants to grow, in turn providing food for our bodies and fuel for our hearths, or in the form of electricity generated by nuclear fission or another source, the details of exactly how it is produced and provided to us have been rendered largely invisible to most. Though over the eons our relationship with energy has evolved and our demands for it have increased, this lack of awareness appears to be a common and increasing feature [2].

In today's world, access to and consumption of energy are largely a function of where one lives. According to data from the International Energy Agency, many living in North America might consume on the order of ten times as much energy per capita compared to those living in parts of Africa [3]. As people around the world strive for and attain higher levels of income and standards of living, their energy consumption tends to rise. To secure a sustainable future, our relationship with energy and the methods and techniques we employ in its sourcing, storage, delivery, and consumption deserve expansive and careful attention.

1.2   Sourcing, Storage, and Delivery of Energy

To be able to consume it when and where it is needed, energy generally requires sourcing, storage, and delivery. Depending on the particular source and the nature of the ultimate application, these three major activities might include subprocesses along the way, such as extraction, refinement, and conversion [4]. Their distinctions may also be blurred, as some energy sources offer features that can be reasonably interpreted as intrinsic storage or delivery.

1.2.1   *Sourcing*

Much attention has been devoted to the primary sources from which our energy originates, especially as the most prevalent of these from the 20th century have proved harmful to the environment: fossil fuels [5]. Despite the significant drawback of altering the climate when consumed in quantity, fossil fuels have proven to be one of nature's most valuable gifts: ancient sunlight, stored in the remains of long-deceased organisms, provides concentrated energy in an enduring form that can be extracted and then transported to where it is needed. Renewable sources like wind, solar, tidal, geothermal, and biomass offer the advantage of being much less damaging for the global climate but typically have geographical limitations that prevent their economically widespread adoption. Nuclear fission, while not renewable, shares a similar climate benefit. It has struggled for years with public perceptions and notable instances of safety challenges. Future sources like nuclear fusion and space solar remain in development.

1.2.2   *Storage*

Most people in the present day have at least a vague sense of the importance of stored energy, whether it manifests as the level of charge of the batteries in their phones or cars, or as a nearby bundle of firewood. Storage can occur in a range of places in the energy chain between sourcing and consumption, and over a range of scales. As indicated above, some primary energy sources present as stored energy, awaiting conversion as demanded. With the advent of the prevalent use of renewable primary sources [6], the urgency behind grid-scale and smaller-scale storage has increased. Similarly, the electrification of transportation and other systems has fed the demand for battery and other electrical energy storage technologies specifically. Whether energy is stored closer to the primary source or closer to the point of consumption, there remains a need to deliver it between those points.

1.2.3 *Delivery*

Delivery is a crucial part of energy systems, though its details are typically unseen to energy consumers. The expectation is that when a light switch is flipped, the lights will come on. How it came from a distant power plant at exactly the moment it was needed is successfully abstracted away, except perhaps to a specialized group of technicians, engineers, and the curious. In the case of the electrical grid, delivery can be broken into transmission and distribution, reminiscent of a tree's trunk dividing into branches. For off-grid applications, delivery might take the form of shipments of fuel. In consumer applications, the ubiquitous power cord often provides the last meter or two of delivery to the devices that consume the energy we use. For each case, traditional solutions feature upsides and downsides.

The energy delivered by electrical grids, cabling, and wiring is notable in that once the infrastructure is emplaced, no additional mass needs to be moved to convey the energy. Though it needs a conductive path for current to flow, it dispenses with the time and additional energy that accompany delivering energy stored in some other form. These conductive paths add some loss, except for those few that are superconductive, but the utility of moving energy without moving mass is enormous. What if the same feat could be accomplished without the wires?

1.3 Wireless Power Transmission

Lurking in the margins for decades has been a historically less-considered form of energy delivery: wireless power transmission

(WPT). In this case, *energy is deliberately sent from a transmitter for capture and utilization at a receiver without using wires or the movement of matter.* WPT is commonly associated with the transmission of electrical energy, though there is no reason to constrain it only to that domain. WPT options include varying electric or magnetic fields, electromagnetic waves as photons, phonons, and perhaps quantum energy teleportation. The scope of this book focuses on the consideration of instances employing transmitted electromagnetic waves beyond the reactive near field for wireless power, and will be referred to as "power beaming". A more detailed definition is developed in the next section. Power beaming is differentiated from other forms of wireless power that utilize inductive means, capacitive means, or resonant forms of these means, as well as from energy harvesting. Energy harvesting depends on taking advantage of extant electromagnetic waves, or energy in the local environment in another form, rather than on those generated specifically for the purpose of transmitting power. Generally, inductive and capacitive approaches are relatively limited in range, whereas there is no theoretical limit to the range of a launched electromagnetic wave. This can be appreciated by merely looking at the night sky and seeing distant stars, whose photons have traveled trillions upon trillions of kilometers, and yet are still readily visible on a clear evening.

WPT is most often associated in the public's mind with Nikola Tesla and Tesla's exploits have been ably documented in both literature [7] and film [8]. Despite Tesla's impressive technical achievements, he was unable to find commercial success for WPT. It was not until 1961 that an early consumer product employing WPT was introduced to the market: the General Electric cordless toothbrush [9]. For consumer electronics, WPT continued to reside largely in the toothbrush charging niche until the advent of magnetically coupled resonance that began in 2006 as a result of work at the Massachusetts Institute of Technology [10,11]. This opened the door for the effective short-range wireless charging of various devices, which today includes the widespread ability to charge personal electronics wirelessly [12,13], as well as electric vehicles [14]. The historical development of WPT and power beaming is explored in greater depth in Chapter 2.

While WPT as a whole has enjoyed a boom since 2006, the power beaming subset of WPT is still emerging. Though power beaming has seen notable research developments and demonstrations since the 1960s [15], it has not yet established itself either in the mind of the general public as a capability, or found suitable application niches in practice. This situation may be changing as both the underlying component technologies and the practicality of integrated systems have begun maturing.

It is instructive to distinguish power beaming from the larger field of WPT and other capabilities with which it may be confused, such as directed energy. In Figure 1.1, power beaming can be seen in context as a subset of WPT and in approximate relation to other applications of electromagnetic waves. This depiction should not be interpreted to discount ways that energy can be moved without employing electromagnetic waves, including transport of mass and acoustic means. It is likewise important to recognize that many of these applications may occur over much wider distance ranges and power densities than the notional regions in Figure 1.1 might suggest. For instance, wireless communications can occur over small distances of much less than a meter or across those of greater than 23 billion km, as in the case of the Voyager 1 spacecraft [16]. In the future, the notional distance limits for many of these applications are likely to increase. The figure's intent is to show some different classifications of how electromagnetic waves are employed and to convey where power beaming falls within this landscape.

One key in distinguishing between the capabilities of the regions in Figure 1.1 can be found by examining the intended application and the technical requirements for satisfying that application's demands. Detection and imaging seek to capture energy to discern information from a region of interest, while radar and lidar enhance this by adding energy into the environment to aid in gathering even more information. Communication links and navigation transmissions have the goal of ensuring the received signal is sufficiently above the noise floor at the receiver to convey information. Energy harvesting's goal is to exploit ambient power of sufficient power density for conversion and application to a task. Directed energy, jamming, and electronic warfare aim to disrupt, disable, destroy, or otherwise co-opt sensors or systems by delivering relatively large or unexpected amounts of energy to unwitting receivers. Industrial equipment, medical devices, microwave ovens, and similar devices impart energy to effect material changes as part of processing, treatment, or cooking. In Bloxton's

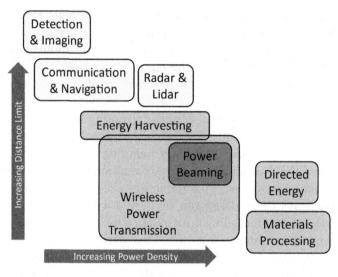

Figure 1.1. Technology capabilities depicted with notional relative distance limits and power densities for some common applications of electromagnetic waves.

paradigm of "bits, atoms, and joules" for quantities of human economic interest [17], the uppermost three categories of Figure 1.1 can be categorized in the realm of "bits" and all others would be in the realm of "joules".

In each capability shown in Figure 1.1, there is an implication of a source of electromagnetic energy and a corresponding sink. For detection and imaging, the source might be a star whose light is reflected by a planet to a sink in the form of an imager, such as ESA's planned ARIEL mission for remote sensing of extrasolar planets [18]. For directed energy or materials processing, the source might be a high-power laser with the sink as an object to be ablated. For energy harvesting, the source might be a natural electromagnetic energy emitter like the sun, with the sink as a solar cell. The source for energy harvesting might also be artificial, such as a television broadcast transmitter [19]. Since energy harvesting can happen with both natural and artificial sources of energy, some instances could fairly be considered as instances of WPT. This is why the domains for energy harvesting and WPT intersect in the figure.

For communication, navigation, and WPT, sources are typically labeled as transmitters whereas the sinks are the receivers. This

Transmitter   Link   Receiver

Figure 1.2. A basic arrangement for communication or energy connections.

implies a deliberate effort to move something from one place to another: for communication, the entity to be moved is information; for WPT, it is energy. In the simplest terms, these situations can be described by the block diagram depicted in Figure 1.2.

For communications, Sklar's seminal communications text has expanded the 3-element depiction of Figure 1.2 to a 22-element depiction, as shown on the cover of [20] and elaborated within. Communications engineering is a very mature field and has a robust ecosystem of textbooks [20–22], technical journals, conferences, standards, and professional societies [23]. Currently, no comparable ecosystem exists for power beaming, though some progress is being made for its parent category of WPT [24,25].

When the elemental transmitter-link-receiver construct of Figure 1.2 is applied to the domain of WPT, a continuum is created based on the degree of coupling between the transmitter and receiver. This continuum yields cases where the transmitter and receiver may be considered strongly coupled, weakly coupled, or uncoupled. Power beaming generally falls into the "uncoupled" category. A discussion examining the implications of both the degree of coupling and coupled mode theory is in Chapter 3.

1.4   Defining Power Beaming

What then defines power beaming? Is it merely uncoupled WPT? A variety of qualitative expressions have been employed to describe things that sound like power beaming. Some of these terms include far-field energy transfer, far-field WPT, long-distance wireless power, free-space power transmission, and other permutations of these. While all of these clearly fall within the larger domain of WPT, they beg the question: what qualifies as "far" or "long"? Different definitions of these could plausibly apply in different situations.

Prior efforts to define power beaming have focused specifically on electromagnetic wave propagation and on the imperative of energy transfer efficiency. For instance, in 1974 William C. Brown posited that "Free-space power transmission by microwave beam is defined as the efficient point-to-point transfer of energy through free space by a highly collimated microwave beam" [26]. He further went on to qualitatively define "free space" but left the threshold as to what constitutes "efficient" unaddressed. However, he did observe that at the time the efficiency range demonstrated was 15%–45% [26, p.12]. Clearly, efficiency must play some role, as an extremely inefficient WPT or power beaming link would likely not satisfy requirements for energy delivery in most situations.

Outlined here are two arbitrary but essentially modality-agnostic guidelines for defining power beaming: (1) a practical guideline for human-scale utility, and (2) a guideline based on the relationship of the distance traversed by the link to the size of the transmit and receive structures.

1.4.1 *A Human-scale Guideline: Satisfying the 1-1-1 Criteria*

A famous quote attributed to Protagoras is often paraphrased as "Man is the measure of all things" [27]. Power beaming can be defined as an instance of WPT in which the following conditions that have an intuitive association with human scales are met simultaneously:

- *at least 1%* of the input source energy is delivered to the output of the link ("end-to-end efficiency") for a user or application,
- a link distance of *at least* 1 m is spanned, and
- the first two conditions are met, averaged over a period of *at least* 1 min.

This guideline can be summarized as the "1-1-1 criteria", where each "1" corresponds to $\geq 1\%$, ≥ 1 m, and ≥ 1 min. Meters and minutes are quantities intuitively familiar to most people, and 1% is likewise an order of magnitude touchpoint with wide accessibility.

Using this 1-1-1 criteria sets no constraints on the size of either the transmit or receive apertures for the establishment of a power beaming link. Another approach is to define power beaming geometrically in terms of the greatest dimensions for the transmit and receive structures in relation to the link distance.

1.4.2 *A Geometric Guideline: Using the Relationship Between the Link Distance and Transmit/Receive Structure Dimensions*

WPT can be performed effectively over distances shorter than a meter by employing relatively large transmit and receive structures, such as in electric vehicle recharging [28]. For power beaming, the usage of the word root "beam" connotes length, which is not evident in these shorter-distance situations. Whether something is "long" or not depends on the object of comparison. Convenient references for comparison to determine whether a given WPT link qualifies as long are the dimensions of the transmit and receive structures. Since these structures may not have projections that are circular, square, or even symmetrical, their effective cross-sectional area may not be easily ascertained. Taking their greatest dimensions simplifies comparisons and provides a reasonable bound on the expectations of link performance. Some systems may have transmit and receive structures of very different sizes. Given these considerations, here is a guideline for distinguishing power beaming from other types of WPT:

- Where the distance that a meaningful amount of energy is transported wirelessly significantly exceeds the sum of the largest dimensions of the transmit and receive structures.

This definition implies an expression for the Beam Aspect Factor or BAF, that conveys a sense of the link's narrowness. This could be expressed as a ratio or a quotient:

$$\frac{\textbf{Link distance}}{\textbf{Sum of the largest dimensions of the transmit and receive structures}}$$

Cases of power beaming can be defined as requiring a quotient value that significantly exceeds one.

For power beaming employing electromagnetic waves, it may make sense to speak in terms of the reactive near-field, radiating near-field, far-field, Fresnel region, or Fraunhofer region. These regions are variously defined by link distance, transmit or receive structure dimensions and focal lengths, and operating wavelength. These are examined in greater depth in Chapter 3.

There is still a point of subjectivity with geometrical guidelines: what constitutes a "meaningful amount of energy"? To address this, we will combine the guideline with the 1% of input source energy being delivered to the output element listed among the three human-scale conditions discussed previously. This results in the following definition of power beaming:

Power beaming: Uncoupled WPT in which the link distance exceeds the sum of the largest dimensions of the transmit and receive structures, and where at least 1% end-to-end energy transmission efficiency is obtained.

Specific contrived examples might fit this definition but wouldn't colloquially be considered power beaming. Regardless, this definition appears reasonable for most situations.

1.5 Motivation for Using Power Beaming

With a provisional definition of power beaming in hand, it now makes sense to determine when, where, and why power beaming might be employed. As other means of energy delivery have served humankind apparently satisfactorily over time, what justification can be cited to introduce a new method?

For a given activity, at a given location, at a given point in time, there may or may not be enough energy immediately available to permit or sustain that activity. This in turn drives a need either for storing a suitable amount of energy locally or for having the means of delivering energy from a source location to the point of consumption. Certain circumstances levy constraints on the amount of energy that can be stored locally or how energy can be delivered. These challenges relate to portable electronics, deployed sensors, vehicles, autonomous or teleoperated systems, and off-grid locations, including those in space. Each of these is explored in greater depth in

Chapter 8. Chapter 8 also introduces a systematic power beaming decision flow, wherein the range of alternatives for energy provision can be examined and weighed against a specific application's requirements. Depending on the application, the power beaming link might span meters or many thousands of kilometers and perhaps require milliwatts or gigawatts delivered to the output. It might be continuous or intermittent. There could be one or more transmitting or receiving nodes, and they might be moving relative to each other. The operational environment might levy requirements, such as the ability to operate in a vacuum or through the atmosphere.

The matching of the specific application's requirements with existing technological capabilities is important. However, the approach to making the proper choices arises from a more basic pair of questions: what is needed, and what resources can be applied to meet those needs? Power beaming has historically fallen short because of both its expense and inefficiency when compared to available energy alternatives. While these shortcomings are lessening, there is another important question: what applications could this new technology enable that have not yet been fully recognized or realized? As more people become aware of power beaming as a means of energy distribution, it is likely that applications will emerge that have not yet been considered.

1.6   Power Beaming Subsystems

1.6.1   *3-Block Diagram*

The scenario depicted in Figure 1.3 resembles a simplified Figure 1.2 link case that is tailored for power beaming.

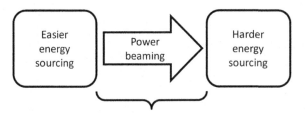

Separation ill-suited for physical or resupply connections

Figure 1.3.   A generic scenario for power beaming.

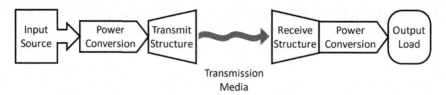

Figure 1.4. Functional blocks for a wireless power or power beaming system.

This depiction includes the source of the energy, the sink for the energy, and the path between them. Note that no assumptions are made about the specific technology to be used. Electromagnetic waves, sound waves, or something else could be employed. The annotation "separation ill-suited for physical or resupply connections" levies the only constraints on how the two places might be connected for energy delivery purposes. A natural next step is to further subdivide this depiction into functional elements, as is accomplished in the next section.

1.6.2   *7-Block Diagram*

For the situation depicted in Figure 1.3, the elements of a power beaming system can be subdivided and functionally portrayed as shown in the block diagram in Figure 1.4.

The "Input Source" may well be given as a constraint in the design of a power beaming system. It could be a wall outlet, generator, or something completely different. The leftmost "Power Conversion" takes all its energy from the Input Source, including not just that which is destined for ultimate delivery to the "Output Load" but also for any supporting elements that might be needed in the implementation of the link, such as thermal management and system control. It is imperative not to neglect the needs and contributions of supporting elements in assessing the link's performance. This is explored in greater detail in Chapter 7. The next block, "Transmit Structure," indicates the subsystem that provides the interface to the media for transmission. This is generically termed "structure" rather than an aperture, array, antenna, lens, or transducer since it might take any of these specific forms or others. The "Transmission Media" must be critically considered for the overall system function. It might consist of a dynamic combination of different mediums, possibly including a vacuum, the atmosphere, and any persistent or transient material that could occur within them. The "Receive Structure" serves the inverse function of the Transmit Structure by collecting the energy from the Transmission Media and presenting it to the receiver-side "Power Conversion" block, after which it is delivered to the "Output Load". The Output Load may often be considered as external to the power beaming system itself. The application employing the energy delivered to the Output Load may set constraints or requirements that influence the design of the entire system.

The functions within each of the blocks can be implemented using a wide range of different technologies. Discussions of some of the options appear in the next section and Chapters 5 and 6.

1.7 Modality Discussion

By approaching electromagnetic spectrum power beaming links from a functional rather than technology-specific perspective, it is easier to develop representative concepts that can be applied broadly. However, specific electromagnetic technologies have notable benefits and drawbacks, and power beaming can be implemented functionally without using the electromagnetic spectrum at all.

In this book, the focus is primarily on applications and technologies that use electromagnetic means of power beaming, but the fundamental concepts of how to characterize links and to determine power beaming suitability should still generally apply to other methods. Similarly, phased-array techniques can be applied to instances of electromagnetic power beaming, acoustic power beaming, and other modalities that employ waves. For a review of acoustic power beaming, also known as acoustic energy transfer, readers should consult [29] and other related literature.

Even within the subset of electromagnetic power beaming, a huge range of potential technologies can be utilized, often making it difficult to determine both the state-of-the-art and the best technology for a given application. The challenge is heightened by the wide range of wavelengths that could be selected, each having advantages and disadvantages for different applications.

For microwave wavelengths, the advantages of lower power density (in most cases) and weather outage resistance are counterbalanced by severe spectrum allocation challenges. For laser wavelengths, the advantages of smaller aperture sizes and avoidance of the radio spectrum and its regulatory challenges are counterbalanced by potential safety and safety perception challenges. For modality selection for any electromagnetic power beaming link, considerations of beam efficiency, device conversion efficiencies, safety, regulations, and atmospheric effects must be assessed. These are addressed in Chapters 4–6. By generalizing comparison criteria for power beaming links and reviewing and comparing notable historical power beaming demonstrations, the most fruitful areas for forward development may be identified and pursued across all candidate modalities.

1.8   Accomplishments to Date

Many compelling demonstrations of power beaming have been accomplished since the middle of the 20th century. Notable power beaming subsystem performance metrics and record-holding power beaming demonstration results have been collected by Dickinson [30] and others [31]. While subsystem development is paramount in creating practical and fieldable power beaming systems, much benefit may be lost if subsystems are not effectively integrated into a full end-to-end operational system. Often, efficiency and other parameters are degraded when high-performance individual subsystems are combined at scale due to thermal and other challenges. For this reason, this section focuses solely on demonstrations that have complete end-to-end power beaming links and that have also met the definition of power beaming as given in Section 1.4.2.

Documentation of previous efforts does not always report values for all the metrics of interest. Sometimes, missing metrics of interest can be estimated, surmised, or otherwise determined. Three particular parameters are of keen interest:

1. The distance over which the link operates.
2. The power (or energy) delivered to the load.
3. The end-to-end efficiency.

Determining the best implementation for a given application relies on a host of other factors, including transmit and receive structure dimensions, the cost per watt or joule delivered to the end user, and the degree of safety of the link. These and other measurable and calculated parameters are explored in greater depth in Chapter 7 and how they might fit in specific applications is explored in Chapter 8.

Previous demonstrations that appear in the literature and which have notable performance in key metrics can be tabulated in a "Power Beaming Leaderboard." In other fields, various parameters and figures of merit have been tracked to show progress toward technological goals. Some examples of this include the US National Renewable Energy Laboratory's "Best Research-Cell Efficiencies" and "Champion Photovoltaic Module Efficiency" charts, available at [32,33], respectively. These are used to keep track of developments in solar cell technology and their corresponding efficiencies. Similar comparisons exist for many other areas, as diverse as Valenta and Durgin's for rectenna element conversion efficiencies [34], Fafard and Masson's for laser power converter efficiencies [35, Fig. 1.3], launch vehicle payload capacity, production automobile acceleration times, record marathon times, and countless others.

Creating a plot of what has been reported to date for the first two parameters from the list above, distance of link operation and power delivered, gives a sense of what has been accomplished to date. Such a plot is shown in Figure 1.5.

This plot uses logarithmic scales for both distance and power to display the operating range of different demonstrations. Points were only included if documentation showed that the demonstration was likely to meet the definition of power beaming from Section 1.4.2. Several of these demonstrations are discussed in greater detail in Chapter 2. The sources consulted for the points are found in Table 1.1.

The well-known 1975 JPL-Raytheon demonstration at Goldstone, California took advantage of the large directivity of NASA's 26 m Deep Space Network communications antenna to execute what has remained both the longest distance and the highest power demonstration for many decades. Other demonstrations that did not meet the definition of power beaming from 1.4.2 but that may have demonstrated an important advance or novel feature, such as the 148 km demonstration by Mankins and Kaya in Hawaii in 2008 [28], are also discussed in Chapter 2.

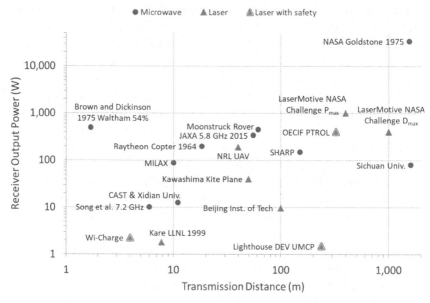

Figure 1.5.   *Receiver power output vs. transmission distance. See Table* 1.1 *for references.*

Table 1.1.   *References for receiver power output vs. transmission plot.*

Label	Cite	Label	Cite
NASA Goldstone 1975	[36]	MILAX	[37]
Brown and Dickinson 1975 Waltham 54%	[38]	Sichuan Univ.	[39]
LaserMotive NASA Challenge P_{max}	[40]	Kawashima Kite Plane	[41]
Moonstruck Rover	[30]	CAST & Xidian Univ.	[39]
OECIF PTROL (PowerLight Technologies)	[42]	Beijing Inst. of Tech	[43]
LaserMotive NASA Challenge D_{max}	[40]	Wi-Charge	[44]
JAXA 5.8 GHz 2015	[45]	Lighthouse DEV UMCP	[46]
NRL UAV	[47]	Raytheon Copter 1964	[48]
SHARP	[49]	Song *et al.* 7.2 GHz	[50]

The third parameter of great interest, efficiency, has not been as widely reported for various demonstrations. This may arise from a reticence among performers to report figures they perceive as

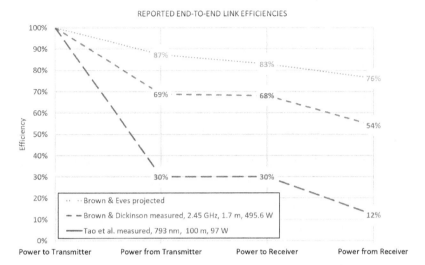

Figure 1.6. Projected and reported link efficiencies. Data from [38,43,51].

disappointingly low to themselves or others, or because of difficulties in quantifying the efficiencies and their contributors. Thus, the pool of demonstrations from which to compare efficiency results is considerably smaller than that for distance and power. A further complication is that measuring efficiency has opportunities for greater subjectivity and error when compared with the measurement of a simpler parameter like distance. A longer discussion of some of the challenges and pitfalls of efficiency reporting can be found in Chapter 7. A selection of three examples of end-to-end efficiency for a projected microwave link, a measured microwave link (over a relatively short distance of 1.7 m), and a measured laser link is plotted in Figure 1.6.

Worth noting in Figure 1.6 is that most of the losses occur in the energy conversion at the transmitter and receiver, and not from the beam propagation itself. This is not always the case, and achieving the high beam collection and transmission efficiencies depicted requires judicious system design and aperture sizing to avoid losses due to diffraction and other sources.

Link efficiencies are perhaps most clearly reported by using a Sankey diagram which maps loss sources clearly to the subsystems of the link implementation, as described in Section 7.5.6.

As of this writing, many ongoing efforts were poised to break new ground in setting records for distance, power, and efficiency. Readers are advised to consult the Further Exploration section in the backmatter of this book for pointers to recent breakthroughs.

1.9   Conclusion

This chapter sets the context for power beaming by reviewing how humanity has depended on the sourcing, delivery, and storage of energy to found and sustain civilization. It introduced the concepts and nomenclature for WPT and power beaming, including the functional subsystems and options for their implementation. Key considerations for when and where power beaming might be used were reviewed, as were a range of notable prior power beaming instances. The balance of this book delves deeper into the history, fundamentals, implementations, metrics, and applications of power beaming.

1.10   Further Reading

For a summary overview of energy, its forms, related topics, and frequently asked questions, the US Department of Energy has developed a well-curated online set of resources available at [52]. A comprehensive treatment of civilization's energy needs, quantitative constraints arising from physical limitations, and exploration of possible solutions can be found in Murphy [53]. A broader range of paths to the exploration of power beaming and related topics can also be found in the back matter of this book.

References

[1] Long Beach, CA. *Innovating to Zero! Bill Gates*, February 20, 2010. [Online Video]. Available at: https://www.youtube.com/watch?v=JaF-fq2Zn7I&ab_channel=TED (accessed September 11, 2020).

[2] International Technology Education Association Gallup Poll 2004. [Online]. Available at: https://www.iteea.org/File.aspx?id=50275 (accessed September 24, 2020).

[3] "Energy Use (kg of Oil Equivalent per Capita)," *Data*. Available at: https://data.worldbank.org/indicator/EG.USE.PCAP.KG.OE?most_recent_value_desc=t3rue (accessed September 21, 2020).

[4] B. Hayes, *Infrastructure: A Field Guide to the Industrial Landscape the Book of Everything for the Industrial Landscape*. New York London: W.W. Norton, 2006.

[5] R. K. Pachauri and L. Mayer, and Intergovernmental Panel on Climate Change, Eds., *Climate Change 2014: Synthesis Report*. Geneva, Switzerland: Intergovernmental Panel on Climate Change, 2015.

[6] Renewable Energy Statistics 2020, p. 408.

[7] W. B. Carlson, *Tesla: Inventor of the Electrical Age*. Princeton, NJ: Princeton University Press, 2013.

[8] *Telsa — Master of Lightning*, 2000. [Online Video]. Available at: https://www.pbs.org/tesla/ (accessed September 28, 2020).

[9] A. Tadinada, "The Evolution of a Tooth Brush: From Antiquity to Present — A Mini-Review," *J. Dent. Health Oral Disord. Ther.*, vol. 2, no. 4, June 2015. doi: 10.15406/jdhodt.2015.02.00055.

[10] "Goodbye Wires!," *MIT News*. Massachusetts Institute of Technology. Available at: https://news.mit.edu/2007/wireless-0607 (accessed September 17, 2020).

[11] A. Karalis, J. D. Joannopoulos, and M. Soljačić, "Efficient Wireless Non-Radiative Mid-Range Energy Transfer," *Ann. Phys.*, vol. 323, no. 1, pp. 34–48, January 2008. doi: 10.1016/j.aop.2007.04.017.

[12] "Qi Wireless Charging," Qi Wireless Charging. Available at: http://www.qiwireless.com/ (accessed September 28, 2020).

[13] "Wireless Power Transfer & Charging Standards — AirFuel Alliance," *AirFuel Alliance*. Available at: https://airfuel.org/ (accessed July 01, 2020).

[14] P. Campbell, "Electric Vehicles to Cut the Cord with Wireless Charging," September 9, 2020. Available at: https://www.ft.com/content/720bc57b-944f-47a5-8d65-12439228571b (accessed September 19, 2020).

[15] B. Strassner and K. Chang, "Microwave Power Transmission: Historical Milestones and System Components," *Proc. IEEE*, vol. 101, no. 6, pp. 1379–1396, June 2013. doi: 10.1109/JPROC.2013.2246132.

[16] "Voyager — Mission Status." Available at: https://voyager.jpl.nasa.gov/mission/status/ (accessed September 19, 2020).

[17] M. Bloxton, "Bits Atoms Joules," *Medium*, April 16, 2020. Available at: https://medium.com/@michael_54806/bits-atoms-joules-f775aadbd910 (accessed October 24, 2020).

[18] E. Gibney, "First Space Mission Dedicated to Exoplanet Atmospheres Gets Green Light," *Nature*, vol. 555, no. 7698, Art. no. 7698, March 2018. doi: 10.1038/d41586-018-03445-5.

[19] R. Vyas, H. Nishimoto, M. Tentzeris, Y. Kawahara, and T. Asami, "A Battery-Less, Energy Harvesting Device for Long Range Scavenging of Wireless Power from Terrestrial TV Broadcasts," in *2012 IEEE/ MTT-S International Microwave Symposium Digest*, Montreal, QC, Canada: IEEE, June 2012, pp. 1–3. doi: 10.1109/MWSYM.2012. 6259708.

[20] B. Sklar, *Digital Communications: Fundamentals and Applications*, Second edition. Upper Saddle River, New Jersey: Prentice-Hall PTR, 2001.

[21] B. P. Lathi, *Modern Digital and Analog Communication Systems*, Third edition, in the Oxford Series in Electrical and Computer Engineering. New York: Oxford University Press, 1998.

[22] S. S. Haykin, *Communication Systems*, Fifth edition. Hoboken, New Jersey: Wiley, 2009.

[23] "IEEE Communications Society — IEEE ComSoc." Available at: https://www.comsoc.org/ (accessed September 20, 2020).

[24] "Wireless Power Transfer," *Cambridge Core*. Available at: http:// www/core/journals/wireless-power-transfer (accessed September 20, 2020).

[25] "IEEE MTT-S Wireless Energy Transfer and Conversion." Available at: https://www.mtt-archives.org/~mtt26/ (accessed September 20, 2020).

[26] W. C. Brown, "The Technology and Application of Free-Space Power Transmission by Microwave Beam," *Proc. IEEE*, vol. 62, no. 1, pp. 11–25, January 1974. doi: 10.1109/PROC.1974.9380.

[27] Joshua J. Mark, "Protagoras," *Ancient History Encyclopedia*, September 02, 2009. Available at: https://www.ancient.eu/protagoras/ (accessed October 10, 2020).

[28] Evatran, "FAQ Gateway," *Plugless Power*. Available at: https://www. pluglesspower.com/faqs/ (accessed June 8, 2021).

[29] M. R. Awal, M. Jusoh, T. Sabapathy, M. R. Kamarudin, and R. A. Rahim, "State-of-the-Art Developments of Acoustic Energy Transfer," *Int. J. Antennas Propag.*, vol. 2016, pp. 1–14, 2016. doi: 10.1155/2016/ 3072528.

[30] R. M. Dickinson and O. Maynard, "Ground Based Wireless and Wired Power Transmission Cost Comparison," presented at the *International Energy Conversion Engineering Conference* (*IECEC*), Vancouver, British Columbia, Canada, August 1999. [Online]. Available at: http://hdl.handle.net/2014/17841 (accessed April 25, 2020).

[31] C. T. Rodenbeck *et al.*, "Microwave and Millimeter Wave Power beaming," *IEEE J. Microw.*, vol. 1, no. 1, pp. 229–259, Winter 2021. doi: 10.1109/JMW.2020.3033992.

[32] "Best Research-Cell Efficiency Chart." Available at: https://www.nrel. gov/pv/cell-efficiency.html (accessed September 13, 2020).

[33] "Champion Photovoltaic Module Efficiency Chart." Available at: https://www.nrel.gov/pv/module-efficiency.html (accessed September 13, 2020).

[34] C. R. Valenta and G. D. Durgin, "Harvesting Wireless Power: Survey of Energy-Harvester Conversion Efficiency in Far-Field, Wireless Power Transfer Systems," *IEEE Microw. Mag.*, vol. 15, no. 4, pp. 108–120, June 2014. doi: 10.1109/MMM.2014.2309499.

[35] S. Fafard and D. Masson, "Vertical Multi-Junction Laser Power Converters with 61% Efficiency at 30 W Output Power and with Tolerance to Beam Non-Uniformity, Partial Illumination, and Beam Displacement," *Photonics*, vol. 10, no. 8, p. 940, August 2023. doi: 10.3390/photonics10080940.

[36] R. M. Dickinson, "Wireless Power Transmission Technology State of the Art," *Acta Astronaut.*, vol. 53, no. 4, pp. 561–570, 2003.

[37] N. Shinohara, "Beam Control Technologies with a High-Efficiency Phased Array for Microwave Power Transmission in Japan," *Proc. IEEE*, vol. 101, no. 6, pp. 1448–1463, June 2013. doi: 10.1109/JPROC. 2013.2253062.

[38] R. M. Dickinson and W. C. Brown, "Radiated Microwave Power Transmission System Efficiency Measurements," Jet Propulsion Lab., California Inst. of Tech., Pasadena, CA, United States, Technical Report NASA-CR-142986, JPL-TM-33-727, May 1975. [Online]. Available at: https://ntrs.nasa.gov/search.jsp?print=yes&R=1975001 8422 (accessed April 25, 2020).

[39] B. Duan, "On New Developments of Space Solar Power Station (SSPS) of China," presented at the *The 36th International Space Development Conference*, St. Louis, MI, May 25, 2017.

[40] Tom Nugent, "Review of Laser Power beaming Demonstrations by PowerLight Technologies (formerly LaserMotive)," presented at the *20th Annual Directed Energy Science and Technology Symposium*, Oxnard, CA, February 2018.

[41] N. Kawashima and K. Takeda, "Laser Energy Transmission for a Wireless Energy Supply to Robots," in *Robotics and Automation in Construction*, 2008, p. 8. [Online]. Available at: https://www.intechopen.com/books/robotics_and_automation_in_construction/laser_energy_tra nsmission_for_a_wireless_energy_supply_to_robots (accessed July 2, 2020).

[42] T. J. Nugent, Jr., D. Bashford, T. Bashford, T. J. Sayles, and A. Hay, "Long-Range, Integrated, Safe Laser Power beaming Demonstration," in *Technical Digest OWPT 2020*, Yokohama, Japan: Optical Wireless

Power Transmission Committee, The Laser Society of Japan, April 2020, pp. 12–13.

[43] T. He *et al.*, "High-Power High-Efficiency Laser Power Transmission at 100 m Using Optimized Multi-Cell GaAs Converter," *Chin. Phys. Lett.*, vol. 31, no. 10, p. 104203, October 2014. doi: 10.1088/0256-307X/31/10/104203.

[44] "WI-CHARGE LTD. Transmitters," *Wi-Charge*. Available at: https://wi-charge.com/product_cat/transmitters/ (accessed July 9, 2020).

[45] "Ground Demonstration Testing of Microwave Wireless Power Transmission. JAXA. Research and Development Directorate." Available at: http://www.kenkai.jaxa.jp/eng/research/ssps/150301.html (accessed July 9, 2020).

[46] R. Winsor and B. Murray, "Optical Wireless Power beaming," October 6, 2013.

[47] P. Sprangle, B. Hafizi, A. Ting, and R. Fischer, "High-Power Lasers for Directed-Energy Applications," *Appl. Opt.*, vol. 54, no. 31, p. F201, November 2015. doi: 10.1364/AO.54.00F201.

[48] W. C. Brown, "The History of the Development of the Rectenna," presented at the *Rectenna Session of the SPS Microwave Systems Workshop*, Lyndon B. Johnson Space Center. Houston, Texas, USA: NASA, January 1980.

[49] J. J. Schlesak, A. Alden, and T. Ohno, "A Microwave Powered High Altitude Platform," *IEEE MTT Dig.*, pp. 283–286, 1988.

[50] K. D. Song *et al.*, "Preliminary Operational Aspects of Microwave-powered Airship Drone," *Int. J. Micro Air Veh.*, vol. 11, p. 1756829319861368, January 2019. doi: 10.1177/1756829319861368.

[51] W. C. Brown and E. E. Eves, "Beamed Microwave Power Transmission and Its Application to Space," *IEEE Trans. Microw. Theory Tech.*, vol. 40, no. 6, pp. 1239–1250, June 1992. doi: 10.1109/22.141357.

[52] "Forms of Energy – U.S. Energy Information Administration (EIA)." Available at: https://www.eia.gov/energyexplained/what-is-energy/forms-of-energy.php (accessed October 14, 2020).

[53] T. W. Murphy, *Energy and Human Ambitions on a Finite Planet: Assessing and Adapting to Planetary Limits.* 2021. [Online]. Available at: https://escholarship.org/uc/energy_ambitions (accessed June 4, 2021).

Chapter 2

History

I do not think that the wireless waves I have discovered will have any practical application.

— Heinrich Hertz

2.1 Introduction

For nearly 150 years, researchers have been able to source power at one location and receive a portion of it somewhere else without any physical connection. In 1886, German physicist Heinrich Hertz observed and documented this phenomenon. While he may have downplayed its usefulness at the time, transmitting power in this manner has had profound impact on our civilization, allowing for such things as wireless communication systems and radar. It only took a few years for Hertz's discovery to be put into practice, as scientists of the time were able to identify application spaces for this newfound ability. Italian inventor Guglielmo Marconi would go on to build transatlantic communication systems, effectively making our world much smaller. At the same time, Serbian-American physicist Nikola Tesla was able to generate and transmit large amounts of power from one spot to another. Tesla essentially laid the groundwork for a whole new field of study: wireless power transmission (WPT).

After the advent of efficient microwave sources, microwave power transmission (MPT), a form of high-frequency WPT, gained popularity due to its reasonably-sized transmit and receive apertures. A variety of MPT prototype systems have been built since

the 1950s using techniques that will be described in this book. Most
of these prototypes over the last 60 years have served to solve some
of the questions surrounding the feasibility of the space solar power
(SSP) concept. Thus, in addition to reviewing power beaming developments, this chapter also discusses many efforts concerning SSP
systems with some details of their performance. In recent decades,
power transmission in the optical regime has also gained substan-
tial traction. Some of those experiments will be covered here as well.
The breadth of experiments and prior research in both microwave
and optical power beaming could each warrant their own separate
books. This text endeavors to summarize some selected and notable
instances.

Figures 2.1–2.3 give a historical timeline of selected WPT experiments from 1880 to the present. Important demonstrations for the
first 100 years are seen chronologically in Figure 2.1. Here WPT's
European origins at the latter part of the 19th century are clearly
noted along with a geographical shift in WPT, namely MPT, research

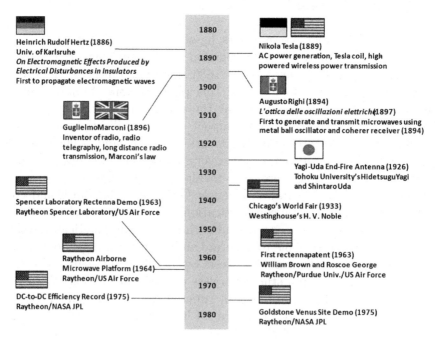

Figure 2.1.   Notable WPT/MPT milestones from 1880 to 1980.

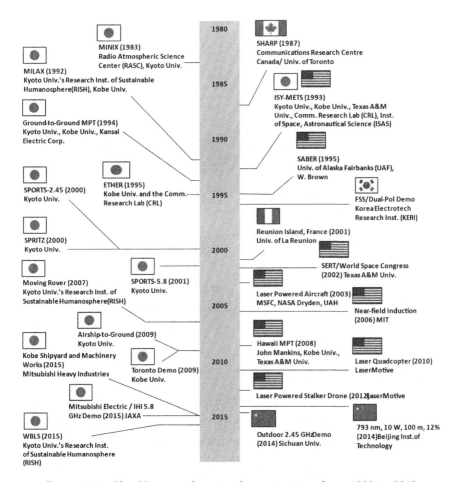

Figure 2.2. Notable power beaming demonstrations from 1980 to 2015.

to the United States after 1930. Early European progress was done primarily by solo researchers or small groups of scientists working within university environments funded by government sponsorship. The later United States contributions came about in cooperative efforts involving corporate, governmental, and academic institutions. This three-pronged collaborative approach led to experimental sophistication, pushing MPT appreciably, with SSP as a driving goal in many cases.

Major WPT demonstrations from 1980 to 2015 are listed chronologically in Figure 2.2. In the early 1980s, Japan became heavily

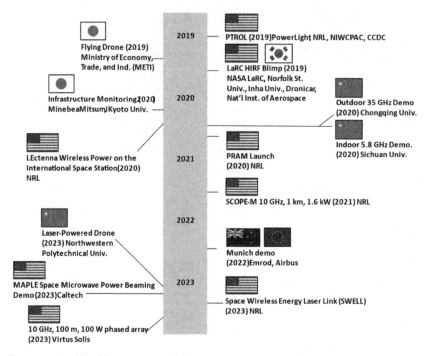

Figure 2.3.   Notable power beaming demonstrations from 2015 to Summer 2023.

involved in MPT research and subsequent demonstrations. In effect, Japan took the lead in promoting MPT to the world. Solo Japanese contributions are charted on the left side of Figure 2.2. Contributions from other countries are listed on the right. Japan has contributed greatly to MPT for more than 40 years with the expressed intent of using space solar as a complement to various other forms of energy generation such as nuclear power. US contributions have also been evident during the same period, albeit less consistently.

Starting in approximately 2014, China began accomplishing significant and noteworthy microwave and laser power beaming demonstrations. Many of these, as well as those resulting from renewed interest in the US and around the world, are shown at the bottom of Figure 2.2 and in Figure 2.3.

The rest of this chapter explores a number of these demonstrations in greater detail. Commercial and consumer-oriented developments are addressed in Chapter 8, whereas the principal focus in this chapter is on research efforts and demonstrations.

By the time this book is published, further advances and demonstrations of power beaming will certainly have occurred and been published and may be included in future editions. Even at the time of this writing, the authors are aware of completed record-breaking demonstrations that have yet to be publicly disclosed.

2.2 Technological Development from Antiquity to 1950

Electromagnetic observations go back to ancient times with such discoveries as amber's electrostatic effects, magnetism, and the compass. The true understanding of the nature of basic electromagnetic principles would have to wait until the latter part of the 1600s and 1700s as higher-level mathematics and the understanding of electromagnetic phenomena began to take shape. The quest to understand the nature of light drove the electromagnetic debate for several hundred years. Scientists such as René Descartes and Isaac Newton put forth the corpuscular theory of light in the 1600s as a means of summarizing previous optical observations. This theory was soon challenged by the wave theory of light championed by researchers such as Dutch physicist Christiaan Huygens in the 1690s and later by British polymath Thomas Young around 1800. The experimental and theoretical work of French physicist Augustin-Jean Fresnel in the 1810s would soon cause widespread acceptance of wave theory as the proper underpinning of electromagnetic behavior.

The 1800s proved to be a period of important foundational discoveries and innovative incremental advances in the areas of electromagnetic device enablers and wireless power transfer. While numerous scientists contributed to the body of knowledge at the time, the fundamental basis for the adaptation of electromagnetic principles into physics was made possible by empirical work conducted by English scientist Michael Faraday in the 1820s and 1830s. Faraday was a creatively gifted experimental researcher who expanded upon previous observations by Danish physicist Hans Christian Ørsted and others through the discovery of numerous interplays between electricity and magnetism. Due to his lack of formal training in higher-level mathematics, Faraday described these interdependences using simple straightforward algebraic relationships which could be understood by a wider audience [1].

It took about 30 more years before Scottish mathematical physicist James Clerk Maxwell provided a higher-level mathematical framework for Faraday's laboratory findings [2]. In 1855, Maxwell published a comprehensive unifying theory on electromagnetism through a set of twenty differential equations described using twenty variables. These equations outlined electromagnetic field relationships based on previous work by scientists such as Faraday, Charles-Augustin de Coulomb, Carl Friedrich Gauss, and André-Marie Ampère. After subsequent revisions by Maxwell and later by both Oliver Heaviside and Heinrich Hertz, employing vector calculus in 1884, the 20 equations evolved into four primary equations which can be described in differential or integral forms. These four equations are famously known as Maxwell's equations. At the same time in 1884, English physicist John Poynting defined the Poynting vector describing the proper orientation of electromagnetic power flow across an area. These mathematical expressions set the stage for accurately understanding wireless power transfer and power beaming.

At the University of Berlin, German physicist Hermann von Helmholtz became interested in proving out in further detail what Maxwell had laid down mathematically. In 1879, Helmholtz suggested to his doctoral student Heinrich Hertz that Hertz's research should center on proving out Maxwell's theories. In 1880, Heinrich Hertz received his Ph.D. from the University of Berlin and stayed for post-doctoral studies, turning his focus to electromagnetic induction. During his time there, Hertz showed greater analytical validity to Maxwell's groundbreaking equations than the accepted "action at a distance" notions held at the time.

In 1883, Hertz became a lecturer at the University of Kiel, and then in 1885 became a full professor at the University of Karlsruhe. Here in 1886, Hertz first noticed the propagation of electromagnetic waves using a Leyden jar and a Riess spiral. The Leyden jar was invented in 1745 by German cleric Ewald George von Kleist and Dutch scientist Pieter van Musschenbroek. The jar could store electricity in large quantities and was credited as the first capacitor. The Reiss spiral, invented by German physicist Peter Reiss in 1873, had a separated pair of spirally wound conductors, each having two metal balls at their respective ends. Placing one conductor above the other formed an induction coil system. When Hertz discharged a Leyden jar into one of the Riess spiral coil's spark gaps, he noticed sparks on the other spiral's coil gap. This validation of wave propagation served as an impetus for Hertz to conduct more complex experiments involving power transmission by radio waves [3, Vol. 6].

One of the devices Hertz used in his experiments was the induction coil. The first induction coil was invented by Irish priest Nicholas Callan in 1836 and a later version was patented and commercialized in 1851 by German inventor Heinrich Ruhmkorff. One of Hertz's most famous experiments from 1888, shown in Figure 2.4(a), used a Rumkorff inductive coil to produce a series of high-voltage pulses that were then introduced across a spark gap centrally located on a dipole antenna. Hertz's dipole had two spherical resonant adjusting balls at either end for tuning and for suppressing electrostatic corona discharge. The oscillating sparks generated at the gap then radiated electromagnetic "Hertzian" waves towards an awaiting loop equipped with its own spark gap. The transmitted energy radiated in the very high frequency (VHF) range near 3 MHz. The setup ultimately generated high-frequency power on one end and detected the same power minus loss on the receiving end. Throughout the 1880s, Hertz created a number of these complete energy transfer systems bringing forth the age of antennas and WPT. These demonstrations served as experimental proof of Maxwell's equations, and the catalysts which propelled the equations to universal acceptance amongst scientists.

Soon after, in the 1890s, Italian physicist Augusto Righi took on the task of investigating adaptations of Hertz's work. Righi along with Indian physicist Jagadish Chandra Bose were the first to generate microwaves. They used a four-metal ball spark oscillator to produce 12 GHz energy and detected them using Hertz's technique of a spark gap dipole antenna. One of Righi's demonstrations shown in Figure 2.4(b) utilized the four-ball oscillator for transmission and a coherer detector for reception. The coherer was a radio signal detection device invented by French physicist Édouard Branly in 1890 that remained popular until 1907. Coherers were eventually replaced by the more sensitive electrolytic and crystal detectors first invented by Karl Braun in 1898. Righi also validated, through a series of experiments, Maxwell's theory that radio waves and light are both electromagnetic waves [7].

(a)

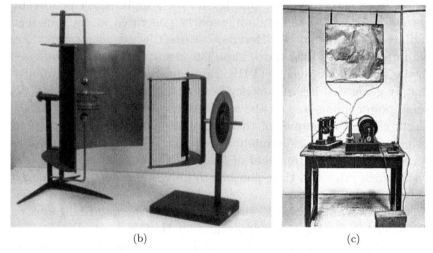

(b)   (c)

Figure 2.4. Spark gap experiments from the 1880s and 1890s: (a) Hertz's 1888 experiment [4], (b) Righi's WPT experiment [5] (CC BY-SA 4.0), and (c) Marconi's first transmitter from 1895 [6, p. 31].

Italian inventor Guglielmo Marconi was intrigued by the electromagnetic propagation work of Hertz and was able to predict its usefulness when applied to the emerging area of wireless telegraphy [8, p. 206]. Marconi visited Righi at his lab for consultations on how to properly adapt Hertz's work to wireless communications. Throughout the 1890s, Marconi developed transmitters capable of sending Morse code over long distances for a variety of experiments.

His first transmitter from 1895 shown in Figure 2.4(c) used an induction coil to power a Righi four-ball spark gap oscillator which in turn generated RF energy that was then sent to a raised copper sheet monopole antenna for transmission. In July 1896, Marconi conducted his first radio transmission demonstration for the British government which led to several follow-on experiments in the English isles. These experiments culminated in the famous 1901 exhibition establishing a link between Poldhu in Cornwall, England, and Clifden in County Galway, Ireland. Marconi was then involved in numerous follow-on radio transmission demonstrations as well as the establishment of commercial services, including transatlantic communication links.

During the time that Marconi was laying the groundwork for wireless radio communication, research scientist and Serbian-American inventor Nikola Tesla was looking at how to transfer higher levels of power effectively. In the 1880s, Tesla looked at conductor-based power transmission. He built several alternating current (AC) power grids and proved their superiority over the then commonly used direct current (DC) systems championed by American entrepreneur Thomas Edison. AC power generation would become the standard for electrical grids throughout the world.

After learning about Hertz's experiments, Tesla then turned some of his focus from conductor-supported power distribution to WPT research [9]. Tesla rightfully recognized that radiated power levels fall off rapidly with increasing distance so the generation of large amounts of source power was necessary for long-distance wireless communications. In his New York City laboratory in 1889, he tried to power a Ruhmkorff induction coil, similar to the one employed by Hertz, with a high-speed alternator to generate oscillating pulses. Unfortunately, the high-frequency current overheated the iron core and melted the insulation between the induction coil's primary and secondary windings. In response to this, Telsa created his own oscillating transformer which employed an air gap instead of insulation material. This transformer could produce high-voltage, low-current, and high-frequency AC energy. It later became known as the Tesla coil and was an integral part of much of his follow on WPT work.

Tesla first attempted to transmit large amounts of power without wires at Colorado Springs, Colorado in 1899. Under a $30,000 grant from Colonel John Jacob Astor, owner of the Waldorf-Astoria Hotel in New York City, Tesla built the huge Tesla coil shown in

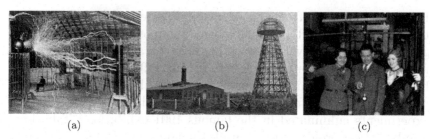

Figure 2.5(a), over which rose a 61-m metallic mast with a 1-m diameter ball positioned at the top. The Tesla coil resonated 300 kW of low-frequency energy at 150 kHz. When Tesla threw the switch, the radio frequency output of the Tesla coil was unleashed into the mast, and 100 MV of RF potential was produced on the sphere, causing man-made lightning discharges, some 135 feet in length. The spectacular experiment was deemed a failure by some since the transmitting structure sent RF energy in all directions creating a situation where the power available at any one location was insignificant.

After the Colorado Springs experiments, Tesla moved back to New York City and managed to obtain financial backing in March 1901 from J.P. Morgan for the construction of a setup like the one in Colorado Springs. The laboratory was to be constructed on about 8 km^2 of land around 100 km east of New York City at Shoreham, in Suffolk County, Long Island. The building plans called for a 47-m wooden tower, called the Wardenclyffe Tower, pictured in Figure 2.5(b), that would support a giant copper electrode 30 m in diameter shaped like a donut at its top. The structure was nearly completed when the financial resources ran dry, forcing a halt to construction. The installation was eventually torn down during World War I by the US Government due to its belief that the structure could be used by the enemy for targeting purposes.

It should be noted that Tesla was decades ahead of his time in his pursuit of transmitting high power from one location for it to be used somewhere else. It wasn't until 1933 that another attempt at high-powered WPT was carried out. This experiment, performed by American researcher H.V. Noble at the Westinghouse Laboratory,

(a)   (b)

Figure 2.6. Yagi-Uda Japanese antenna inventors: (a) Shintaro Uda [15] and (b) Hidetsugu Yagi [16].

consisted of identical transmitting and receiving 100-MHz dipoles separated by about 7 m. No attempts to focus the energy were made, but several hundred watts of power were transferred between the two dipoles. This experiment was demonstrated again to the general public at the Chicago's World Fair of 1933–1934 as seen in Figure 2.5(c).

After taking notice of the successes of Marconi, Tesla, and others, a few researchers in Japan started their own antenna development programs. One of the most famous Japanese contributions to the wireless transmitting realm during this period was the Yagi-Uda end-fire antenna array invented at the University of Tohoku in the 1920s. The arrays pictured in Figures 2.6(a) and 2.6(b) were devised by Shintaro Uda with support from Hidetsugu Yagi. At the time of the discovery, Uda served as an assistant professor to Yagi at the University of Tohoku in the city of Sendai within the Miyagi Prefecture of Japan. The array used a dipole-driven element, a reflective dipole parasitic element, and a collection of parallel parasitic dipole directive elements to send the RF energy along the end-fire direction [13,14]. The novel array received the Japanese patent number 69,115 in 1926, and the US patent number 1,860,123 later in 1932. The Yagi-Uda array has been used heavily in broadcast TV reception as well as some radar applications. It can also be adapted to higher power transmit/receive functions.

The primary reason that WPT received little interest in the first part of the 20th century was that knowledgeable engineers and scientists knew that to achieve efficient point-to-point transmission of power the electromagnetic energy had to be concentrated into a narrow beam, reducing what is referred to today as spillover loss. It was theorized at this time that the only way to obtain such confined energy would be to utilize energy at high frequencies and use radiating elements of relatively large size. The other problem was that the existing sources that created high-frequency energy could output only a few milliwatts of energy, which was generally not enough for a feasible WPT system.

In the late 1930s, two inventions were made that enabled the generation of high power at high frequencies. The first was the velocity-modulated beam tube, first described by German engineer Oskar Heil and his wife Russian physicist Agnesa Arsenjewa-Heil, which was modified into the klystron tube. The second enabler was the improvement of the microwave cavity magnetron by English physicists John Randall and Harry Boot at the University of Birmingham in Great Britain in 1940 and passed to the US during World War II [17]. These developments allowed WPT to transition to higher frequencies for what became known as MPT. The advent of radar, made possible by the introduction of both the klystron and magnetron, helped to bring about significant improvements in both antenna development and microwave generation. The US Government took notice and started proposing programs based on the new capabilities.

MPT possibilities also started to capture the imaginations of science fiction writers. In Isaac Asimov's fictional short story "Reason," published in April of 1941 [18], two astronauts residing at a space station oversee the supplying of energy to other planets via power beams. Asimov, with artistic embellishment, references the possibility of misguided beams and their potentially harmful effects while exploring deep questions about logic, knowledge, and artificial intelligence.

2.3 Modern US Contributions (1950–1980)

Prior to 1950, many of the developments in MPT had come from the efforts of individual researchers often working in isolation. Now that the fundamental technologies existed for conducting MPT, corporate entities started to step forward with the goal of developing complete MPT systems. Special focus was given to realizing efficient performance, both in an overall sense as well as at the subsystem component level. Notable MPT contributions during this 30-year period were achieved largely in the United States, as it had already become a primary center of MPT research 50 years prior.

Successful developments in tube-based microwave energy sourcing motivated the Raytheon Company to propose the Raytheon Airborne Microwave Platform (RAMP) concept in 1959 to the US Department of Defense as a solution to surveillance and communication shortcomings. The proposed platform was a large helicopter positioned above the jet stream at 15 km (about 50,000 ft) where atmospheric winds can be weaker. To fly at this altitude, the helicopter could be powered from Earth by an Amplitron having an output of 400 kW of energy at 3 GHz with an efficiency of over 80%. This high-powered Amplitron was developed in 1960 at Raytheon's Spencer Laboratory in Burlington, MA by American engineer William Brown, who is largely regarded as the principal pioneer of practical MPT [19]. The only capability missing was the ability to convert microwave energy to DC power to drive motors attached to the rotor blades. The US Air Force awarded several contracts to study this rectification problem. One of the studies carried out by American researchers Roscoe George and Elias Sabbagh at Purdue University showed that a semiconductor diode could be used as an effective rectifier [20]. At the same time, Brown at Raytheon carried out research on using a thermionic diode rectifier [21]. With both high-powered sources and efficient rectifiers available, researchers could now explore MPT's usefulness.

The Air Force continued to partner with Brown and Raytheon during the early 1960s in the pursuit of emerging MPT possibilities. In 1963, the first modern MPT system was constructed at Raytheon's Spencer Lab. This system, seen in Figure 2.7(a), used a DC-fed magnetron and reflector to send microwave energy at 3 GHz to a horn antenna located 5.5 m away [22]. The horn was connected to a single close-spaced thermionic diode rectifier placed within the horn's waveguide section. DC-rectified power of 100 W, corresponding to 15% DC-to-DC conversion efficiency, was seen across the diode's terminals. This demo resulted in a US Air Force contract for remotely powering an airborne communications platform.

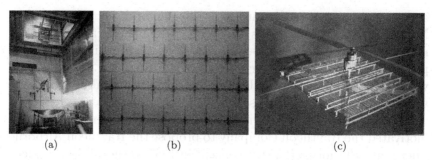

(a)   (b)   (c)

Figure 2.7.   Raytheon/Air Force MPT Experiments: (a) Spencer Laboratory [26], (b) first rectenna array [27], and (c) MPT-powered helicopter [27].

In 1963, Brown and George discovered that many solid-state diode rectifiers, operating in the 2–3 GHz range, could be combined within a waveguide and could output reasonable amounts of DC power. To facilitate the test, they placed a dense array of diodes inside a waveguide that was directly attached to a receiving horn [23]. This setup resulted in low-efficiency numbers. To improve the efficiency, the full-wave rectifiers were then placed outside of the waveguide over a reflecting plane. It was the first demonstration of a rectifying antenna ("rectenna") array in which each half-wave dipole antenna element was assigned its own semiconductor diode. Brown, along with George, received a patent for the rectenna array [24] shown in Figure 2.7(b), and their use of a reflecting plane has been utilized in many of the most efficient rectenna arrays to date.

One of the most well-known examples of MPT was conducted by Brown on July 1, 1964 inside Raytheon's Spencer Laboratory. There, a microwave-powered helicopter, much smaller than the one proposed in RAMP, was flown a few inches off the ground. It was the first heavier-than-air vehicle to be flown using power beaming and was sustained by a 2.45 GHz microwave beam. This experiment was demonstrated again to the mass media on October 28, 1964. The presentation was covered by Walter Cronkite's CBS news program, providing the world with a glimpse of MPT's potential. The helicopter, with a 1.83-m (6-ft) rotor as shown in Figure 2.7(c), flew for 10 h at an altitude of 15 m (50 ft) [25]. Dipole antennas were used to collect the incoming microwave energy, and the DC energy that powered the propeller was obtained using 4,480 semiconductor diodes. A grating of rods was placed in front of the diode plane to impedance match the diodes to free space. Groups of diodes were connected in series to form diode modules, with each module being the equivalent of a half-wave dipole. After the helicopter flight, the US Air Force decided that their objectives were met and elected to discontinue their MPT endeavors. Thus in 1967, Brown began to court German-American engineer Dr. Werner von Braun and his staff at NASA's Marshall Space Flight Center (MSFC) on MPT alternative possibilities in space.

One such application of MPT in space was the idea of SSP using solar power satellites (SPSs). While Asimov imagined accessing remotely generated microwave power, Czechoslovakian-born American scientist Peter Glaser of the Arthur D. Little Company in 1968 described a constellation of SPSs that could move the science fiction concept into reality [28]. SSP was envisioned as an MPT system utilizing solar cells tied to magnetrons for microwave power generation. The idea called for a constellation of satellites to be placed in geosynchronous orbit (about 36,000 km above Earth) to capture the sun's energy, convert it to microwave energy, and then beam this energy to Earth. Terrestrial "farms" of rectenna arrays would then collect the incoming energy and convert it to DC power [29]. The DC power would then be stored, consumed, or perhaps converted back to low-frequency AC for distribution to consumers on the electrical grid. The architectural implementation that would be best suited for space solar remains unsettled even today.

In 1970, with NASA now looking at MPT's potential role in space, MSFC awarded Raytheon a contract to improve the overall DC-to-DC efficiency of MPT systems. This DC-to-DC efficiency includes the conversion from DC to RF in the magnetron at the transmitter, the aperture-to-aperture link transfer efficiency, and the RF-to-DC conversion of the receiving rectenna array. By multiplying these three efficiencies, overall system efficiency can be determined. The MSFC contract led to advances in solid-state rectifying diodes, improving the RF-to-DC conversion significantly. Another technological breakthrough at this time enabling more efficient MPT was the design of the dual-mode horn by Philip Potter at the Jet Propulsion Laboratory (JPL) [30]. This modified horn emanated a Gaussian beam with negligible sidelobes from its circular aperture. The beam was comprised of TE_{11} and TM_{11} in-phase energy with equal beamwidths in both the E- and H-planes. This type of horn would be used in numerous follow-on MPT experiments, allowing for the improvement of aperture transfer efficiencies.

In 1971, W. Brown and P. Glaser, along with members of Northrop Grumman and the photovoltaic company Textron, conducted a 6-month study concluding that SPS was worth pursuing. A letter was then sent to the Director of NASA requesting funding [31]. This led to NASA's Lewis Research Center (LeRC) awarding a small contract to Brown and his Raytheon colleagues for improving the overall efficiency of existing MPT systems.

During the early 1970s, NASA and Raytheon began to shift more focus to space solar. NASA's JPL now became center stage for SSP research under the guidance of American engineer Richard Dickinson. This collaboration focused on the construction of more robust rectenna designs such as the one illustrated in Figure 2.8(a) with its accompanying Figure 2.8(b) equivalent circuit. This type of rectenna was used in the 1975 MPT setup shown in Figure 2.9(a) which produced an overall DC-to-DC conversion efficiency of around 54% [33].

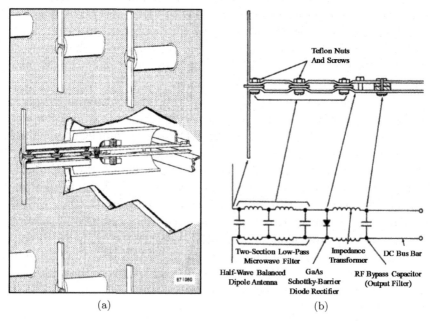

(a)   (b)

Figure 2.8. Raytheon/JPL rectenna design: (a) array implementation [32] and (b) rectenna illustration with equivalent circuit [27].

(a) (b) (c)

Figure 2.9. Raytheon/JPL MPT experiments: (a) Raytheon setup that achieved 54.18% DC-to-DC system efficiency [33], (b) JPL Goldstone facility experiment [29], and (c) RXCV rectenna array [29].

This high efficiency was made possible by each rectenna's being terminated into its own matched resistive load. Individual loading compensated for power density variability on the rectenna array's surface that was created from the horn's Gaussian pattern distribution. The operating frequency was 2.446 GHz, and the rectenna array's output DC power level was 495 W. The end-to-end efficiency was certified by JPL's quality assurance organization and to this day stands as the highest MPT DC-to-DC efficiency. The breakdown of the 54.18% ± 0.94% overall efficiency is 68.87% ± 1.0% for the DC-to-RF conversion and 78.67% ± 1.1% for the RF-to-DC conversion [33]. This demonstration is particularly notable for the high level of detail and transparency given in the reported specifics regarding what was accomplished and is explored further in Chapter 7.

Also in 1975, another important milestone was shown at the Venus Site of JPL's Goldstone Facility. In this demonstration shown in Figure 2.9(b), microwave energy at 2.388 GHz was sent over a 1.54-km (0.957-mile) distance of the Mojave Desert to the awaiting microwave power Reception-Conversion (RXCV) rectenna. This rectenna array, displayed in Figure 2.9(c), was the largest constructed. The 24-m^2 (3.4 m × 7.2 m) RXCV array consisted of 17 subarrays with each subarray having 270 of Figure 2.8(a) rectenna elements arranged in an equilateral triangular grid. It was designed by W. Brown at Raytheon and outputted 30 kW of DC power [34]. This DC output power was the highest recorded. Both the JPL-certified and Goldstone experiments gave NASA confidence regarding MPT viability and its possible use in Glaser's SPS concept.

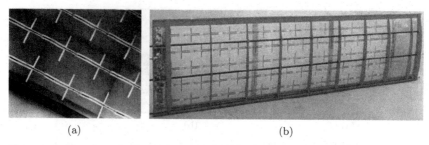

(a)   (b)

Figure 2.10. Linearly polarized 2.45 GHz half-wave dipole rectenna arrays: (a) W.C. Brown's "thin film" [27] and (b) SHARP's wing [27].

Even with the success of Goldstone, LeRC continued to push Brown for improvements in both transmitting antenna arrays and rectennas. In 1980, Brown improved the design of rectenna arrays by introducing linearly polarized thin film-etched rectennas, such as the one seen in Figure 2.10(a), in which the DC bussing is achieved in the plane of the antennas [31]. Before etched rectennas, the DC networks were attached behind the antennas making previous rectenna arrays more complex, bulkier, much heavier, and more costly. Another similar rectenna used a few years later in a Canadian remotely powered plane is pictured in Figure 2.10(b). Many currently designed rectennas are etched.

Between 1977 and 1980, NASA teamed up with the US Department of Energy (DOE) to further evaluate SSP's potential in providing affordable energy to consumers on Earth. The resulting mass of documents, including a nearly 700-page program summary, determined that SSP was feasible and should be pursued in the future [35]. One concept from the study was the use of retro-directivity to keep the microwave energy beam focused on a target receiver that simultaneously sent a pilot beam back to the transmitter. However, the NASA-sponsored program ended in 1980, causing the US to lose its SSP leadership role.

2.4   International Involvement (1980–1995)

After 1980, MPT research and development noticeably shifted from the United States to Japan, with Canada and countries in Europe also making contributions. In 1980, a program to develop a long-endurance, high-altitude platform called the Stationary High-Altitude Relay Program (SHARP) was proposed in Canada [36]. The airplane was designed by the Communications Research Centre (CRC) Canada and built by the University of Toronto Institute for Aerospace Studies (UTIAS). On September 17, 1987, the 1/8-scale prototype SHARP with a 4.5-m wingspan seen in Figure 2.11(a) flew on beamed microwave power for 20 min at an altitude of 150 m. A 2.45 GHz microwave beam was transmitted by a parabolic dish antenna, providing a power density at the airplane of 400 W/m^2. The dual-polarized rectenna array, a portion of which is shown in Figure 2.10(b), received enough microwave energy to generate 150 W of DC power to the electric motor to lift and fly the 4.1 kg airplane.

While demonstrations such as Goldstone and SHARP showed MPT's utility, the automatic beam alignment between the RF transmitter and the rectenna array remained an issue. In 1987, researchers at Kyoto University, Kobe University, and Mitsubishi Electric Corporation developed the first retrodirective transmitter for MPT [37]. The seven-dipole 90-W transmitter could intentionally send energy in the direction of incoming pilot signals. It used a phase conjugation circuit to compare two asymmetric pilot frequencies to remove the $2n\pi$ ambiguity in the phase comparison. Another retrodirective transmitter, developed later in 1996 by Kyoto University and Nissan Motor Co. Ltd., used pilot signals that were one-third the transmitting frequency. This simplified the system design at the expense of a larger antenna for receiving the pilot signals. The Nissan system used a pilot signal of 815 MHz to keep 80 W of energy at 2.45 GHz on target [37].

Another example of driving a small aircraft using microwave power was the 1992 MIcrowave Lifted Airplane eXperiment (MILAX) demonstration conducted by Kyoto University's Research Institute of Sustainable Humanosphere (RISH) and Kobe University in Japan in 1992. The experiment used an electronically scanned phased array to focus a 2.411 GHz microwave beam on the aircraft shown in Figure 2.11(b). Two charge-coupled device (CCD) cameras recognized the airplane's outline, revealing the location to a computer which scanned the array to the appropriate location. The transmitting array was located on a sports utility vehicle which was also in motion during the tests. MILAX received nationwide media coverage in Japan and endeared SSP favorably to the Japanese public [38].

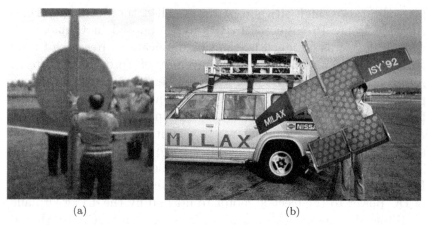

(a)   (b)

Figure 2.11. MPT applied to unmanned remotely powered model aircraft: (a) SHARP [27] and (b) MILAX [27].

Japan's status of being a large energy consumer with few natural energy resources likely facilitated funding for programs exploring SSP as an alternative renewable energy source. A couple of these programs have involved in-space experiments. The first of Japan's in-space experiments was the Microwave Ionosphere Nonlinear Interaction eXperiment (MINIX) conducted by Matsumoto and colleagues at the Radio Atmospheric Science Center (RASC), Kyoto University in 1983. MINIX focused on how the plasma wave dynamic spectrum is created and morphs as high-powered microwave energy is transmitted into ionospheric plasma [39,40]. The RF energy at 2.45 GHz was sent from the rocket seen in Figure 2.12(a) to the elevated detachable daughter portion.

The second in-space experiment was the International Space Year–Microwave Energy Transmission in Space (ISY-METS) conducted by Kyoto University, Kobe University, Texas A&M University, the Communication Research Laboratory (CRL), and the Institute of Space and Astronautical Science (ISAS). ISY-METS was launched from the Kagoshima Space Center in February 1993 to study microwave-generated plasma waves. It also represented the first example of MPT in space [42]. At an altitude of 270 km, the rocket is divided into two units. Approximately 800 W of microwave energy was transmitted from the S-520-16 parent rocket. An electronically steered phased array was used to focus this microwave energy into a

(a)   (b)   (c)

Figure 2.12. Japanese plasma-wave rocket experiments: (a) MINIX [27], (b) ISY-METS folded for launch [41], and (c) ISY-METS expanded for demonstration [41].

narrow beam to a smaller detached rocket section equipped with the receiving arrays seen in Figures 2.12(b) and 2.12(c). A narrow beam was necessary for accurately calculating the angle between the magnetic field and the excited plasma wave caused by the high-power microwave beam. This angle was used to get an accurate assessment of the excited plasma wave's magnitude. The plasma wave observations were conducted in the HF range and indicated that natural plasma waves are enhanced by microwave energy nonlinear interactions within the ionospheric plasma environment. ISY-METS gave researchers insight into the effects that high-powered microwave transmission from space may have on broadcasting and communications systems here on Earth.

Another Japanese MPT program, conducted in 1994, was the aptly named Ground-to-Ground MPT program [43] seen in Figure 2.13(a). The system's development was a joint design effort between Kyoto University, Kobe University, and Kansai Electric Corporation. Kansai's interests were rooted in how MPT could be adapted to a power grid. Here, 5 kW of magnetron-generated microwave energy at 2.45 GHz was fed to a 3-m parabolic antenna. The parabola then transmitted the power to a 3.2 m × 3.54 m rectenna array (2,304 elements) located 42 m away. The beam efficiency was about 74%.

In 1995, Japanese researcher Nobuyuki Kaya of Kobe University and the CRL led the project that resulted in the construction of

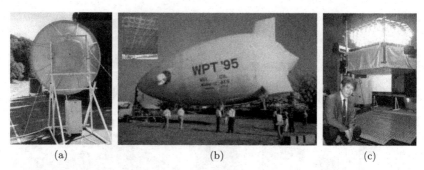

(a)　　　　　　　　　　(b)　　　　　　　　　　(c)

Figure 2.13. Japanese MPT experiments in the 1990s: (a) Ground-to-Ground MPT [27], (b) ETHER airship [27], and (c) Hiroshi Matsumoto and the SPRITZ system [27].

the airship seen in Figure 2.13(b). This airship, constructed for the ETHER (Energy Transmission toward High-altitude long endurance airship ExpeRiment) program, received 10 kW of microwave power at 2.45 GHz [44]. The energy was transmitted from a parabolic antenna/dual magnetron system on the ground to the blimp's 3 m × 3 m (1,200 elements) rectenna array allowing it to fly at a height of 50 m for 4 min. This experiment was exhibited in 1995 at the international WPT conference (WPT'95) in Kobe, Japan. Such floatable systems could be used for long-term loitering allowing for such things as temporary communication links and reconnaissance.

In the latter part of the 1990s, Kyoto University developed the fully integrated Solar Power RadIo Transmitter (SPRITZ) under the guidance of Hiroshi Matsumoto. This system, seen in Figure 2.13(c), was exhibited at several places including the 2002 World Space Congress in Houston, TX. The DC-to-DC system used 133 75-W halogen lamps to provide illumination for the solar cells. These solar cells output around 166 W at 15% efficiency. The microwave transmitter had 3 bits of phase control and operated at 5.77 GHz while outputting 25 W. The rectenna array had 1,848 individual rectennas providing the DC power for illuminating LEDs [45].

Raytheon's 1964 helicopter experiment inspired researchers at the University of Alaska Fairbanks (UAF), along with Brown, to design and build a similar version pictured in Figure 2.14(a). The program known as the Semi-Autonomous BEam Rider (SABER) used a JPL/Raytheon-designed slotted antenna array to transmit 2.45 GHz energy 3 m to the helicopter's rectenna array to drive a

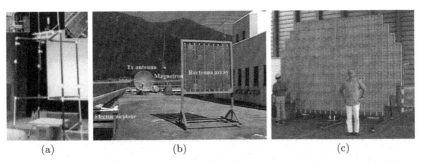

Figure 2.14. MPT programs developed in the 1990s outside Japan: (a) SABER [27], (b) KERI MPT setup [27], and (c) Reunion Island rectenna [27].

DC motor [46]. The motor caused the helicopter's propeller to rotate rapidly as the aircraft ascended two guideposts. This experiment was exhibited at WPT'95 in Kobe, Japan.

During the 1995–2000 period, South Korea also instituted its own power beaming programs. Dr. K.H. Kim of the Korea Electrotechnology Research Institute (KERI) demonstrated the power beaming arrangement shown in Figure 2.14(b). The transmitter used a magnetron-driven parabolic dish antenna to beam microwave energy to a rectenna array located 50 m away. The rectenna was composed of 2,016 dual-polarized microstrip patch antennas arrayed on a plane with a frequency-selective surface (FSS) directly in front of the radiators. The transmitter power and receiver DC output were 2.3 kW and 1.02 kW, respectively, for a total power beaming efficiency of 44% [27, p. 237].

At the May 2001 WPT conference held in Reunion Island, France, researchers from the University of La Reunion unveiled their industrial prototype rectenna array pictured in Figure 2.14(c). This array, first proposed in 1994, was designed as a means for one day supplying isolated villages such as Grand-Bassin on Reunion Island with remotely derived energy [47]. For the demo, 800 W of microwave power was transmitted to the rectenna array yielding 65 W of DC rectified output power, corresponding to an efficiency of 5%. While the rectifying efficiency was poor, the rectenna array did succeed in turning on three low-consumption light bulbs. Numerous design improvements were noted by the La Reunion team for improving the RF-to-DC conversion efficiency. The conference island location was chosen to reveal to researchers the challenges associated with delivering SSP-derived power to remote locations like Grand-Bassin without connective infrastructure, such as pipes or power lines.

2.5   Renewed WPT Focus in the USA (1995–2001)

From 1995 to 1997, NASA under guidance from John Mankins at the Office of Advanced Concepts and Technology conducted its "Fresh Look" study on the feasibility of SSP systems to deliver energy into terrestrial power grids. Enabling technologies, such as power beaming, and system concepts were identified. The study focused on several SSP architecture types: a sun-synchronous low Earth orbit (LEO) constellation, a middle Earth orbit (MEO) multiple-inclination constellation, and different stand-alone geostationary Earth orbit (GEO) SPSs. The study also examined the market factors that might make SSP economically viable.

In 1999, NASA started its Space Solar Power Exploratory Research & Technology (SERT) program as a follow-on to the "Fresh Look" study. SERT meetings were held at the University of Alabama in Huntsville (UAH) to bring the community of researchers together. The program broadened the scientific community's involvement and generated successful demonstrations on a variety of system-level components. Large system SPS concepts such as the integrated symmetrical concentrator (ISC) SPS, affectionately referred to as "clamshells," were also drawn up [48]. An artistic rendering of "clamshells" is presented in Figure 2.15(a). The scheme called for two opposing lightweight arrays of solar-sail concentrator mirrors, contained within space frame structures, to illuminate photovoltaic arrays positioned on a centrally located transmitting body. The mirrors were to always face the Sun with very little, if any, shadowing. The transmitting body would then generate microwave energy and beam it to Earth.

SERT pushed further into SSP technologies and funded advanced component development. One such advance was the integration of circular polarization (CP) into rectenna design, culminating in Figure 2.15(b) rectenna array introduced by Texas A&M University researchers in 2000. This array was composed of 18 parallel DC-combined columns with each column having four DC-series-connected 1×4 element CP rectenna arrays. Each of these 1×4 CP arrays accepted relatively low CP power densities at 5.8 GHz and

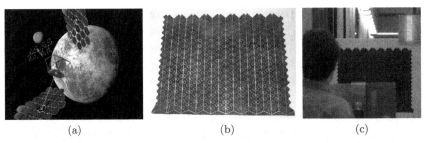

Figure 2.15. SERT MPT systems: (a) "Clamshells" SPS concept [50], (b) Texas A&M University CP rectenna array [27], and (c) CP demonstration [27].

output rectified DC at 82% efficiency [49]. This high efficiency was made possible by flip-chip Schottky diodes with low parasitic losses. The interrogating 8×8 element CP patch array held in Figure 2.15(c) was focused on the CP rectenna array's center, resulting in the illumination of 72 GaN diodes. CP was chosen because of atmospheric depolarization effects such as Faraday rotation that could be an issue for future SSP propagation links. Linearly polarized (LP) systems used for SSP would experience significant degradations in system efficiency due to polarization mismatch. The energy at 5.8 GHz propagates through the atmosphere with relatively low loss, and the receiving and transmitting antennas have smaller footprints than those at the traditional SSP frequency of choice 2.45 GHz. The individual CP elements of Figure 2.14(c) array were placed approximately $\lambda_0/4$ over a ground plane, providing gains 3 dB higher than single CP patches without it. These higher gains reduced the number of rectifying diodes needed within the 2D array by at least a factor of two. For very large rectenna arrays that might employ billions of individual rectenna elements, this reduction could result in significant cost savings. A joint demonstration of this system, orchestrated by Texas A&M University and NASA's Johnson Space Center (JSC) researchers at the 2002 World Space Congress in Houston, TX, was witnessed by media, international officials, and 10,000 members of the public.

During this same era, optical power beaming began moving from component and concept development to demonstration. Spurred by concepts, analyses, and designs in the 1980s for the use of lasers (1) to launch rockets [51], (2) to power satellites [52], and (3) a range of other power beaming in space applications [53], researchers in the USA began hardware investigations for laser power beaming. Olsen *et al.* developed gallium arsenide photovoltaic cells to match the bandgap of 800–840 nm diode lasers [54]. They measured PV efficiencies as high as 53% at 0.4 W/cm^2 light intensity. By 2002, PV efficiency greater than 50% was measured by van Riesen *et al.* at much higher incident light intensities, up to 6.5 W/cm^2, and still achieving 44% efficiency at 42 W/cm^2 [55]. Kare, Mitlitsky, and Weisberg from Lawrence Livermore National Laboratory performed a comprehensive end-to-end hardware demonstration of optical power beaming. Their 1999 paper details their effort to beam 850 nm infrared light over 10 m [56]. Though "operating constraints and unexpected effects" affected the laser power conversion efficiency and kept them from realizing high end-to-end efficiency, their effort is notable for its rigor. It also showed that expectations for achievable end-to-end efficiency had risen to 13%, more than an order of magnitude higher than the 1% anticipated to be realizable several decades prior [57, p. 427].

2.6 New Millennium Brings New Demonstrations of Power Beaming (2001–2010)

While the last century laid the technological groundwork for power beaming, the new millennium has featured demonstrations and concepts of increasing sophistication that build on previous efforts. In 2001, the Japanese Aerospace Exploration Agency (JAXA) of Japan announced plans for an SPS that could beam 1 GW of RF power at 5.8 GHz [58]. In 2002, the European Network on Solar Power from Space was created to evaluate the possibilities of SPSs and to support the European Space Agency's (ESA) activities in that arena [59]. In 2004, EADS Astrium and the European Space Agency revisited the idea of SPS systems powering remote bases on the Moon and Mars using microwave or laser beaming [60]. With the new century also came new terminology, as SSP is often referred to as space-based solar power (SBSP) or simply as space solar.

Early in the new millennium, Kyoto University unveiled two MPT systems known as SPORTS-2.45 and SPORTS-5.8. Semiconductor technology has been unable to provide high-efficiency beamforming at the power levels required by MPT. The Space POwer

Radio Transmission System (SPORTS) demonstrations showed that phased-controlled magnetrons (PCMs), an idea originally from W. Brown, could circumvent this problem by exhibiting frequency stability using injection locking and phase-locked loop (PLL) feedback [61]. Additionally, the PCMs were purposely limited to a couple hundred watts, eliminating the need for power division between the PCMs and the radiating elements. Certain SPORTS array configurations allowed for each array element to have its own PCM which decreased system losses and the amount of heat generated. SPORTS-2.45 was a full-terrestrial SBSP system equipped with an outdoor solar cell panel producing 8.4-kW DC. This DC power then powered 12 PCMs combining to output 4 kW of 2.45-GHz energy. The RF power was then inputted into a 12-element horn or into the 12×8 dipole retro-directive (400 MHz pilot) electronically steered phased array (2 kW at 2.45 GHz) seen in Figure 2.16(a). The RF power was then transmitted to a planar 2,692 element Yagi-Uda rectenna array for conversion back to DC.

SPORTS-5.8 used magnetrons developed by Matsushita Co. to drive the 144 (12×12) microstrip antennas shown in Figure 2.16(b) along with their 144 four-bit phase shifters to transfer 7 W of power at 5.77 GHz to the rectenna array arrangements. Unlike SPORTS-2.45, SPORTS-5.8 used a collection of rectenna panels that could be formed into different geometries [62].

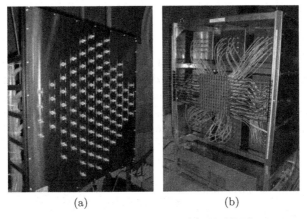

(a)   (b)

Figure 2.16.   SPORTS PCM-based systems: (a) SPORTS-2.45 transmitter [27] *and (b) SPORTS-5.8 transmitter* [27].

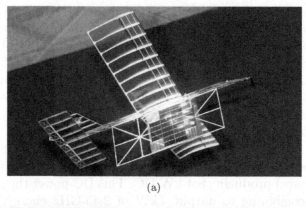

(a)

(b)   (c)

Figure 2.17. Recent US WPT systems: (a) NASA laser-powered aircraft [64], (b) MIT near-field [65], and (c) Hawaii MPT experiment.

Two wireless power experiments of note that fall above and below the microwave region are NASA's 2003 first-ever laser-powered aircraft and MIT's 2006 resonant induction demonstration. In September 2003, NASA's MSFC, NASA's Dryden Flight Research Center, and UAH successfully demonstrated the first-ever small-scale aircraft to be powered solely from power delivered by a ground-based laser [63]. The 1.5-m wingspan plane seen in Figure 2.17(a), weighing about 312 g (just over 11 ounces) was flown by an engineer who manually focused a laser beam on an infrared-sensitive photovoltaic cell panel hanging from the plane's underside. The energy received was enough to power a DC motor which rotated a propeller on the nose of the craft, allowing the plane to fly in circles inside a large building. Optical power beaming also has the potential for powering space probes using space-based laser systems, as there are no detrimental atmospheric attenuation effects in such environments.

The 2006 MIT WPT experiment, carried out by physicist Croatian-American physicist Marin Soljacic and his colleagues, involved resonant inductive near-field coupling of power from a primary AC-excited coil to a secondary coil located two meters away. A 60-W light bulb attached to the secondary coil seen in Figure 2.17(b) was successfully lit. The MIT researchers quantified the power transfer efficiency at 15%, and the measurable power levels at 10 MHz were 14 times higher than the safety standards. Regardless, MIT researchers correctly saw this induction work as a precursor to the non-contact charging of personal electronic devices [66], albeit over relatively short distances. The explosion in short-range wirelessly charged devices based on Qi [67] and other standards [68] has validated their prescience.

Another experiment that garnered attention was the Hawaii MPT experiment carried out in 2008 by American physicist John Mankins and Professor Nobuyuki Kaya of Kobe University in conjunction with researchers at Texas A&M University. The Hawaii demo was sponsored and documented by Discovery Communications for the program Project Earth [69]. The experiment transmitted power over a 148-km distance from the Maui-based array, seen in Figure 2.17(c), to the big island of Hawaii [70], but the amount of power received at the big island was less than 1/1,000th of 1% of the power transmitted from Maui [71]. This very low efficiency was expected since the transmitting and receiving arrays were far too small to establish appreciable beam collection efficiency over 148 km at the operating wavelength.

Over the last two decades, both the Kobe University and the Kyoto University have been involved in the development of microwave-powered rovers for a variety of applications. A Japanese consortium including the Research Institute for Sustainable Humanosphere (RISH) at Kyoto University, the Institute of Space and Astronautical Science (ISAS), and the former Institute for Unmanned Space Experiment Free-flyer (USEF) (now Japan Space Systems) has been successful in building working microwave powered rovers as well as the transmitting arrays for moving them. Some of these transmitting arrays are seen in Figure 2.18. Two

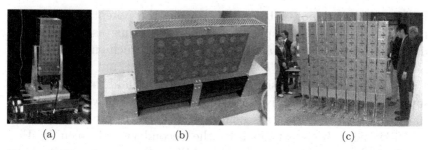

(a)   (b)   (c)

Figure 2.18. Japanese Rover MPT Experiment Transmitting Arrays: (a) 5.8 GHz 4×8 active element gimballed transmitting array with 120 W 5.8 GHz CW [75], (b) 5.8 GHz 8 × 4 element electronically steered transmitter [75], and (c) 2.45 GHz 72-element transmitting array for the rover demonstration at the SPS 2009 Toronto conference, photo credit Paul Jaffe.

Kyoto University transmitters that have been used several times with different rovers are seen in Figures 2.18(a) and 2.18(b) [72,73]. These arrays both operated at 5.8 GHz with one offering a single-axis gimballed option with 120 W CW, and the other implementing electronic steering for maintaining rover illumination. Figure 2.18(c) shows a Kobe University retrodirective phased array that operated at 2.45 GHz. This array was used in a public demonstration to power a rover at the SPS 2009 conference at the Ontario Science Center (OSC) in Toronto, Canada [74, p. 60].

Some of the Kyoto University rover designs are presented in Figure 2.19. The rovers in Figures 2.19(a) and 2.19(b) were illuminated by a 700-W CW transmitter operating at 2.45 GHz [76]. Non-ideal tracking alignments still delivered between 5 and 35 W to the motors which was enough to move the vehicles. In 2007, a rover operated at 5.8 GHz, and was driven by the transmitting arrays seen in Figures 2.18(a) and 2.18(b) [77]. Its rectenna array was subdivided into three series-connected concentrically shaped partitions. These partitions properly accommodated the power density profiles that illuminated the rectenna array's surface. The rectenna elements within each section were wired in parallel to their respective resistive loadings. Powering rovers remotely could be useful for applications such as the assessment of contaminated areas where humans can't safely enter.

In 2009, Kyoto University fielded another experiment using the proven PCM technology. Unlike ETHER, here power was beamed

(a)   (b)

Figure 2.19.   *Japanese rovers: (a) 2.45 GHz rectenna arrays on all four sides* [75] *and (b) single-sided 42-element 2.45 GHz rectenna array* [75].

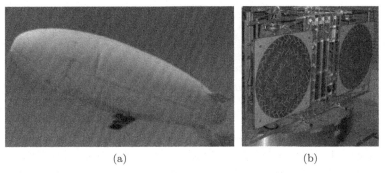

(a)   (b)

Figure 2.20.   *Airship-to-ground MPT experiment: (a) blimp* [27] *and (b) radial slot array* [27].

from the airship to the ground, representative of a small-scale SSP system. Two 110-W PCMs generated 2.46-GHz microwave power which was then fed individually to two radial slot antennas located on the underside of a blimp as seen in Figure 2.20(a). Each of the 72-cm diameter slot antennas pictured in Figure 2.20(b) had a gain and aperture efficiency of 22.7 dB and 54.6%, respectively. The system employed retrodirective tracking at 5.8 GHz, making the pilot frequency higher than the MPT frequency [41]. This differed from previous designs where the retrodirective frequency was lower, and thus the pilot antenna size could be made smaller.

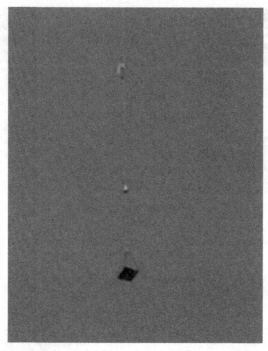

Figure 2.21.   LaserMotive's tether climber ascends a 1-km cable suspended from a helicopter.

Advances in optical power beaming were also evident, spurred in large part by NASA's "Space Elevator Power Beaming Challenge Games" [78]. To support technology development geared toward clarifying the feasibility of space elevators [79], NASA's Centennial Challenges program and the Spaceward Foundation hosted the competition, which saw entrants employing a diverse range of different approaches, including microwave power beaming, optical power beaming, and reflected sunlight [80]. Three teams made it to the final competition in 2009: the University of Saskatchewan, the Kansas City Space Pirates, and LaserMotive. LaserMotive ultimately took first place, winning $900,000 US [81]. The climber is shown ascending the cable in Figure 2.21.

This demonstration was also notable in that is showed the establishment of a laser power beaming link over a distance of 1 km, as well as a link that delivered at least one kilowatt, though these conditions were not met simultaneously.

The Space Elevator Challenge spurred further activity in power beaming development via subsequent activities from members of the teams [82], including the formation of PowerLight Technologies, which has gone on to further mature laser power beaming [83].

2.7 Selected Global Activities from 2010 to 2023

Since around 2010, interest in both space solar and power beaming technologies has spread across the globe. Numerous nations, along with the historical principal contributors, have outlined futuristic political goals and cross-border collaborations for bringing such systems online. For instance, the Indian Space Research Organization and the US's National Space Society launched the Kalam-NSS initiative in 2010 to lay the groundwork for a SBSP program between the two countries [84]. Similarly, in 2012 during a visit by former Indian President Dr. APJ Abdul Kalam, China proposed that India and China work to jointly develop an SPS [85]. These types of cooperative agreements have been on the rise over the last decade. During this time, there has also been an increase in the number of private companies conducting experiments that attempt to move the WPT idea into practice. Notable among these are the efforts of the New Zealand company Emrod, which in September 2022 performed a demonstration in Munich, Germany in partnership with Airbus, as described in press releases from both companies and elsewhere. Other major power beaming demonstrations over the last decade or so have come from four countries, including Japan, South Korea, China, and the United States. Some of these experiments are described below.

2.7.1 *Japan*

Japan has continued to lead the world in power beaming and space solar both technically and politically. In 2009, the Japanese Ministry of Economy, Trade, and Industry (METI) initiated an SPS research committee chaired by Naoki Shinohara of Kyoto University. Since then, the committee has fostered working relationships between Japanese government agencies, universities, and industrial partners for the development of SPS and power beaming systems.

(a)   (b)

Figure 2.22.   METI/MHI 2015 experiment: (a) PCM transmit array [27] and (b) rectenna array [27].

These collaborations have resulted in several successful demonstrations of some of the most sophisticated MPT systems to date. For more than a decade, Mitsubishi Heavy Industries, Ltd. (MHI) has teamed up with its university partners to push the state-of-the-art in MPT research. On February 24, 2015, METI and MHI announced that they had set an MPT record in Japan by transmitting 10 kW of 2.45-GHz RF power over a 500-m distance [86]. The test was conducted on a pier at the Kobe Shipyard and Machinery Works in Hyogo Prefecture. The magnetron and rectenna arrays used in the test are shown in Figures 2.22(a) and 2.22(b), respectively. Both are 8 m × 8 m. The power received was used to light up the MHI logo as seen in Figure 2.22(b).

An *IEEE Spectrum* magazine article published in 2014, written by JAXA Professor Emeritus Susumu Sasaki, described JAXA's intent to place a 1-GW commercial SBSP system in space by the 2030s [87]. The 1-GW system would provide the same level of power as a typical nuclear power plant. In the article, JAXA provides the two concepts illustrated in Figure 2.23. The first, simpler design has a huge square panel of photovoltaic elements on its topside and transmission antennas underneath. Here, the sunlight that hits directly varies greatly due to the fixed solar panel array's orbital mechanics. The second concept, which operates functionally like NASA's clamshell-style SPS, differs in that it has free-floating mirror panels that focus light

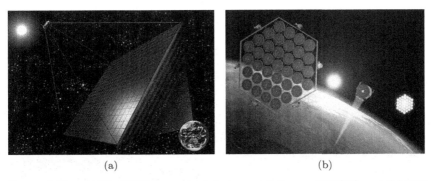

(a) (b)

Figure 2.23. JAXA SBSP conceptual designs: (a) Square SPS [88] *and (b) SPS with formation flying* [87].

to solar panels located on the transmission unit. The three objects would invoke formation flying to align with the sun and each other to increase the amount of power collected and avoid any shadowing.

On March 8, 2015, the METI SPS team successfully carried out an advanced horizontal ground microwave power beaming experiment at the Mitsubishi Electric Facility in Hyogo Prefecture. The novel phased array pictured in Figure 2.24(a) was designed to operate at 5.8 GHz, took 6 years to complete, and was equipped with 304 state-of-the-art GaN amplifiers [89]. Its four phased array panels combined to transmit 1.8 kW of 5.8-GHz microwave energy over a 55-m distance to the rectenna array shown in Figure 2.24(b). The phased array was composed of four 2.5 cm × 60 cm × 60 cm panels, each weighing 16.1 kg. Each individual panel was composed of 76 2 × 2 subarrays arranged along a 9 × 9 rectangular lattice. Five subarrays are noticeably absent in each of the panels' centers to accommodate their corresponding five-element 2.45 GHz retrodirective direction-finding pilot arrays [90–92]. Each panel was also equipped with 76 GaN class-F amplifiers, whose average amplifier efficiency was 60.3%. Each individual amplifier was connected to a host 2 × 2 antenna element subarray. The total DC-to-RF efficiency was 35.1%.

The 2.6 m × 2.3 m rectenna array developed for the field experiment used 36 rectenna modules with a centrally located retrodirective pilot array. Schottky diode rectifiers contained within the rectenna had RF-to-DC conversion efficiencies varying from 59% to 62%. The total receive efficiency, including the 96% efficiency of the power distribution unit, was 42%. The output rectified power was

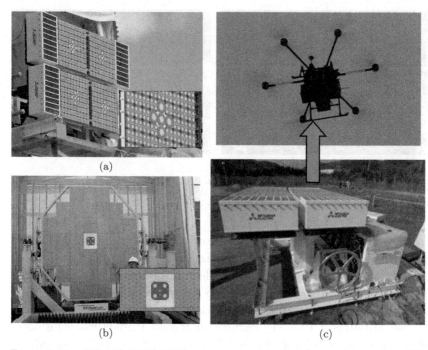

Figure 2.24.   METI/JAXA 2015 experiments: (a) phased array transmitter [27], *(b) rectenna array* [27], *and (c) 2019 METI demonstration of a flying drone* [27].

about 335 W DC. Prior to the outdoor experiment, exhaustive laboratory testing was done in which the four phased array panels were slightly misaligned orthogonally from their ideal planar configuration. The retrodirective subsystem was able to maintain the beam pointing at the rectenna. This analysis is directly applicable to an SPS which may experience subtle moment-generated deformations in space. The R&D project was conducted by Japan Space Systems. The phased array and the rectenna array were developed by Mitsubishi Electric Corp. and IHI Aerospace Co. Ltd., respectively. JAXA played a role as well in overseeing the experiment.

In May 2019, another METI experiment successfully beamed power to the flying drone shown in Figure 2.24(c) [63]. This experiment reused Figure 2.24(a) phased array for transmitting power to a newly developed small and lightweight rectenna array located on the underside of the drone. This new rectenna array, equipped with 17 circular microstrip antennas, measured 200 mm × 186 mm and was used to add additional DC supply power to that contributed by an onboard battery [93]. The phased array transmitted power over a 10-m distance to produce a microwave power density of 4 kW/m^2 at the rectenna array and subsequently a DC output power of 60 W at the rectenna array's output. At a transmit distance of 30 m, 42 W of DC power was produced by the rectenna array. Although this received power was insufficient to fly the drone without a battery, the beamed power was able to extend the battery life and therefore the drone's flight time. Based on this and other prior successes, the METI SPS committee in 2019 began a 5-year development period for the creation of a planar sandwich structure that marries a microwave-phased array to a panel of solar cells.

In other related work, Tsukuba University developed a 303-GHz rectenna in 2018 [94] using the MACOM MA4E1317 GaAs Schottky diode shown in Figure 2.25. The rectenna was able to output 17.1 mW of DC power at an RF-to-DC conversion efficiency of 2.17%. A custom GaN Schottky diode is currently in development with the potential to increase the efficiency in sub-THz rectennas [95]. Power beaming in this frequency range could be suitable for future exoatmospheric applications where propagation losses do not play a significant factor, though the component efficiencies do not yet compete with those for optical power beaming.

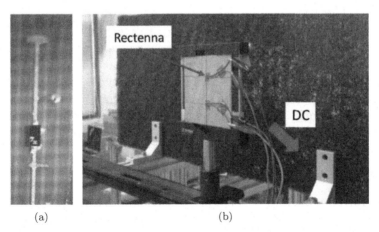

(a)   (b)

Figure 2.25. (a) 303 GHz rectenna with GaAs Schottky diode, notch filter, and antenna [27] and (b) experiment measuring the efficiency of rectennas using a 303-GHz gyrotron [27].

(a) (b)

Figure 2.26. Tunnel Infrastructure deterioration monitoring: (a) interrogating vehicle and (b) identifying sensor.

Over the course of five days in October 2020, MinebeaMitsumi, Inc. and the Kyoto University conducted an infrastructure monitoring experiment inside the evacuation tunnel of the Takiba Jizo Tunnel, Miyazu City, Kyoto Prefecture [96]. The array seen in Figure 2.26(a) sent 10 W of transit power to the batteryless bolt axial force sensor seen in Figure 2.26(b) while the vehicle was in motion. The 10 W of directed microwave energy was enough to turn the sensor on and have it return an identifying waveform. This link operated much like passive RFID does but at higher power levels to cover the distance. The goal is to put passive sensors of this type on sections of tunnels, bridges, sewers, and other types of infrastructure to track their deterioration movements over time using signal processing techniques. A drive-by monitoring system of this type would provide a quick way to check the health of such structures.

2.7.2 *South Korea*

In 2019, Korean and US researchers collaborated to beam power at 10 GHz to the 2.2 m long by 1.1 m diameter helium-filled 1-kg airship seen in Figure 2.27(a) [97]. The floating craft used three DC-driven small propellers to maneuver at 7 mph within NASA

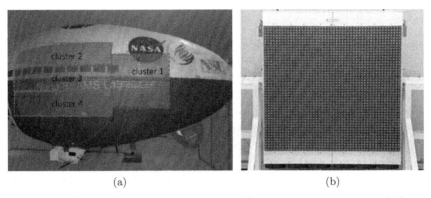

(a)   (b)

Figure 2.27. Korean X-band contributions: (a) Joint US–Korean airship [27] and (b) RFcore X-band transmitter [27].

Langley Research Center's High Intensity-Radiated Fields (HIRF) chamber. Two propellers were used for forward/backward movements while another propeller shifted the craft laterally. The three DC-driven propellers collectively needed 7 V and 8 W to operate properly. This DC power was output by 16 etched flexible-circuit-membrane rectenna array sheets, as seen in the four clusters in Figure 2.27(a), cascaded on either side of the craft, adding to 32 in total. Each sheet contained 90 linearly polarized half-wavelength dipole rectenna elements subdivided into 10 columns with nine rectennas in each column. The use of LP rectennas required proper orthogonal alignment of the interrogating LP X-band horn antenna for maximum RF reception. These rectenna sheets each measured 18 cm × 19 cm and weighed 3 g. This minuscule weight was made possible by the ultra-thin 20-μm polyimide used as the supporting substrate material. A long-term goal of this research is to explore the integration of rectenna technology into notional heavy-lift airships, which might be able to transport large amounts of material with a dramatically reduced fuel burden.

Another Korean contribution to the MPT arena is the X-band high-power microwave transmit array pictured in Figure 2.27(b). This array was manufactured by the Korean company RFCore [98]. The design uses a 2,304-element GaN MMIC-based phased array with 1.5 kW transmit power. Such an array could be used to drive rectenna arrays such as the ones in Figure 2.27(a) airship.

2.7.3 *China*

Since 2010, China has become increasingly active in power beaming. Vice President Z. Pang of the China Academy of Space Technology (CAST) made a salient statement to the media in 2019 that "China is expected to become the first country to build a SSP station with practical value" [99]. To facilitate this goal, the Chinese SPS Promotion Committee was founded on January 17, 2018. This committee is a voluntary non-government think tank composed of professors and engineers from research institutions interested in SPS. On December 6, 2018, it was announced that an SPS experiment base was being established in Chongqing. Its main stated purpose will be to demonstrate power beaming technology and a complete SPS system based on an airship. The experiment's base footprint was proposed to cover an area of 130,000 m^2 and cost \$30 million to construct. Future goals include the facilitation of major scientific and technological-related infrastructure through various programs.

Chinese researchers have had several successful indoor and outdoor power beaming demonstrations since 2010. In 2014, Professor C. Liu's group at Sichuan University, Chengdu, transmitted microwave energy over a 4.5-m distance to the 2.45-GHz rectenna array seen to the left in Figure 2.28(a). The rectenna output 7.1 W of DC power and achieved an overall efficiency of 14.2% [27, p. 240]. In 2016, Sichuan University proposed a "subarray decomposition" rectenna technology, akin to JPL's 1974 individual loading technique and the Japanese 2007 rover rectenna subarray method. In experimental results at 5.8 GHz, they verified that the technique could improve the overall rectification efficiencies by more than 10% [100]. In 2020, Prof. X. Chen's research group, also at Sichuan University, designed the 5.8-GHz microwave power beaming system pictured in Figure 2.28(b). The setup operated over a 10-m distance with an overall efficiency of 18.5% [101].

In 2018, Chinese researchers based in Xi'an designed a 5.8-GHz LHCP rectenna array system shown in Figure 2.28(c). An 8 × 8 truncated patch transmitting array, divided into 16 2 × 2 subarrays, was designed to focus on the near field. All 2 × 2 subarrays were connectorized to receive the necessary inputs for constructing a parabolic phase front across the transmitting array's surface. On the receiving end, a "focused" rectenna array used an 8 × 8 array

Figure 2.28. Chinese microwave power beaming experiments: (a) Chengdu 2.45 GHz, (b) Sichuan 5.8 GHz, (c) Xi'an focused 5.8 GHz CP experiment, (d) Chonqing phase-controlled source and rectenna array, (e) 35 GHz spatiotemporal transmitting antenna, and (f) receiving system and 35 GHz spatiotemporal receiving antenna (inset) in Wuhan in 2020. All images from [27].

of subwavelength radiating elements to provide harmonic filtering, decreased mutual coupling, and reduced backscatter. The rectenna array was able to achieve 57.74% RF-to-DC conversion efficiency in the near-field [102]. In December 2018, another SPS project named ZhuRi (Chase the Sun) was announced with the goal of demonstrating a complete ground-based SPS system modeled on OMEGA SPS, one of the overarching Chinese SPS concepts [27, p. 240]. Professor D. Baoyan of Xidian University was named as the project lead. In 2020, Chongqing University unveiled the phase-controlled source, transmit antenna, and power sub-system shown in Figure 2.28(d). The university stated its intent to demonstrate a 5.8-GHz, 640-W power beaming experiment at somewhere between 60 m and 100 m using a 64-channel phase-adjustable microwave source [27, p. 240].

In Wuhan, Dr. L. Xiao of the China Ship Development and Design Center disclosed a far-field "diffraction-free" millimeter-wave power beaming. This concept beams power to the far-field using a spatiotemporal beam where each spatial frequency is assigned to a single wavelength. In a 2020 experiment, a custom 35-GHz tube delivered 1 kW of power to Figure 2.28(e) spatiotemporal transmitting antenna. The energy was then radiated 300 m to the rectenna array seen in Figure 2.28(f). This rectenna was fitted with a custom GaN diode and achieved a total system efficiency of 9.89% measured from the 220 V AC at the transmitter input to the 36 V DC at the receiver output. This work was supported under the State Grid Corporation of China project "Research on Ten Meter-Level Microwave Radio Wireless Power Transmission Technology." Before this millimeter-wave experiment, in 2019 the group had already succeeded in a 10 GHz microwave power beaming experiment covering 100 m. This prior experiment had a total efficiency of 19.5%, converting 220 V AC on the transmit side to 24 V DC at the rectenna's output. This 10 GHz experiment also featured a custom microwave tube design and a GaN diode rectenna [27, p. 241].

2.7.4   *The United States*

The United States has also exhibited an uptick in power beaming and space solar activity since 2010. Government institutions and private companies have come together to push the technology ahead. In the summer of 2010, the private company LaserMotive (now PowerLight Technologies) made a portable laser power beaming demonstrator shown in Figure 2.29(a). It powered a commercially available toy helicopter, enabling it to remain aloft indefinitely, and had automatic interlocks as well as optical aspects to render reflections safe. The unit helped educate the public at many events, starting at the AUVSI trade show in 2010. Only a few months later, on October 28, 2010, they used laser illumination of specialized photovoltaics to remotely power the quadcopter shown in Figure 2.29(b) for 12.5 h. This test, conducted at Boeing's Future of Flight Center in Mukilteo, WA, set an endurance record for power beaming and for remotely powering a drone [103]. The demonstration was also notable due to it being safe for observers on the ground without needing laser eye protection.

Two years later, on July 12, 2012, PowerLight further announced that it had used a laser system to fly Lockheed Martin's 6 kg (13.2 lb) Stalker drone in a wind tunnel for 48 h straight. The company revealed a month later that the outdoor flight testing, as seen in Figure 2.29(c), demonstrated that recharging the drone's battery while in flight left it with more electricity stored than when the testing

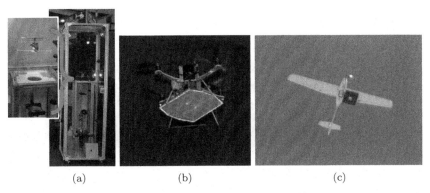

(a) (b) (c)

Figure 2.29. PowerLight laser beaming experiments: (a) mobile, safe toy helicopter demonstrator, (b) quadcopter, and (c) stalker drone. Images courtesy PowerLight Technologies.

began [104]. The flight tests of the Stalker also demonstrated the integration of the laser transmitter with the Laser Clearinghouse, turning off the laser whenever it might be pointing near a satellite while still tracking the aircraft.

In 2016, the D3 Space Solar proposal advocated the expressed goal of the United States becoming a leader in the development of SSP. The proposal recognized that other nations had active SSP programs while the US did not. The proposers had both technical and political backgrounds and represented such entities as the US Department of State, US Department of Defense, and private industry [105].

In 2018, a collaboration between Caltech and Northrop Grumman Corporation developed an ultralight, high-efficiency photovoltaic phased-array system. The design distributed power dynamically and featured ultralight deployable space structures [106,107]. The photovoltaic portion is seen in Figure 2.30(a), and the flexible phased array is shown in Figure 2.30(b). In 2020, the Caltech team announced several innovations that advance SPS power beaming, including flexible RFIC-based phased arrays with dynamic calibration [108,109] and timing devices for large-scale phased array synchronization [110]. The flexible nature of the panels would allow for continued directive beam formation in a situation where planarity is disturbed.

Optical power beaming continued to make strides during this time as well, with a PowerLight demonstration in May 2019 that was approved by the Navy's Laser Safety Review Board and the

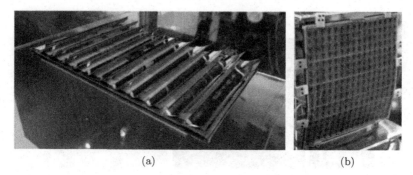

(a)   (b)

Figure 2.30.   (a) Caltech/Northrup Grumman photovoltaic phased array [27] and (b) Caltech flexible RFIC-based phased array [27].

demonstration site's Laser Safety Officer at the Naval Surface Warfare Center Carderock. The demonstration established a link of 325 m delivering more than 400 watts of output and was attended by nearly 100 VIPs attending over the course of multiple days [111]. A nearly identical demonstration in preparation had been conducted at the Port of Seattle, as pictured on the back cover of this book.

Late in 2019, Northrop Grumman announced that it was working with the US Air Force Research Laboratory (AFRL) on the Space Solar Power Incremental Demonstrations and Research (SSPIDR) project to develop an SSP system that can beam power from space to expeditionary forces on Earth [112]. SSPIDR will develop critical technologies and use incremental demonstrations to validate them. SSPIDR has three planned demonstrations called: (1) Arachne, (2) SPINDLE, and (3) SPIRRAL. Arachne is slated to fly in 2025 and would be the world's first space-to-ground solar-to-RF beaming demonstration. SPINDLE and SPIRRAL are follow-on experiments designed to test certain metrics of Arachne's performance. SPINDLE will test the onboard structural dynamics of the operational system to optimize beam formation for an aperture with dynamic variations in planarity. SPIRRAL will focus on the thermal management of Arachne and ways to maintain high-performance reliability over the long term. One of the critical technologies developed by the US Naval Research Laboratory (NRL) for Arachne is the 1-m^2 rectenna array of Figure 2.31(a). This array has been designed to achieve highly efficient rectification over a wide range of very low power densities from 1 to 1,000 mW/m^2 [27].

Figure 2.31. NRL MPT experiments: (a) low-power density rectenna array for SSPIDR [27], (b) PRAM [116], and (c) an X-37B orbital test vehicle [117].

On May 17, 2020, the NRL launched its Photovoltaic Radio-frequency Antenna Module Flight Experiment (PRAM FX) seen in Figure 2.31(b) aboard an Air Force X-37B orbital test vehicle such as the one pictured in Figure 2.31(c). PRAM FX was a 30-cm^2 tile that collected solar energy and converted it to RF microwave power [113]. The PRAM FX experiment was the first demonstration of SPS hardware, operating through the conclusion of the X-37B Orbital Test Vehicle's 6th mission on November 12, 2022. The NRL has also authored reports citing space solar as a potential means of providing energy to locations such as humanitarian and disaster response areas and forward operating bases [114,115].

Apart from space solar, the span from 2019 to 2023 has also yielded compelling demonstrations of power beaming capabilities worldwide. In early 2020, Astronaut Jessica Meir demonstrated wireless power in orbit for the first time on the International Space Station with the NRL's LEctenna project (https://www.nrl.navy.mil/STEM/LEctenna-Challenge/). In 2021, NRL also led compelling microwave and optical power beaming demonstrations, achieving power levels exceeding 1.6 kW and distances of 1 km, as described in a paper by Rodenbeck *et al.* in [27,118] and in videos on NRL's YouTube channel (https://www.youtube.com/@USNRL).

From the early 2020s, there was an increase in power beaming from mainstream consumer electronics companies. In October 2021, Ericsson announced the world's first wirelessly-powered 5G radio

Figure 2.32. Laser power beaming receiver used for Ericsson 5G base station demonstration.

base station in collaboration with PowerLight Technologies [119], safely delivering "hundreds of watts over hundreds of meters through the air." The receiver for this demonstration is shown in Figure 2.32.

In September 2022, the lock company Alfred and the Israeli-based optical wireless charging company Wi-charge announced a partnership to release smart locks using the latter's laser power beaming technology [120]. Many of the other companies making forays into power beaming are listed in Section 8.5.

In January 2023, the California Institute of Technology launched a novel space experiment to test microwave power beaming, solar cells, and space deployable structures to advance those technologies for SPSs. The effort and its predecessors were underwritten by the billionaire Donald Bren in an amount of "at least $100 million" [121]. The *Wall Street Journal* reported in June of 2023 that the microwave power beaming experiment, titled Microwave Array for Power-transfer Low-orbit Experiment (MAPLE) had been successful. It lit up LEDs onboard the spacecraft, with power transmitted from about 30 cm (one foot) away [122].

March 2023 also saw the first demonstration of laser power beaming in space with the Space Wireless Energy Laser Link (SWELL) project, also conducted by NRL on the International Space Station [123]. It set a new record for a distance of 144 cm for a power beaming link in orbit of any type, microwave or optical. In September

2023, the US Defense Advanced Research Projects Agency (DARPA) announced that it had selected performers to develop airborne laser power beaming relays as part of its Persistent Optical Wireless Energy Relay (POWER) program, seeking to significantly extend the range of optical power beaming [124].

2.8 Conclusion

Development of power beaming and space solar technology appears to be accelerating in a global sense. The work of Tesla, Brown, Dickinson, Matsumoto, Kaya, Kare, and other researchers has laid the groundwork, and now those currently working in the field are pushing further. The political backing from a handful of countries appears to be making such a push possible. Future collaborations in the mold of SERT or those coming out of RISH, ESA's Solaris effort, or the UK's Space Energy Initiative may serve to focus attention on improving space solar subsystem components through a variety of means and metrics, making the tasks of their emplacement into space more realizable. Joint efforts between countries, similar to the International Space Station or International Thermonuclear Experimental Reactor (ITER) may offer a potential path forward as more nations take part in the ambitious effort to produce a working SPS system and exploit its potential benefits.

At the time of this writing, several compelling additional power beaming demonstrations had been executed, but their results had not yet been published or publicly acknowledged. Many further power beaming demonstrations were in the planning stages and are likely to have occurred by the time you read this. The history of power beaming continues to unfold and promises to have a forward path for some time.

References

[1] S. P. Israelsen, "The Scientific Theories of Michael Faraday and James Clerk Maxwell," *The Purdue Historian*, vol. 7, 2014. [Online]. Available at: http://docs.lib.purdue.edu/puhistorian/vol7/iss1/1 (accessed March 21, 2023).

[2] J. C. Maxwell, *On Faraday's Lines of Force*. 1855. [Online]. Available at: https://en.wikisource.org/wiki/On_Faraday%27s_Lines_of_Force.

[3] C. C. Gillispie, Ed., *Dictionary of Scientific Biography*. New York, NY: Charles Scribner's Sons, 1980.

[4] H. Hertz, "Photo Taken by Heinrich Hertz of His Laboratory." [Online]. Available at: https://www.nutsvolts.com/uploads/wygwam/ NV_0119_Steber_Fig10_Hertz_lab.jpg (accessed September 24, 2023).

[5] Museo Nazionale della Scienza e della Tecnica "Leonardo da Vinci," *oscillatore di Righi con riflettore parabolico*. 1984. [Rame]. Available at: https://commons.wikimedia.org/wiki/File:Oscillatore_di_ Righi_con_riflettore_parabolico_-_Museo_scienza_tecnologia_Milano_ 08757_1.jpg (accessed September 16, 2023).

[6] "Radio-Broadcast-1926-11.pdf." [Online]. Available at: https://wor ldradiohistory.com/Archive-Radio-Broadcast/Radio-Broadcast-1926- 11.pdf (accessed September 16, 2023).

[7] A. Righi, *L'Ottica Delle Oscillazioni Elettriche*. 1897. [Online]. Available at: http://archive.org/details/righi-ottica-delle-oscillazioni-elet triche (accessed September 16, 2023).

[8] G. Marconi, "Nobel Lectures, Physics 1901–1921," in *Wireless Telegraphic Communication*. Amsterdam: Elsevier Publishing Company, 1909, pp. 196–222. [Online]. Available at: https://www.nobelprize. org/uploads/2018/06/marconi-lecture.pdf (accessed September 16, 2023).

[9] M. Cheney, *Tesla: Man out of Time*. Englewood Cliffs, NJ: Prentice-Hall, 1981.

[10] *English: Nikola Tesla, with His Equipment*, 1899. [Online]. Available at: https://commons.wikimedia.org/wiki/File:Nikola_Tesla,_with_his_ equipment_Wellcome_M0014782.jpg (accessed September 16, 2023).

[11] *Tesla's Tower at Wardenclyffe*. 2016. [Online]. Available at: https:// commons.wikimedia.org/wiki/File:Wardenclyffe_Tower_-_1904.jpg (accessed September 16, 2023).

[12] H. J. Visser, "A Brief History of Radiative Wireless Power Transfer," in *2017 11th European Conference on Antennas and Propagation (EUCAP)*, March 2017, pp. 327–330. doi: 10.23919/ EuCAP.2017.7928700.

[13] S. Uda, "On the Wireless Beam of Short Electric Waves," *Journal of the Institute of Electrical Engineers of Japan*, 1927. [Online]. Available at: https://books.google.com/books?id=nmmYAQAACAAJ.

[14] H. Yagi, "Beam Transmission of Ultra Short Waves," *Proc. Instit. Radio Eng.*, vol. 16, no. 6, pp. 715–740, June 1928. doi: 10.1109/ JRPROC.1928.221464.

[15] *Shintaro Uda*. [Online]. Available at: https://dxnews.com/forum/ upload/18620-uda.jpg (accessed September 16, 2023).

[16] *H. Yagi.* [Online]. Available at: https://dxnews.com/forum/upload/ 18619-yagi.jpg (accessed September 16, 2023).

[17] H. A. H. Boot and J. T. Randall, "Historical Notes on the Cavity Magnetron," *IEEE Trans.Electron Dev.*, vol. 23, no. 7, pp. 724–729, Jul. 1976. doi: 10.1109/T-ED.1976.18476.

[18] I. Asimov, "Reason," in *I, Robot.* Street & Smith Publications Inc., 1941. [Online]. Available at: http://addsdonna.com/old-website/ ADDS_DONNA/Science_Fiction_files/2_Asimov_Reason.pdf (accessed September 10, 2020).

[19] W. C. Brown, J. F. Skowron, G. H. MacMaster, and J. W. Buckley, "The Super Power CW Amplitron," in *1963 International Electron Devices Meeting*, October 1963, pp. 52–52. doi: 10.1109/ IEDM.1963.187384.

[20] R. H. George and E. M. Sabbagh, "An Efficient Means of Converting Microwave Energy to DC Using Semiconductor Diodes," *Proc. IEEE*, vol. 51, no. 3, pp. 530–530, March 1963. doi: 10.1109/ PROC.1963.2119.

[21] W. C. Brown, "Thermionic Diode Rectifier," in *Microwave Power Engineering: Generation, Transmission, Rectification*, in Electrical science series, v. 1. Academic Press, 1968, pp. 295–298. Available at: https://books.google.com/books?id=Obw3BQAAQBAJ.

[22] W. C. Brown, "The History of the Development of the Rectenna," presented at the *Rectenna Session of the SPS Microwave Systems Workshop, Lyndon B. Johnson Space Center.* Houston, Texas, USA: NASA, January 1980.

[23] W. C. Brown and R. H. George, "Rectification of Microwave Power," *IEEE Spectrum*, vol. 1, no. 10, pp. 92–97, October 1964. doi: 10.1109/MSPEC.1964.6501196.

[24] W. C. Brown, R. H. George, N. I. Heenan, and R. C. Wonson, "Microwave to DC Converter," 3,434,678, March 25, 1969. [Online]. Available at: https://patentimages.storage.googleapis.com/81/c6/ bd/58c6b606e1bf1c/US3434678.pdf (accessed September 16, 2023).

[25] W. C. Brown, "Experimental Airborne Microwave Supported Platform," *Griffiss Air Force Base.* New York, RADC-TR-65-188, December 1965. [Online]. Available at: https://apps.dtic.mil/sti/pdfs/AD0 474925.pdf (accessed September 16, 2023).

[26] W. C. Brown, "Electronic and Mechanical Improvement of the Receiving Terminal of a Free-Space Microwave Power Transmission System," NASA Lewis Research Center, NASA-CR-135194, August 1977. [Online]. Available at: https://space.nss.org/wp-content/ uploads/1977-Receiving-Terminal-For-Free-Space-Microwave-Power. pdf (accessed September 24, 2023).

[27] C. T. Rodenbeck *et al.*, "Microwave and Millimeter Wave Power beaming," *IEEE J. Microwaves*, vol. 1, no. 1, pp. 229–259, winter 2021. doi: 10.1109/JMW.2020.3033992.

[28] P. E. Glaser, "Power from the Sun: Its Future," *Science*, vol. 162, no. 3856, pp. 857–861, November 1968.

[29] R. M. Dickinson, "Evaluation of a Microwave High-Power Reception-Conversion Array for Wireless Power Transmission," NASA-CR-145625, JPL-TM-33-741, 1975. [Online]. Available at: https://ntrs.nasa.gov/archive/nasa/casi.ntrs.nasa.gov/19760004119. pdf (accessed August 2, 2020).

[30] W. C. Brown, "Free-Space Microwave Power Transmission Study Combined Phase Ill and Final Report," Marshall Space Flight Center, Huntsville, AL, PT-4601, September 1975. [Online]. Available at: https://ntrs.nasa.gov/api/citations/19760009531/downloads/197600 09531.pdf (accessed September 16, 2023).

[31] W. C. Brown, "Design Definition of a Microwave Power Reception and Conversion System for Use on a High Altitude Powered Platform," NASA Wallops Flight Center, Wallps Island, VA, 156866, May 1981. [Online]. Available at: https://ntrs.nasa.gov/api/citations/ 19810018863/downloads/19810018863.pdf (accessed September 16, 2023).

[32] W. C. Brown, "Rectenna Technology Program: Ultra Light 2.45 GHz Rectenna and 20 GHz Rectenna," NASA-CR-179558, March 1987. [Online]. Available at: https://ntrs.nasa.gov/api/citations/19870010 123/downloads/19870010123.pdf (accessed September 24, 2023).

[33] R. M. Dickinson and W. C. Brown, "Radiated Microwave Power Transmission System Efficiency Measurements," Jet Propulsion Lab., California Inst. of Tech., Pasadena, CA, Technical Report NASA-CR-142986, JPL-TM-33-727, May 1975. [Online]. Available at: https:// ntrs.nasa.gov/search.jsp?print=yes&R=19750018422 (accessed April 25, 2020).

[34] "Reception - Conversion Subsystem (RXCV) for Microwave Power Transmission System," ER75-4386, September 1975.

[35] "The Final Proceedings of the Solar Power Satellite Program Review," July 1980. [Online]. Available at: https://space.nss.org/ wp-content/uploads/1981-DOE-SPS-Final-Proceedings-Of-The-Solar-Power-Satellite-Program-Review.pdf .

[36] J. Schlesak, A. Alden, and T. Ohno, "SHARP Rectenna and Low Altitude Flight Trials," presented at the *IEEE Global Telecommunications Conference, New Orleans*, December 1985. [Online]. Available at: https://www.friendsofcrc.ca/Projects/SHARP/sharp. html (accessed September 16, 2023).

[37] H. Matsumoto, "Microwave Power Transmission," *J. Aerospace Soc.*, vol. 32, pp. 120–127, 1989.

[38] Y. Fujino *et al.*, "A Rectenna for MILAX," in *Proceedings of the 1st Wireless Power Transmission. Conference*, Texas, February 1993, pp. 273–277.

[39] H. Matsumoto, N. Kaya, I. Kimura, S. Miyatake, M. Nagatomo, and T. Obayashi, "MINIX Project toward the Solar Power Satellite-Rocket Experiment of Microwave Energy Transmission and Associated Nonlinear Plasma Physics in the Ionosphere," presented at the *ISAS Space Energy Symposium*, 1982, pp. 69–76.

[40] M. Nagatomo and N. Kaya, "Engineering Aspect of the Microwave Ionosphere Nonlinear Interaction Experiment (MINIX) with a Sounding Rocket," *Acta Astronautica*, vol. 13, no. 1, pp. 23–29, 1986.

[41] N. Shinohara, "Beam Control Technologies with a High-Efficiency Phased Array for Microwave Power Transmission in Japan," *Proc. IEEE*, vol. 101, no. 6, pp. 1448–1463, June 2013. doi: 10.1109/JPROC.2013.2253062.

[42] N. Kaya, H. Kojima, H. Matsumoto, M. Hinada, and R. Akiba, "ISY-METS Rocket Experiment for Microwave Energy Transmission," *Acta Astronautica*, vol. 34, pp. 43–46, October 1994. doi: 10.1016/0094-5765(94)90241-0.

[43] N. Shinohara and H. Matsumoto, "Dependence of DC Output of a Rectenna Array on the Method of Interconnection of Its Array Elements," *Elect. Eng. Jpn.*, vol. 125, no. 1, pp. 9–17, October 1998. doi: 10.1002/(SICI)1520-6416(199810)125:1<9::AID-EEJ2>3.0.CO;2-3.

[44] N. Kaya, S. Ida, Y. Fujino, and M. Fujita, "Transmitting Antenna System for Airship Demonstration (ETHER)," *Space Energy Transport.*, vol. 1, no. 4, pp. 237–245, 1996.

[45] H. Matsumoto, "Research on Solar Power Satellites and Microwave Power Transmission in Japan," *IEEE Microwave Magaz.*, vol. 3, no. 4, pp. 36–45, December 2002. doi: 10.1109/MMW.2002.114 5674.

[46] J. Hawkins, S. Houston, M. Hatfield, and W. Brown, "The SABER Microwave-Powered Helicopter Project and Related WPT Research at the University of Alaska Fairbanks," in *AIP Conference Proceedings*. Albuquerque, NM: AIP, 1998, pp. 1092–1097. doi: 10.1063/1.54725.

[47] A. Celeste, P. Jeanty, and G. Pignolet, "Case Study in Reunion Island," *Acta Astronautica*, vol. 54, no. 4, pp. 253–258, February 2004. doi: 10.1016/S0094-5765(02)00302-8.

[48] J. O. McSpadden and J. C. Mankins, "Space Solar Power Programs and Microwave Wireless Power Transmission Technology," *IEEE Microwave Magaz.*, vol. 3, no. 4, pp. 46–57, December 2002. doi: 10.1109/MMW.2002.1145675.

[49] B. Strassner and K. Chang, "5.8-GHz Circularly Polarized Dual-Rhombic-Loop Traveling-Wave Rectifying Antenna for Low Power-Density Wireless Power Transmission Applications," *IEEE Trans. Microwave Theory Techniq.*, vol. 51, no. 5, pp. 1548–1553, May 2003. doi: 10.1109/TMTT.2003.810137.

[50] NASA, *English: SPS in Front, Solar Clipper (Space Tug) Top Left.* 2011. [Online]. Available at: https://commons.wikimedia.org/wiki/File:Solar_power_satellite_sandwich_or_abascus_concept.jpg (accessed September 24, 2023).

[51] J. Kare, "Program and Applications for a Near-Term Laser Launch System," UCID-21718, 6907065, ON: DE90011760, June 1989. doi: 10.2172/6907065.

[52] D. Young, G. H. Walker, and G. L. Schuster, "Preliminary Design and Cost of a 1-Megawatt Solar-Pumped Iodide Laser Space-to-Space Transmission Station," NASA Technical Memorandum 4002, September 1987.

[53] "Space Laser Power Transmission System Studies," NASA Conference Publication 2214, 1982. [Online]. Available at: https://ntrs.nasa.gov/api/citations/19820010704/downloads/19820010704.pdf (accessed October 26, 2023).

[54] L. C. Olsen, G. Dunham, D. A. Huber, F. William Addis, N. Anheier, and E. P. Coomes, "GaAs Solar Cells for Laser Power Beaming," presented at the *Space Photovoltaic Research and Technology Conference*, NASA Lewis Research Center, August 1991. [Online]. Available at: https://ntrs.nasa.gov/api/citations/19910020915/downloads/19919910020915.pdf.

[55] S. van Riesen, U. Schubert, and A. W. Bett, "GaAs Photovoltaic Cells for Laser Power Beaming at High Power Densities," 2002, [Online]. Available at: https://publica.fraunhofer.de/handle/publica/341879 (accessed October 26, 2023).

[56] J. T. Kare, F. Mitlitsky, and A. Weisberg, "Preliminary Demonstration of Power Beaming with Non-Coherent Laser Diode Arrays," in *AIP Conference Proceedings*, Albuquerque, NM: AIP, 1999, pp. 1641–1646. doi: 10.1063/1.57492.

[57] W. C. Brown, *The Private Journals of William C. Brown. Book 1: Father of Microwave Power Transmission*, Apollo 11 50th anniversary edition. Alpharetta, GA: BookLogix: Space Solar Power Institute, 2018.

[58] M. Mori, H. Kagawa, and Y. Saito, "Summary of Studies on Space Solar Power Systems of Japan Aerospace Exploration Agency (JAXA)," *Acta Astronautica*, vol. 59, no. 1–5, pp. 132–138, July 2006. doi: 10.1016/j.actaastro.2006.02.033.

[59] "ESA Work on Solar Power from Space: Concluded and Ongoing Activities," *Advanced Concepts Team — ESA*, January 2008. [Online]. Available at: https://www.esa.int/gsp/ACT/doc/POW/ACT-RPT-NRG-2209-SPS_concluded_and_ongoing_activities_reduced_size.pdf (accessed September 17, 2023).

[60] C. Cougnet, "Solar Power Satellite — SPS-REPOSE STUDY," September 2004. [Online]. Available at: https://www.esa.int/gsp/ACT/doc/POW/GSP-RPT-SPS-0501%20Executive%20Summary%20REPOSE-EADS%20Astrium.pdf (accessed September 17, 2023).

[61] N. Shinohara, H. Matsumoto, and K. Hashimoto, "Phase-controlled magnetron development for SPORTS: Space power radio transmission system," *URSI Radio Science Bulletin*, vol. 2004, no. 310, pp. 29–35, September 2004. doi: 10.23919/URSIRSB.2004.7909435.

[62] H. Matsumoto, K. Hashimoto, N. Shinohara, and T. Mitani, "Experimental Equipments for Microwave Power Transmission in Kyoto University," in *Proceedings of the 4th International Conference on Solar Power from Space*, European Space Agency, July 2004.

[63] Y. Gibbs, "NASA Dryden Fact Sheets — Beamed Laser Power," *NASA*. [Online]. Available at: http://www.nasa.gov/centers/armstrong/news/FactSheets/FS-087-DFRC.html (accessed August 26, 2022).

[64] M. Conner, *Power Beaming Flight Demonstration (2003-09-18, 1 of 2 images)*. 2015. [Online]. Available at: http://www.nasa.gov/centers/dryden/multimedia/imagegallery/Power Beaming/ED03-0249-18.html (accessed November 11, 2022).

[65] http://taminelectriccom/%D9%85%D9%82%D8%A7%D9%84%D8%A7%D8%AA-%D8%B9%D9%84%D9%85%DB%8C/Id/90?title=%D8%A7%D9%86%D8%AA%D9%82%D8%A7%D9%84-%D8%A8%D8%B1%D9%82-%D8%A8%DB%8C-%D8%B3%DB%8C%D9%85, *English: Stanford scientists*. [Online]. Available at: https://commons.wikimedia.org/wiki/File:Intelwirelesselectricity.jpg?uselang=fa (accessed September 24, 2023).

[66] A. Kurs, A. Karalis, R. Moffatt, J. D. Joannopoulos, P. Fisher, and M. Soljačić, "Wireless Power Transfer via Strongly Coupled Magnetic Resonances," *Science*, vol. 317, no. 5834, pp. 83–86, July 2007. doi: 10.1126/science.1143254.

[67] "Qi Wireless Charging |Wireless Power Consortium." [Online]. Available at: https://www.wirelesspowerconsortium.com/qi/ (accessed September 25, 2023).

[68] "RF Wireless Power & Radio Frequency Charging," *AirFuel Alliance*. [Online]. Available at: https://airfuel.org/airfuel-rf/ (accessed April 27, 2022).

[69] "Researchers Beam 'Space' Solar Power in Hawaii," *Wired*. [Online]. Available at: https://www.wired.com/2008/09/visionary-beams/ (accessed November 24, 2020).

[70] N. Kaya, M. Iwashita, F. Little, N. Marzwell, and J. C. Mankins, "Microwave Power Beaming Test in Hawaii," in *Proc. 60th Int. Astronaut. Congr.*, 2009, pp. 6128–6132.

[71] "Solar Power Beamed rom Space within A Decade?," *New Atlas*. [Online]. Available at: https://newatlas.com/solar-power-space-satellite/11064/ (accessed September 19, 2023).

[72] Shoichiro Mihara, "WPT Technology Demonstration Options at USEF," presented at the *International Symposium on Solar Energy from Space*, Toronto, Canada, September 10, 2009.

[73] S. Kawasaki, "Microwave WPT to a Rover Using Active Integrated Phased Array Antennas," in *Proceedings of the 5th European Conference on Antennas and Propagation (EUCAP)*, April 2011, pp. 3909–3912.

[74] J. C. Mankins, "SPS-ALPHA: The First Practical Solar Power Satellite Via Arbitrarily Large Phased Array," 2012. [Online]. Available at: https://www.nasa.gov/sites/default/files/atoms/files/niac_2011_phasei_mankins_spsalpha_tagged.pdf.

[75] A. Massa, G. Oliveri, F. Viani, and P. Rocca, "Array Designs for Long-Distance Wireless Power Transmission: State-of-the-Art and Innovative Solutions," *Proc. IEEE*, vol. 101, no. 6, pp. 1464–1481, June 2013. doi: 10.1109/JPROC.2013.2245491.

[76] A. Oida, H. Nakashima, J. Miyasaka, K. Ohdoi, H. Matsumoto, and N. Shinohara, "Development of a New Type of Electric Off-Road Vehicle Powered By Microwaves Transmitted through Air," *J. Terramech.*, vol. 44, no. 5, pp. 329–338, November 2007. doi: 10.1016/j.jterra.2007.10.002.

[77] N. Shinohara *et al.*, "Experiment of Microwave Power Transmission to the Moving Rover," in *Proceedings of ISAP2007*, Niigata, Japan, 2007, pp. 648–651.

[78] "2009 Space Elevator Power Beaming Contest Reset for Nov. 4 — NASA." [Online]. Available at: https://www.nasa.gov/news-release/2009-space-elevator-power beaming-contest-reset-for-nov-4/ (accessed October 25, 2023).

[79] P. Swann, *Space Elevators: An Assessment of the Technological Feasibility and the Way Forward*. Virginia Edition Publishing Co., 2014.

[80] J. T. Kare, T. J. Nugent, and A. V. Pakhomov, "Laser Power Beaming on A Shoestring," in *AIP Conference Proceedings*, Kailua-Kona (Hawaii): AIP, 2008, pp. 97–108. doi: 10.1063/1.2931935.

[81] N. Administrator, "LaserMotive Wins $900,000 from NASA in Space Elevator Games," *NASA*. [Online]. Available at: http://www.nasa. gov/centers/dryden/status_reports/power_beam.html (accessed October 2, 2022).

[82] R. Winsor, A. Bakos, B. Murray, and T. Stone, "Power Beaming techniques for NASA's 2009 Centennial Challenge," presented at the *Government Microcircuit and Applications Conference* (GOMAC), 2010, p. 5.

[83] Tom Nugent, "Review of Laser Power Beaming Demonstrations by PowerLight Technologies (formerly LaserMotive)," presented at the 20th Annual Directed Energy Science and Technology Symposium, Oxnard, CA, February 2018.

[84] R. Gopalaswami, "An International Preliminary Feasibility Study on Space Based Solar Power Stations." Kalam-National Space Society Energy Technology Universal Initiative, August 16, 2010. [Online]. Available at: https://space.nss.org/wp-content/uploads/KALAM-NSS-Initiative.pdf.

[85] "China Proposes Space Collaboration with India," *The Times of India*, November 2, 2012. [Online]. Available at: https://timesofin dia.indiatimes.com/india/china-proposes-space-collaboration-with-india/articleshow/17066537.cms (accessed September 19, 2023).

[86] T. Nishioka and S. Yano, "Mitsubishi Heavy Takes Step toward Long-Distance Wireless Power," *Nikkei Asia*. [Online]. Available at: https://asia.nikkei.com/Business/Biotechnology/Mitsubishi-Heavy-takes-step-toward-long-distance-wireless-power (accessed September 19, 2023).

[87] Susumu Sasaki, "How Japan Plans to Build an Orbital Solar Farm — IEEE Spectrum." Available at: https://spectrum.ieee.org/how-japan-plans-to-build-an-orbital-solar-farm (accessed September 19, 2023).

[88] USEF, *SSPS.jpg (399× 282)*. [Online]. Available at: http://1.bp. blogspot.com/_fSvarQSvbd0/SvZfrL6wXhI/AAAAAAAArY/LSD7 A4i8Or8/s1600/SSPS.jpg (accessed September 24, 2023).

[89] S. Mihara *et al.*, "The Result of Ground Experiment of Microwave Wireless Power Transmission," in *Proceedings of the 2015 International Astronautical Congress*, International Astronautical Federation, 2015. [Online]. Available at: https://iafastro.directory/iac/archive/browse/IAC-15/C3/2/28587/ (accessed September 19, 2023).

[90] T. Takahashi *et al.*, "Phased Array System for High Efficiency and High Accuracy Microwave Power Transmission," in *2016 IEEE International Symposium on Phased Array Systems and Technology (PAST)*, October 2016, pp. 1–7. doi: 10.1109/ARRAY.2016.783 2563.

[91] Shoichiro Mihara *et al.*, "The Current Status of Microwave Power Transmission for SSPS and Industry Application," in *Proceedings of the 68th International Astronauts Congress*, Adelaide, Australia: International Astronautical Federation, 2017.

[92] S. Mihara *et al.*, "The Plan of Microwave Power Transmission Development for SSPS and Its Industry Application," in *2018 Asia-Pacific Microwave Conference (APMC)*, November 2018, pp. 443–445. doi: 10.23919/APMC.2018.8617218.

[93] N. Shinohara, N. Hasegawa, S. Kojima, and N. Takabayashi, "New Beam Forming Technology for Narrow Beam Microwave Power Transfer," in *Proceedings of the 8th Asia-Pacific Conference Antennas Propag. (APCAP)*, 2019.

[94] S. Mizojiri and K. Shimamura, "Wireless Power Transfer via Subterahertz-Wave," *Applied Sciences*, vol. 8, no. 12, p. 2653, December 2018. doi: 10.3390/app8122653.

[95] S. Mizojiri *et al.*, "Demonstration of Sub-Terahertz Coplanar Rectenna Using 265 GHz Gyrotron," in *2019 IEEE Wireless Power Transfer Conference (WPTC)*, June 2019, pp. 409–412. doi: 10.1109/WPTC45513.2019.9055555.

[96] "Kyoto University and MinebeaMitsumi Started Social Demonstration Test with Wireless Power Supply, Using the Joint National Strategy Special Zone — MinebeaMitsumi." [Online]. Available at: https://www.minebeamitsumi.com/english/news/press/2020/11 99445_13882.html (accessed March 23, 2021).

[97] K. D. Song *et al.*, "Preliminary Operational Aspects of Microwave-Powered Airship Drone," *International Journal of Micro Air Vehicles*, vol. 11, p. 1756829319861368, January 2019. doi: 10.1177/1756829319861368.

[98] L. Jeon, "Transmitter for High Power Microwave Wireless Power Transmission," *Mag. Korea Inst. Elect. Eng. (KIEE)*, pp. 11–14, September 2019.

[99] K. Needham, "Plans for First Chinese Solar Power Station in Space Revealed," The Sydney Morning Herald. [Online]. Available at: https://www.smh.com.au/world/asia/plans-for-first-chinese-sol ar-power-station-in-space-revealed-20190214-p50xtg.html (accessed June 1, 2020).

[100] Hexin Zhang and Changjun Liu, "A High-Efficiency Microwave Rectenna Array Based on Subarray Decomposition," *Appl. Sci. Technol.*, vol. 43, no. 4, pp. 57–61, 2016. doi: 10.11991/yykj.201602001.

[101] Q. Chen, "Research on High-Performance Receiving and Rectifying Technology for Microwave Wireless Power Transmission," Sichuan University, 2020.

[102] Y. Dong *et al.*, "Focused Microwave Power Transmission System with High-Efficiency Rectifying Surface," *IET Microwaves, Antennas & Propagation*, vol. 12, no. 5, pp. 808–813, 2018. doi: 10.1049/iet-map.2017.0530.

[103] Alan Boyle, "Copter Sets a Laser-Powered Record," NBC News. [Online]. Available at: http://www.nbcnews.com/science/cos mic-log/copter-sets-laser-powered-record-flna6C10403609 (accessed September 19, 2023).

[104] "Laser Powers Lockheed Martin's Stalker UAS For 48 Hours," Media — Lockheed Martin. [Online]. Available at: https://news.lockheed martin.com/2012-07-11-Laser-Powers-Lockheed-Martins-Stalker-UAS-For-48-Hours (accessed October 27, 2022).

[105] "Space Solar Power Dominates the D3...Is Now the #1 Idea in the Federal Government!," Space Development Steering Committee. [Online]. Available at: https://spacedevelopmentsteeringcommittee. org/space-solar-power-dominates-the-d3is-now-the-1-idea-in-the-federal-government/ (accessed September 19, 2023).

[106] "Space-Based Solar Power Project Funded," California Institute of Technology. [Online]. Available at: https://www.caltech.edu/ about/news/space-based-solar-power-project-funded-46644 (accessed September 19, 2023).

[107] E. Gdoutos *et al.*, "A Lightweight Tile Structure Integrating Photovoltaic Conversion and RF Power Transfer for Space Solar Power Applications," in *2018 AIAA Spacecraft Structures Conference*, Kissimmee, FL: American Institute of Aeronautics and Astronautics, January 2018. doi: 10.2514/6.2018-2202.

[108] M. Gal-Katziri, A. Fikes, F. Bohn, B. Abiri, M. R. Hashemi, and A. Hajimiri, "Scalable, Deployable, Flexible Phased Array Sheets," presented at the *IEEE/MTT-S International Microwave Symposium*, 2020, pp. 1085–1088.

[109] A. C. Fikes, A. Safaripour, F. Bohn, B. Abiri, and A. Hajimiri, "Flexible, Conformal Phased Arrays with Dynamic Array Shape Self-Calibration," in *2019 IEEE MTT-S International Microwave Symposium (IMS)*, June 2019, pp. 1458–1461. doi: 10.1109/MWSYM.2019.8701107.

[110] M. Gal-Katziri and A. Hajimiri, "A Sub-Picosecond Hybrid DLL for Large-Scale Phased Array Synchronization," in *2018 IEEE Asian*

Solid-State Circuits Conference (A-SSCC), November 2018, pp. 231–234. doi: 10.1109/ASSCC.2018.8579340.

[111] Thomas J. Nugent, Jr., David Bashford, Thomas Bashford, Thomas J. Sayles, and Alex Hay, "Long-Range, Integrated, Safe Laser Power Beaming Demonstration," in *Technical Digest OWPT 2020*, Yokohama, Japan: Optical Wireless Power Transmission Committee, The Laser Society of Japan, April 2020, pp. 12–13.

[112] Matt Jorgenson, "Northrop Grumman and US Air Force Research Laboratory Partner to Provide Critical Advanced Technology in Space Solar Power," Northrop Grumman Newsroom. [Online]. Available at: https://news.northropgrumman.com/news/features/northrop-grumman-and-us-air-force-research-laboratory-partner-to-provide-critical-advanced-technology-in-space-solar-power (accessed September 19, 2023).

[113] "NRL Conducts First Test of Solar Power Satellite Hardware in Orbit." [Online]. Available at: https://www.nrl.navy.mil/news/releases/nrl-conducts-first-test-solar-power-satellite-hardware-orbit (accessed June 26, 2020).

[114] N. W. Johnson *et al.*, "Space-based Solar Power: Possible Defense Applications and Opportunities for NRL Contributions," NRL/FR/7650–09-10,179, 2009.

[115] P. Jaffe *et al.*, "Opportunities and Challenges for Space Solar for Remote Installations," U.S. Naval Research Laboratory, Washington, DC, Memo Report NRL/MR/8243–19-9813, October 2019. [Online]. Available at: https://apps.dtic.mil/sti/pdfs/AD1082903.pdf (accessed April 21, 2020).

[116] "CHIPS Articles: Solar Power When It's Raining: NRL Builds Space Satellite Module to Try." [Online]. Available at: https://www.doncio.navy.mil/Chips/ArticleDetails.aspx?ID=4992 (accessed September 24, 2023).

[117] *Encapsulated X-37B Orbital Test Vehicle for United States Space Force-7 Mission (Courtesy of Boeing)*. [Online]. Available at: https://www.spaceforce.mil/Multimedia/Photos/igphoto/2002295322/ (accessed September 24, 2023).

[118] C. T. Rodenbeck *et al.*, "Terrestrial Microwave Power Beaming," *IEEE Journal of Microwaves*, vol. 2, no. 1, pp. 28–43, January 2022. doi: 10.1109/JMW.2021.3130765.

[119] "Ericsson and Powerlight Base Station Wireless Charging Breakthrough." [Online]. Available at: https://www.ericsson.com/en/news/2021/10/ericsson-and-powerlight-achieve-base-station-wireless-charging-breakthrough (accessed October 4, 2021).

[120] Wi-Charge, "Alfred and Wi-Charge Deliver the First Wirelessly-Charged Smart Locks." [Online]. Available at: https://www.prnewswire.com/news-releases/alfred-and-wi-charge-deliver-the-first-wirelessly-charged-smart-locks-301637165.html (accessed October 26, 2023).

[121] "$100M gift from Irvine Co.'s Bren Powers Caltech Space Electricity Idea," *Press Telegram*. [Online]. Available at: https://www.ocregister.com/2021/07/30/how-donald-brens-100-million-gift-is-powering-caltech-space-power-concept (accessed August 2, 2021).

[122] C. S. Powell, "Beaming Solar Energy From Space Gets a Step Closer," *Wall Street Journal*, June 7, 2023. [Online]. Available at: https://www.wsj.com/articles/beaming-solar-energy-from-space-gets-a-step-closer-fc903658 (accessed June 12, 2023).

[123] Mary E. Hamisevicz, "First In-Space Laser Power Beaming Experiment Surpasses 100 Days of Successful On-Orbit Op," U.S. Naval Research Laboratory. [Online]. Available at: https://www.nrl.navy.mil/Media/News/Article/3457014/first-in-space-laser-power beaming-experiment-surpasses-100-days-of-successful/https%3A%2F%2Fwww.nrl.navy.mil%2FMedia%2FNews%2FArticle%2F3457014%2Ffirst-in-space-laser-power-beaming-experiment-surpasses-100-days-of-successful%2F (accessed September 19, 2023).

[124] "POWER Program Selects Teams to Design Power Beaming Relays," DARPA. [Online]. Available at: https://www.darpa.mil/news-events/2023-09-07a (accessed September 24, 2023).

Chapter 3

Survey of Fundamentals and Conceptual Tools

Most of the fundamental ideas of science are essentially simple, and may, as a rule, be expressed in a language comprehensible to everyone.

— The Evolution of Physics
by Albert Einstein and Leopold Infeld [1, p. 29]

3.1 Introduction

The ability to design, implement, and assess power beaming systems is enhanced by an understanding of the applicability of different models for different situations, the regimes that define the behavior of electromagnetic (EM) waves, and how to make meaningful comparisons between the different modalities of power transfer. Background topics such as energy, power, coupling, optics, waves, and electromagnetism are likewise helpful in designing and understanding power beaming systems and applications. This chapter provides an overview of each of these areas, with additional topics being addressed as warranted. References for further study are identified for those wishing to delve deeper.

3.1.1 *Energy and Power*

Though often used interchangeably, energy and power are distinct. *Energy* is frequently defined as "the capacity to do work." It can be thought of as a directionless quantity that can be used to effect changes. To date, energy has always been shown to be conserved, which is to say that it is neither created nor destroyed, but only converted into different forms.[1] Energy can be measured in joules (J) or kilowatt-hours (kWh), each of which is derived from three of the seven metric base units for mass, length, and time [3,4], which are measured in kilograms (kg), meters (m), and seconds (s), respectively:

$$J = \frac{kg \cdot m^2}{s^2} \tag{3.1}$$

$$kWh = 1,000\,W \cdot 3,600\,s = 3.6 \times 10^6\,J \tag{3.2}$$

Power measured in watts (W) relates to how quickly energy is delivered, used, or otherwise transferred, and thus requires an additional factor of 1/s to the base unit definition of joule to become the base unit definition of power:

$$W = \frac{kg \cdot m^2}{s^3} = J/s \tag{3.3}$$

A situation involving a small amount of energy could exhibit high power if the time needed to transfer that energy is relatively short. Conversely, a situation involving a very large amount of energy can still exhibit low power if the amount of time needed to transfer that energy is extremely long.

The energy unit kilowatt-hour is more familiar to many than the joule by virtue of kilowatt-hours being a feature of electricity bills. There may sometimes be occasions where energy consumption is described in terms of kilowatt-hours used during a time period. This may be intended to convey a total amount of energy consumed, rather than as a conventional expression of power, though in both cases, a quantity of energy for a given unit of time is presented.

For power beaming, it is important to consider the peak power transfer level that a system can achieve, as well as the total energy that can be transferred within a given time period. The duty cycle,

---

[1]It should be noted that it is possible to convert energy into mass and vice versa per Einstein's famous $E = mc^2$, but this does not violate the principle of conservation of energy, as mass is a form of energy [2].

or ratio of the system's time operating in an "on" state compared to its being in an "off" state, will affect how much energy is delivered. This duty cycle may be constrained by the operating scenario, system thermal performance, or other limitations. For systems that can operate at a range of power levels, as opposed to simply being "on" or "off", it may be necessary to create a more sophisticated model and to sum or integrate fractional or instantaneous operating power levels over time periods of interest.

3.1.2 *Light*

Power beaming can be accomplished with different modalities and implementations. Most cases to date have utilized EM waves, which is the focus of this book. Many of the concepts described herein are extensible to modalities that do not involve EM waves but that still employ waves or means that exhibit wave-like characteristics.

The EM spectrum is continuous. Whether one is concerned with power beaming links that operate in the microwave, visible, or any other region of the EM spectrum, the key difference between them is the wavelength of the EM waves. Alternatively, this difference can be thought of in terms of frequency, with the relationship between wavelength and frequency described as

$$\lambda f = \frac{c}{n} \tag{3.4}$$

where λ is the wavelength (in meters, m), f is the frequency (in Hertz, 1/s), c is the speed of light (in meters per second, m/s), and n is the index of refraction of the transmission medium. By definition, the value of n in a vacuum is 1.0, which is the value typically taken for propagation calculations even in air (where n is ~1.00029). The value of c is approximately 300 million meters per second [5].

While radio waves or X-rays may not always be thought of as "light", they are both EM waves that have wavelengths occurring in different parts of the EM spectrum and are still subject to Equation (3.4). Waves occurring at any portion of the EM spectrum can be considered as light, and this book uses the term light to refer to EM waves across the spectrum. A wide range of the EM spectrum is shown in Figure 3.1.

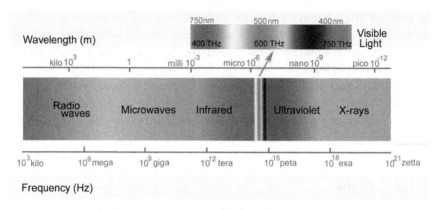

Figure 3.1. The electromagnetic spectrum [6].

In one sense, microwave and optical power beaming are the same thing. Both project a beam of EM radiation from an emitting transmitter to a collecting receiver which converts that beam into electricity. In another sense, the two are drastically different by virtue of the wavelengths involved. Microwave power beaming uses wavelengths generally ranging from about 3 mm up to about 300 mm, whereas optical power beaming uses wavelengths typically in the range of about $0.4\,\mu$m up to about $2\,\mu$m. These wavelength ranges differ by factors of around 1,500× to around 750,000×. For instance, the wavelength ratio of 5 GHz microwave to $1\,\mu$m laser is 60,000×. This means that the mechanisms and components used to convert electricity to EM radiation and vice versa can be substantially different, with implications for efficiency, cost, size, safety, and more.

One practical implication of the different wavelength size scales is that emitters and beam-shaping apertures for optical wavelengths can easily have sizes that are thousands of times the wavelength, whereas, for microwave wavelengths, single emitter apertures in a phased array might be smaller than a single wavelength and shaping aperture sizes might only be a small multiple of the wavelength. This ratio of aperture to wavelength has further implications for beam shaping and intensity profiles. Some examples of relevant dimensions and ratios:

(1) A "small" 2.5 cm lens has a diameter that is around 25,000× the size of its typical optical operating wavelength.

(2) A "massive" 70-m antenna, such as those used in the Deep Space Network, is around 600× as large as its 12 cm operating wavelength.

Having a large shaping aperture relative to the wavelength affects the degree of beam divergence and diffraction.

3.1.3 *Coupling*

In Chapter 1, power beaming was defined as "uncoupled wireless power transmission in which the link distance exceeds the sum of the largest dimensions of the transmit and receive structures, and where at least 1% end-to-end energy transmission efficiency is obtained". A key element within this definition is the concept of coupling, which was introduced in brief at the end of Section 1.3. From the power beaming definition, what does it mean for a transmitter and receiver to be uncoupled? At a basic and intuitive level, an uncoupled link can be thought of in these terms: *a change at the receiver will not substantively affect the power drawn from the transmitter.* This stands in contrast to coupled forms of wireless power transmission, such as inductive and capacitive resonance, where matching the impedances on the transmitter and receiver sides can be critical in establishing a resonance condition that ensures energy is transferred effectively [7]. Authors have categorized the degree to which components or systems couple as weak, strong, over, under, critical, etc. This degree exists on a continuum and will be considered as negligible between the transmitter and receiver in the case of power beaming. For an in-depth treatment of the mathematical underpinnings of coupling and coupled mode theory and the considerations involved, readers can consult Haus [8, Ch. 7] and Loisell [9].

An everyday example of an uncoupled link can be found by considering a solar-illuminated photovoltaic panel. It's clear that whether all or none of the energy available from the panel is used, it will have no effect on how much energy the sun produces. Similarly, the shining of a flashlight across a room onto an absorptive or reflective surface will not affect how long the flashlight's battery lasts. Finally, it's likewise evident that the act of tuning into a particular radio station doesn't increase the power consumption for a distant broadcaster.

Though power beaming doesn't involve meaningful direct coupling between the transmitter and receiver, there is utility in applying the concept of coupling in another sense. The transmitter can be thought to couple energy into an emanated wave that propagates in free space or the medium that exists between the transmitter and receiver. The transmitter's design and implementation will influence how effectively it couples or launches[2] the energy into free space or the transmission medium for reception by the receiver. Likewise, how effectively the receiver couples to free space or the transmission medium for the wavelength of interest will affect how easily it will be able to capture the energy. Impedance matching of rectenna elements in the microwave region and the use of anti-reflection coatings in the optical region are examples where coupling to free space or the transmission medium is improved to reduce the reflection of incoming energy.

Depending on the operating wavelength and dimensions of the transmit structure, energy may be stored near the transmit structure in the form of oscillating fields prior to their coalescence into propagating waves. This allows for the same transmitter and receiver to be coupled with each other when in close proximity but uncoupled when their separation exceeds a certain distance.

What boundary determines the distance at which a transmitter and receiver transition from being coupled to uncoupled with each other? The answer is not always straightforward, as it depends on the particulars of the system's operating wavelength and the characteristics of the transmitter and receiver. In general, the maximum distance from a transmitter to where effective coupling to a receiver can occur decreases with decreasing wavelength, as shown graphically in [10, p. 81] by Ohira.

3.2 Optical Models

In most power beaming scenarios, the transmitters and receivers will be uncoupled as described previously. As a result, power beaming

---

[2] "Launching" can be a useful way to think about the energy sent from a power beaming transmitter, as it evokes the mental image of a rocket traveling skyward from a pad. Once launched and sufficiently distant, what the rocket does (unless it returns) no longer affects what happens at the pad, and vice versa.

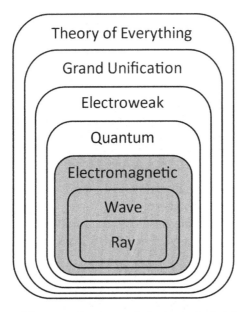

Figure 3.2.   Theories of optics and physics, adapted from [11,12].

links can be approached from the perspective of optics. A hierarchy of theoretical paradigms that have evolved over time for optics and physics is depicted in Figure 3.2.

The sophistication of non-classical theories generally isn't necessary for an adequate practical treatment of power beaming links. The three theories in the shaded region of Figure 3.2 will almost always suffice and are summarized next from innermost to outermost, in an order reflecting increasing complexity.

3.2.1   *Ray Optics*

In ray optics, light is modeled as a ray that travels in a straight line unless it impinges on a surface or is affected by a change in the medium, such as via refraction (Snell's law) [11, p. 54]. Ray optics is an approximately accurate model for power beaming if the wavelength of the light is much shorter (for instance, by at least a factor of 100) than the dimensions of the transmit and receive structures and the distance between them. If the distance between the transmit and receive structures becomes very large relative to the aperture diameters, divergence (which is inherent to the light

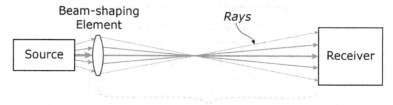

Figure 3.3. Depiction of a power beaming link using ray optics.

source) plus diffraction effects result in the beam spreading out as it traverses the distance to the receiver.

A power beam can be modeled by a collection of rays between the transmitter and receiver, or emitter and absorber, as shown in Figure 3.3.

While this depiction shows an optical element for focusing the beam, this should be thought of as generalized to apply to longer wavelengths as well, where the same function might be realized via a parabolic reflector, phased array, or other means. It is also assumed that the receive structure is perfectly absorptive and that none of the energy carried by the rays is reflected, although this assumption is not required for general analysis in this model. Many ray-tracing models attribute differing amounts of optical power in each ray, enabling an approximation to the beam intensity profile. Defining an axis between the centers of the transmitter and receiver permits the introduction of a helpful tool: the paraxial approximation, which holds that the rays are taken to make a small enough angle θ with the axis such that $\sin \theta \approx \theta$ [11, p. 9]. The paraxial approximation can often be applied to simplify the analysis of power beam shaping.

3.2.2 Scalar Wave Optics

Ray optics has utility but falls short for most real-world power beaming situations because it does not account for diffraction or interference. Scalar wave optics, or simply wave optics, is a better descriptor in which the wave nature of light is accounted for, and the concept of wavelength is introduced.

Waves can be found nearly everywhere. They exist not just as EM radiation, but in mechanical, acoustic, and other systems as well.

In each case, the wave behavior can be described by a mathematical relationship between a vector quantity, the speed of the wave, and time.

For electromagnetism, the differential forms of Faraday's law and the Ampère–Maxwell law in conjunction with vector identities and the assumption of a current- and charge-free region can be used to derive the wave equation [13, p. 122]. This goes with the idea that one field can induce the other, and vice versa. The general form of the wave equation is

$$\nabla^2 \mathbf{A} = \frac{1}{c^2} \frac{\partial^2 \mathbf{A}}{\partial t^2} \qquad (3.5)$$

In this form, $\mathbf{A}$ is a vector quantity which can represent either the electric or magnetic field's amplitude and direction, c is the speed of light in a vacuum, and t is time. The wave equation is a linear, second-order, homogeneous partial differential equation with various possible solutions; further discussion can be found in Fleisch and Kinnaman [14]. Irradiance (power per unit area, W/m^2) is the square of the wave function.[3] Commonly referenced solutions include plane, spherical, and Gaussian (an approximate solution [11, p. 106]). For power beaming, it is often useful to consider waves as having the solution in a form that is essentially Gaussian because it can closely match the beam profile from single/low-mode lasers and highly directive microwave and millimeter wave apertures.

A refined depiction of a power beaming link using the (scalar) wave optics model is shown in Figure 3.4.

Note that the arrows depicting rays in Figure 3.3 are orthogonal to the wavefronts represented by the shadowed bars in Figure 3.4. Lateral variations in the amplitude of the wave are represented with changes from light gray to dark. The wavelength is represented by the distance between the center of two adjacent peaks.

In addition to wavelength and amplitude, waves have a propagation speed and a phase. These qualities permit the description of

---

[3]This can also be called intensity, but it may introduce confusion because the term "intensity" is used in different ways across various disciplines. Irradiance and intensity are both used is used in this book depending on the context. Background, disambiguation, and explanation are presented by McCluney [15, p. 20] and Paschotta [16].

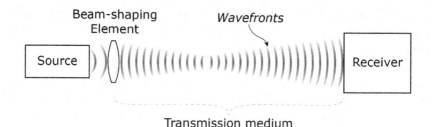

Figure 3.4.   *Depiction of a power beaming link using scalar wave optics.*

interference and diffraction, which have relevance for power beaming as they may affect the amount of energy delivered to the receiver. For power beaming scenarios involving EM radiation, scalar wave optics don't address polarization, which can likewise affect the amount of power recovered at the receiver for microwaves and the transmissivity of visible and near-IR light through certain optics. It also doesn't adequately address scattering and absorption. For those topics, an exploration of EM waves is needed. Though an overview of this transition is provided in the next section, readers are directed to Saleh and Teich for a more in-depth treatment of the distinction between scalar wave optics and EM vector optics [11, p. 180].

3.2.3   *Electromagnetic (Vector Wave) Optics*

Regardless of where on the EM spectrum waves appear, they are comprised of coupled oscillating electric and magnetic fields that induce each other. This interaction results in a propagating wave. The behavior of the electric and magnetic fields, the wave of which they are a part, and the power that the wave delivers are described respectively by Maxwell's equations, the wave equation, and the Poynting vector. These can account for effects including polarization, nonlinear behavior, and some aspects of reflection and refraction at interfaces.

3.3   Maxwell's Equations

Maxwell's equations are named for Scottish mathematician James Clerk Maxwell, and while they are taught today as four equations

(Gauss' law for electric fields, Gauss' law for magnetic fields, Faraday's law of induction, and the Ampère-Maxwell law), Maxwell actually formulated 20 equations to describe the behavior of electric and magnetic fields. It was years after Maxwell's death that FitzGerald, Heaviside, Lodge, and Hertz synthesized from those original equations the four used today [17].

Of these four equations, two are of principal interest: Faraday's law of induction and the Ampère-Maxwell law. The first states that a magnetic field that changes with time produces a circulating electric field, and the second states that an electric current or time-varying electric field produces a circulating magnetic field [13, pp. 75, 101]. Taken together, it can be intuited that changing electric and magnetic fields could induce each other indefinitely when their energy doesn't go elsewhere.

Deeper historical [17] and mathematical [18,19] explorations of Maxwell's equations may be found in other texts and are not discussed further here. Readers with limited technical backgrounds who wish to understand Maxwell's equations are advised to consult Fleisch's short book [13], which dwells solely on them and the required background material. This stands in contrast to most texts which instead place them in the larger context of classical EMs and electrodynamics.

A depiction of power beaming employing EM optics with a Gaussian irradiance distribution and phase focusing is shown in Figure 3.5. This arrangement results in the beam waist appearing between the transmitter and receiver. Compared with the ray and wave optics depictions of Figures 3.3 and 3.4, respectively, Figure 3.5 starts to

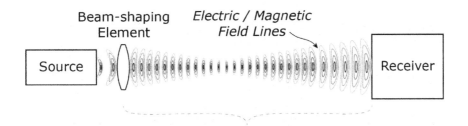

Figure 3.5.   Depiction of a power beaming link using electromagnetic (vector wave) optics.

resemble a real power beaming system more accurately. The contour lines represent electric or magnetic field lines but note that in Figure 3.5 that the wavelength shown is comparable in size to the system elements. For shorter wavelengths, the groupings of field lines will become more densely packed.

The defining parameters and subtleties concerning Gaussian beams receive full treatments in [11, Sec. 3.1, 20]. Additional beam-like solutions to the wave equation (including Ince-Gaussian beams, Bessel beams, and Airy beams) are introduced in [11], and are beyond the scope of this book.

3.4 The Poynting Vector

The Poynting vector is of interest for power beaming, as it expresses power flow per unit area. It can be stated as

$$\mathbf{S} = \mathbf{E} \times \mathbf{H} \tag{3.6}$$

where $\mathbf{S}$ is the power density vector of the EM field in watts per square meter (W/m^2), $\mathbf{E}$ is the electric field intensity vector in volts per meter (V/m), and $\mathbf{H}$ is the magnetic field intensity vector in amperes per meter (A/m) [21]. Since the electric and magnetic field intensities can vary with time, $\mathbf{S}$ can likewise vary with time and may be most useful when time-averaged [22, p. 384]. $\mathbf{E}, \mathbf{H}, \mathbf{S}$, the direction of wave propagation $\mathbf{R}$, and the wavelength λ are shown in Figure 3.6. The fields E and H oscillate in their respective planes perpendicular to the direction of the Poynting vector. The oscillations, and the relationship between oscillations, define the polarization of the propagating wave. Power travels in the same direction as the propagating wave. In media, the wave may be subject to scattering and absorption, as discussed in Chapter 4.

3.5 Diffraction Regimes

Depending on the distance from the aperture or Gaussian beam waist, diffraction effects can be modeled with different approximations. For power beaming, a sense of the appropriate approach to use can be used by finding the Fresnel number. The Fresnel number N_F

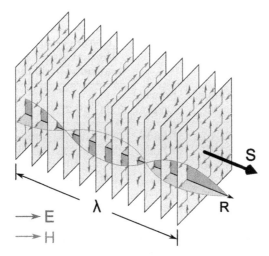

Figure 3.6. A linearly polarized electromagnetic plane wave showing fields and power propagation [23].

of a circular aperture of radius r can be found for a given wavelength λ at distance l [11, p. 121]:

$$N_F = \frac{r^2}{l\lambda} \tag{3.7}$$

When the Fresnel number is approximately one or larger, Fresnel diffraction will apply [24, p. 512]. At this distance, for a uniformly illuminated circular aperture, the irradiance will have a maximum [24, Fig. 10.63], making it a potential point of interest for power beaming, although power beaming may use Gaussian instead of uniform illumination. This distance might also be called the Fresnel range, but this introduces the possibility of confusion with the terms Fresnel length and Fresnel distance, which are defined and used differently.[4] Table 3.1 shows the distance in meters at which the Fresnel number is one for transmit apertures with diameters of 5, 20, 100, and 500 cm for a variety of laser and microwave wavelengths. This shows why microwave and radio-frequency power beaming require relatively larger apertures to stay within the Fresnel diffraction regime.

---

[4] "Fresnel distance" and "Fresnel length" have been used interchangeably, and have been defined independently of aperture size as $\sqrt{l\lambda}$, see [25,26].

Table 3.1.   Comparison of distances where the fresnel number is one.

	Wavelength (m)	Source type	Distance (m) at which Fresnel # = 1 for given transmitter diameter (m)			
			0.05	0.2	1	5
Laser	8.08E-07	Diode	774	12,376	309,406	7,735,149
	9.76E-07	Diode	640	10,246	256,148	6,403,689
	1.08E-06	Fiber	581	9,302	232,558	5,813,953
	1.05E-05	CO_2	60	952	23,810	595,238
Microwave	0.0030	100 GHz	0.2	3.3	83	2,083
	0.030	10 GHz	0.021	0.33	8.3	208.3
	0.052	5.8 GHz	0.012	0.19	4.8	120.8
	0.1224	2.45 GHz	0.005	0.082	2.0	51.0

The implication is that for most single aperture cases where there is a hard limit on the size of the transmit aperture for a given wavelength, it will be necessary either to have a comparatively large receiver or to constrain the distance the receiver is located from the transmitter. The converse is true if there is a hard limit on the size of the receiver aperture. This tradeoff is also captured in the Goubau relationship approximation described in Section 3.9.5.

At distances where the Fresnel number is much greater than one, then Fraunhofer diffraction is a better model for beam intensity profile. See [11, p. 130] for more details.

The terms Rayleigh length, Rayleigh distance, and Rayleigh range are also used but may be defined differently depending on the context. In antenna theory, the Rayleigh distance has been defined as the distance to the far field [27, p. 42]

$$l_{ff} = \frac{2r^2}{\lambda} \qquad (3.8)$$

where r is the maximum dimension of the antenna [28, p. 39]. However, for Gaussian beams, the Rayleigh range is defined as [29, p. 668].

$$z_R = \frac{\pi W_0^2}{\lambda} \qquad (3.9)$$

where W_0 is the beam waist radius. The Rayleigh range can also vary depending on the beam's level of coherence [30]. To avoid ambiguity in this text, the distance at which the Fresnel number is one has been used, as seen in Table 3.1. A further discussion of the subtleties of defining field regions appears in Section 3.9.4.

3.6   Beam Profiles

An approximate way to think about beam profiles and propagation is that when the Fresnel number is one or higher, a beam profile can be meaningfully shaped at the receiver. For a Gaussian beam, the depth of focus is twice the Rayleigh range [11, p. 85] as it was defined in Equation (3.9). At a great enough distance, beams devolve to an Airy or Gaussian-like profile due to diffraction. This far-field intensity distribution can be found by applying a Fourier transform to the aperture's intensity distribution, as predicated by Fourier optics [31,32]. At extremely great distances a localized sampling of the irradiance distribution will give results resembling a plane wave [19, p. 560].

A Gaussian beam has an intensity profile J with this form [33]:

$$J = Ae^{\left(\frac{-2r^2}{w^2}\right)} \tag{3.10}$$

where J is the beam intensity in W/cm^2, A is the peak beam intensity, r is radial distance from the center of the beam, and w is the radius association with the characteristic width. The total captured power is defined by radially integrating the intensity:

$$P(r_0) = \int_0^{r_0} \int_0^{2\pi} Ae^{\left(\frac{-2r^2}{w^2}\right)} r\, dr\, d\theta = \pi A \left(1 - e^{\left(\frac{-2r_0^2}{w^2}\right)}\right) \tag{3.11}$$

From this equation, the power within a given diameter r_0 is linearly proportional to the intensity at a point.

As part of conveying the beam profile, beam divergence angle and beam parameter product (BPP) are usually characterized in measurements and product specification sheets for lasers at one of a variety of power points such as the full width at half maximum (FWHM), $1/e^2$, 95%, $1/e^4$, and possibly others. There are technical definitions for these points for non-Gaussian beams, and the system

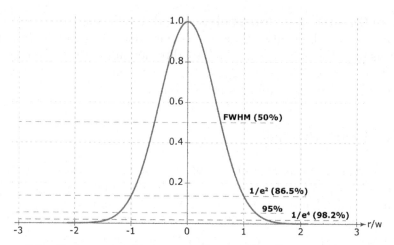

Figure 3.7.   Various points for measuring beamwidth, relative to a Gaussian beam profile.

designer should be certain that they are accounted for. The graph in Figure 3.7 shows where these levels fall for a Gaussian profile ($w = 1$ and $A = 1$). Though similar specifications are not typically given for microwave transmitters, the principles for a highly directive transmitter are the same.

ISO Standard 11146 "Lasers and laser-related equipment — Test methods for laser beam widths, divergence angles, and beam propagation ratios" uses the second moment of the beamwidth as a function of the beam intensity $I(x, y)$. This method is called the $D4\sigma$ method because the diameter is 4× the standard deviation of the intensity profile.

$$w_x = 2\sqrt{\frac{\int x^2 I(x, y)dxdy}{\int I(x, y)dxdy}} \tag{3.12}$$

This method is not as accurate when the tails of the intensity distribution are large.

Figure 3.8 compares an Airy pattern to a Gaussian profile. The first zero of the Airy is at $r = 3.83$, and the Gaussian is down to the 1.8% ($1/e^4$) point at $r = 3.73$. At $r = 3.83$, the Gaussian is down to 1.5%. The difference between the two along their radial direction varies by nearly ±3% of the peak intensity.

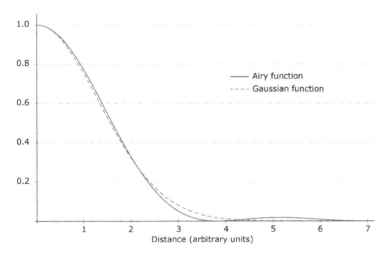

Figure 3.8.   Comparison of an Airy pattern with a Gaussian profile.

In many cases, the beam profile is flatter near the middle and falls off more steeply at the edges. The Gaussian equation is often a reasonable way of approximating some of these profiles:

$$J = Ae^{-\left(\left(\frac{r}{w}\right)^2\right)^{SG}} \tag{3.13}$$

The super-Gaussian number SG determines the deviation from a simple Gaussian. Figure 3.9 compares a standard Gaussian (SG = 1) to examples where SG is set to 2, 3, and 8. As SG goes to infinity, the profile becomes uniform and is known as a "top hat".

Other profiles are possible and may have applicability for particular situations. Hansen created a single-parameter approach in a microwave context for determining aperture distributions and the resulting beam characteristics [34], and others have explored flat-top and pseudo-Bessel beams [35]. Different varieties of annular beams are possible as well [36]. While this discussion has centered on optical concepts and terminology, they are largely extensible to the longer wavelength scenarios of the millimeter and microwave regimes where the apertures are comparatively large, as has been done in texts on quasioptics like [20]. The use of diverse forms of phased arrays makes both the range of possible apertures and possible beam profiles effectively limitless.

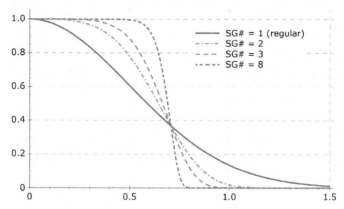

Figure 3.9.   Super-Gaussian profiles with factors of 1, 2, 3, and 8 from left to right nearest the y-axis.

3.7   Radiance and Étendue

Modeling a power beam mathematically using exact or approximate solutions to the wave equation may not always be necessary for developing engineering designs. Presented here are a pair of concepts intended to be useful for the engineering design process, especially in the earlier stages of scoping and design.

A way of estimating system performance limits based on major design parameters such as aperture sizes, wavelength, range, and power is with radiance. Informally, radiance is sometimes referred to as "brightness" but this usage should be confined to non-quantitative characterizations [37]. Radiance is related to étendue, which is depicted conceptually in Figure 3.10. Here A_1 is a transmitter with radius r_1, and A_2 is a receiver with radius r_2.

The solid angle Ω_1 in steradians is the surface area of the target divided by the area of a sphere with a radius equal to that distance. Étendue is the product of the area of the transmitter and the solid angle of the receiver as seen from the transmitter (shown as the shaded region, cone-like structure, of Figure 3.10), modified by the index of refraction n between the two points and the angle ϕ_1 that the beam centerline connecting the two apertures make relative to each other:

$$\text{Étendue} = n^2 \partial\Omega_1 \partial A_1 \cos\phi_1 \approx A_1 \Omega_1 \tag{3.14}$$

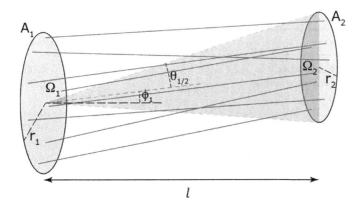

Figure 3.10. Depiction of étendue. Solid lines are light rays.

Étendue represents the "collecting power" of an optical system, and it is a measure of how much light can be efficiently collected. Étendue is conserved, in that it can only stay constant or increase. Increasing étendue represents a worsening of the ability to collect power. In real-world systems, étendue will only increase, and often the system will be designed to minimize this increase as much as possible. An entertaining layperson's explanation of the conservation of étendue is provided by Munroe [38]. Further details are provided by Koshel [39].

Radiance L is power P divided by étendue, shown here for the transmitter aperture:

$$L_1 = n^2 \frac{\partial^2 P_1}{\partial \Omega_1 \partial A_1 \cos \phi_1} \approx n^2 \frac{P_1}{A_1 \Omega_1} \tag{3.15}$$

It will be assumed that the medium of operation, typically air or vacuum for power beaming, will have a refractive index n that is approximately 1. Radiance is also conserved because it is effectively the conservation of energy combined with the conservation of étendue.

While using radiance and étendue in this manner is often associated with optical systems, [40, App. C] shows its application to microwave power beaming as well.

Using some simple assumptions, we can use radiance to characterize light sources. Table 3.2 shows relative approximate radiance values for different sources. Lasers have radiance values ranging

Table 3.2.   Comparison of radiance values for a variety of light sources.

Light source	Radiance (W/m^2*str)	Radiance (TBr)
Sun (6,000 K blackbody)	$2 * 10^7$	0.000020
Microwave oven magnetron (2.45 GHz)*	$5 * 10^7$	0.000050
LED, 810 nm, 0.6 W	$7 * 10^6$	0.000007
1 mW red laser pointer, near-single mode	$2.5 * 10^9$	0.0025
Fiber-coupled diode laser, 808 nm, 130 W, 200 μm fiber ($\sim$0.19NA)	$4 * 10^{10}$	0.04
Fiber-coupled diode array, 976 nm, 220 W, 105 μm fiber ($\sim$0.20NA)	$2 * 10^{11}$	0.2
CO$_2$ laser, 1 kW @ 10.6 μm, M^2 nearly 1	$9 * 10^{12}$	9
Fiber laser, 10 kW, 1.07 μm, in 100 μm fiber (8 mm*mrad)	$1.6 * 10^{13}$	16
Fiber laser, near-single-mode ($M^2 = 1.3$) 10 kW, 1.07 μm	$5 * 10^{15}$	5,000

Note: *Value for microwave oven magnetron is estimated.

from 10^{10} to 10^{16} W/m^2*str, so a handy shorthand notation is to shorten the old laser-industry word for radiance, "brightness", down to "bright" to replace W/m^2*str, and to use 10^{12} ("tera" in exponential notation) as the default scale to reduce the need for exponential notation and simplify radiance values to "terabrights" ("TBr"). In this nomenclature, a 10-kW multimode beam from a fiber laser with a 100 μm core diameter fiber and a 0.16 NA would have a radiance of about 16 TBr. The Sun, in comparison, has a radiance of only 0.00002 TBr, which is why sunlight can't be concentrated usefully for long-distance projection. The approximately 10 orders of magnitude representing the range are an important factor when designing power beaming systems.

The BPP described in Section 3.3 provides a measure of both the spatial size and angular divergence of the laser beam and is commonly expressed in units mm-mrad (millimeter-milliradians). A smaller BPP value indicates a smaller beam waist radius and/or lower divergence, resulting in a more tightly focused and collimated beam. For the same power, a smaller BPP indicates more of the power can be concentrated in a smaller area at a distance. BPP is defined as the beam radius times the beam divergence half-angle, or equivalently the beam quality factor M^2 times the minimum possible BPP:

$$\text{BPP} = r_i \theta_{\frac{1}{2}} = \frac{M^2 \lambda}{\pi} \tag{3.16}$$

where $\theta_{1/2}$ is the half-angle of the beam divergence measured in radians and M^2 is a parameter that is measured in accordance with ISO Standard 11146 [41]. M^2 is widely used to assess the quality of a laser beam and can be thought of as a measure of how closely that beam is to the ideal Gaussian shape. The M^2 value is dimensionless, and when equal to 1 it describes a theoretically ideal Gaussian. That case also defines the minimum possible BPP $= \lambda/\pi$. For all other beams, M^2 is larger than 1. BPP is normally defined at the $1/e^2$ point, but if defined at a different point, such as the $1/e^4$ point, the relation still holds. Radiance is then equivalent to

$$L_1 = n^2 \frac{P_1}{A_1 \Omega_1} = \frac{P_1}{(\pi \text{BPP})^2} \tag{3.17}$$

Beam divergence is nominally defined at distances much larger than the Rayleigh range as it was defined in Equation (3.9) (see more in [11, p. 81]), but a collimated beam might have a Rayleigh range much longer than the distance to the receiver.

For $M^2 = 1$, the minimum BPP for a 1,070 nm fiber laser is about 0.34 mm mrad. At 808 nm it is 0.26 mm mrad, and for a 10 μm CO_2 laser it is 3.3 mm mrad. The minimum BPP for 2.45 GHz microwave is 38,977 mm mrad, whereas for 10 GHz it is 9,549 mm mrad.

The radiance equation can be rewritten to compare the limits of how much power can be delivered versus distance for a given set of aperture diameters and beam source types:

$$P = \frac{L A_{\text{TX}} A_{\text{RX}}}{l^2} \tag{3.18}$$

where l is the separation distance between the transmitter and receiver. Using the 16 TBr value for the 10 kW multi-mode fiber laser example from above and assuming a transmit beam aperture diameter of 30 cm and a receive beam aperture diameter of 40 cm, then $Pl^2 = 1.4 \cdot 10^{11}$ W·m^2. For that specific laser power of 10 kW, the maximum distance to theoretically capture all the power (not accounting for atmospheric losses) within the two apertures is about 3.77 km.

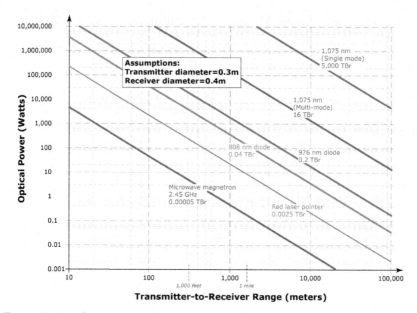

Figure 3.11.   *Comparison of maximum possible power vs. distance for various source types, assuming a transmitter diameter of 30 cm and a receiver diameter of 40 cm, neglecting atmospheric losses.*

If the fact that multi-mode fiber lasers are made at different power levels is used, and it is assumed they can be combined while maintaining brightness, then the equation above defines a curve showing the possible trade-offs between the amount of optical power transmitted and the maximum separation between transmitter and receiver. Given the many orders of magnitude of radiance separating different emitter source types and possible aperture sizes, we can graph that curve for many source types (using the values from Table 3.2) on a log-log plot, which is shown in Figure 3.11. As one example, one could send 10 W of red laser pointer light to a maximum distance of just 1.49 km with the 30 and 40 cm apertures, if one optically combined 10,000 laser pointers (outputting 1 mW each) perfectly. Equation (3.18) could be rearranged to hold optical power and receiver aperture constant, and instead compare source types on transmit aperture size versus distance.

Because the radiance values of source types depend on the amount of power as well as beam quality, the values presented may improve in the future. Single-mode lasers have come out with increasing output power. Microwave sources generally aren't characterized using radiance, but improving device thermal management to emit more power with the same beam quality could also improve this quantity.

In general, power beaming with smaller apertures will result in more favorable beam collection efficiencies when the source's radiance is high. Because BPP depends on wavelength, higher radiance is associated with shorter wavelengths. This isn't to say that beam collection efficiency can't be high at longer wavelengths, but it does mean the apertures need to be larger.

3.8 Aperture Sizing

With the concepts described previously, a key question can be addressed: within the systems' constraints, what sizes should the apertures be to maximize the power transferred?

As with many instances in engineering, the answers to this question may be found iteratively.

The different models described so far could produce different results. To decide the minimum required fidelity for the situation, care must be exercised in parameter definitions that properly model the outcome. For example, when using a Gaussian beam model, the beamwidth is usually defined at the point where the intensity is $1/e^2$ (about 13.5%) of the peak intensity. This approach makes calculations simpler but may also lead to the discarding of a lot of power.

Many treatments introduce the diffraction limit to resolving a spot in the context of imaging. These treatments describe the Airy disk and the first zero in intensity as

$$\theta_{\frac{1}{2}} \approx 1.22 \frac{\lambda}{d_1} \tag{3.19}$$

where $\theta_{1/2}$ is the half-angle of the beam divergence and is assumed to be small enough to make the small-angle approximation, and d_1 is the transmitter diameter. This model is used in imaging to distinguish between two points, and arises from the uncertainty principle as applied to a photon [42, p. 234]. Projecting a power beam, on the other hand, is different for multiple reasons: (1) it may not be operating in the "far field" (where Fraunhofer diffraction applies), which is assumed for the above equation; (2) it is likely important how much power is captured, and so the extent and shape of the beam profile needs to be understood. Note that when estimating beam spot size at the receiver, the $1.22\lambda/d_1$ factor frequently referenced in the literature is the point at which it is hard to distinguish two spots, typically for an imaging application, and not the point at which they overlap.

The equation for beam divergence can be used to derive the required transmit aperture size for a given receiver size in consideration of the size of the beam at the receiver:

$$d_1 = \frac{4l\text{BPP}}{d_2} \qquad (3.20)$$

where d_1 is the transmit aperture diameter, and d_2 is the receive aperture diameter. An important point to remember is that while Gaussian expressions can be used to describe power beams and their propagation, actual beam profiles are not precisely Gaussian. Non-single mode beams have different profiles, such as super-Gaussian, top hat, donut, and others as described in Section 3.3. Far from the transmitter, diffraction effects can further increase the effective required transmit aperture diameter, favoring instead the use of this expression:

$$d_1 = \frac{l(4\text{BPP} + 2.44\lambda)}{d_2} \qquad (3.21)$$

3.8.1   *Design Example*

Assume usage of a 976 nm CW fiber-coupled diode array, with a maximum output power of 2 kW. The diode array output is coupled into a fiber with 0.22 numerical aperture (NA), and a 400 μm core diameter. The beam is expanded through a telescope system to about 20 cm diameter and can be focused by imaging the fiber tip at distances from approximately 40 m to infinity. At 325 m, which is the expected beam size? Rayleigh range? Fresnel number?

The beam parameter product is BPP $= (1/20.4\,\text{mm})(220\,\text{mrad}) = 44\,\text{mm} * \text{mrad}$ which implies M^2 is about 142.

The Rayleigh range is

$$z_R = \frac{\pi W_0^2}{M^2 \lambda} = \frac{\pi (0.1\,\text{m})^2}{142 * 976\,\text{nm}} = 227\,\text{m} \tag{3.22}$$

Thus, at 325 m the receiver is not too much farther than the Rayleigh range.

The Fresnel number is

$$N_F = \frac{r^2}{l\lambda} = 32 \gg 1 \tag{3.23}$$

The beam is also well within the result from the radiative near-field region often used in the radiofrequency regime [43, p. 15], which is double the distance where the Fresnel number is one:

$$l < \frac{2r^2}{\lambda} = 81{,}967\,\text{m} \tag{3.24}$$

What is the significance of these calculations? Saleh and Teich point out that within the Fresnel regime, the profile is the "shadow of the aperture" [11, p. 135] and slowly morphs into the Fraunhofer diffraction pattern at much longer distances. Since the distance in this example is also within the Rayleigh range, with reasonable optics the beam profile should be relatively focused.

To determine the beam size, the simple estimate from using Equation (3.20) would show:

$$d_R = l \left(\frac{4\text{BPP}}{d_T} \right) \cong 29\,\text{cm} \tag{3.25}$$

This example is drawn from an actual implementation, which is described in [44,45]. The actual, near-top-hat-profile beam diameter was measured to be slightly over 30 cm.

3.9 Other Tools and Topics

In the discussion and exploration of power beaming theory and models, a range of topics beyond those already covered in the chapter may be encountered. This section addresses some of these in brief.

3.9.1   Nomenclature Differences

With many fields of study concerned with propagating EM waves in different parts of the spectrum, there has sometimes been a divergence of nomenclature for similar concepts. For instance, in antenna design, changing the amplitude and phase distributions across an aperture to reduce off-axis radiation is often termed sidelobe reduction. In optics, a method to achieve a similar goal might be referred to as apodization [46, p. 122]. In this text, the nomenclature convention observed varies but generally follows a rationale based on the wavelength and aperture sizes of the particular modality.

3.9.2   Misapplication of the Inverse Square Law

A concern sometimes raised about power beaming is that it can never be done efficiently because of the inverse square law. This law states that a quantity, such as electric field intensity, will decrease with the square of distance [47]. However, this law applies to isotropic point sources or situations where sources can be accurately modeled as points. While it is true that at a great enough distance, this law applies to every practical source, power beaming occurs within the distance range where beam divergence is minimal, as outlined earlier in this chapter.

3.9.3   Misapplication of the Friis Transmission Formula

Engineers with a background in wireless communications often misconceive the feasibility of power beaming because they are used to modeling links using the Friis transmission formula [48]:

$$P_{\text{RX}-\text{in}} = \frac{P_{\text{TX}-\text{out}} A_{\text{RX}} A_{\text{TX}}}{\lambda^2 d^2} \tag{3.26}$$

where power received $P_{\text{RX}-\text{in}}$ is a function of the power transmitted $P_{\text{TX}-\text{out}}$, areas of the transmit and receive apertures, the operating wavelength λ, and their separation d.

In his 1946 paper, Friis clearly conveys limitations on the use of this formula, including that it assumes plane waves, which may not be the case for power beaming scenarios. Furthermore, by inspection, it can be seen that if the wavelength λ is short enough, the formula will produce an anomalous result showing that more power can be received than was transmitted. For these reasons, it is typically more instructive to use the Goubau relationship described in Section 3.9.5 for approximating power beaming link theoretical performance. Shinohara shows graphically where the Friis formula and Goubau relationship diverge in [49].

3.9.4 *Field Regions*

In microwave and RF communities, apertures are often thought of as having a reactive near field, radiative near field, and far-field regions. These are sometimes equated with Fresnel and Fraunhofer regions, but this is discouraged for the reasons described by Hansen [50, pp. 31–32]. The IEEE Standard for Definitions of Terms for Antennas [43] rightly defines each of these five terms separately. An examination of the differing approaches to defining these field regions is found in Capp [51]. For power beaming and scenarios with focused beams, optical and quasioptical approaches are preferred [20].

3.9.5 *Goubau Relationship*

The Goubau relationship [52,53] can provide a quick, straightforward approximation of beam collection efficiency for power beaming links. It addresses geometric factors only and thus neglects inefficiencies of conversion or sources of loss in the transmission medium.

For planar apertures that are widely separated, perfectly efficient, and optimally illuminated without losses from misalignment or mismatched polarization, an expression for beam collection efficiency can be derived using methods outlined by Borgiotti in [54]. The parameter τ [52] is expressed as[5]

$$\tau = \frac{\sqrt{A_{\mathrm{TX}}A_{\mathrm{RX}}}}{\lambda d} \tag{3.27}$$

---

[5]The parameter τ has been defined differently by different authors [55–57]. This text uses the definition from [55].

where A_{TX}, A_{RX}, λ, and d are the transmit aperture area, receive aperture area, the wavelength of operation, and the distance between the apertures, respectively.

Then τ can be used in the following expression to find the beam collection efficiency, as in [58, p. 24]:

$$\eta_{BC} = e^{-\tau^2} \tag{3.28}$$

It is critical to recognize that η_{BC} is not a hard limit on nor a guarantee of the efficiency of a power beaming link, but rather a useful approximation. In practice, neither the transmitter nor the receiver will be perfectly aligned and completely efficient. Other sources of loss will be present.

The condition of "optimal illumination" means that the phase distributions of the antennas are focused on each other and that the amplitude distributions are "equal to the field patterns of the mode with minimum diffraction loss in two confocal resonator mirrors, whose separation, sizes, and shapes are equal to those of the apertures in question" [54]. In practice, the amplitude distribution typically closely resembles a Gaussian distribution. As the distance between the transmitter and receiver increases greatly, the beneficial effects of optimal phase and amplitude distributions are lessened [59].

As the Goubau relationship shows intuitively, the sum of aperture areas is minimized for a given link distance and beam collection efficiency when the apertures are equal in size. Robert Winsor has called this "the equal aperture rule."

3.9.6   Unfilled Apertures

Since the directivity of an aperture is increased with its increasing maximum dimension, using unfilled apertures has sometimes been suggested for power beaming. Unfilled apertures, also called sparse or thinned arrays, will provide a narrower beam if they are mutually coherent. However, much of the transmitted energy will be in the resulting grating lobes, away from the central axis of the beam. A derivation of why this happens can be found in [60] and further considerations concerning the use of thinned arrays can be found in [61,62, pp. 2–27].

3.9.7 *Superdirectivity*

Though perhaps counterintuitive, it is true that "A fixed aperture size can achieve (in theory) any desired directivity value" [63, p. 102]. This finding arises from Bouwkamp and de Bruijn's landmark 1946 paper [64]. However, important practical constraints on the implementation of such apertures limit their potential to improve power beaming link performance. Hansen suggests several constraints in optimizing a superdirective antenna design: antenna quality factor Q, sidelobe level, tolerance (both electrical and mechanical), and efficiency [63, pp. 110–111].

3.9.8 *Sub-diffraction Methods*

Though yet to be demonstrated in a practical power beaming context, proposed methods for overcoming the diffraction limit have been elucidated. These include using far-field superlenses to convert evanescent fields into propagating ones [65]. The use of metamaterials and superlenses may be able to enhance power beaming systems in the future.

3.9.9 *Speculative Technologies for Moving Energy*

The emergence of interest in orbital angular momentum (OAM) in the 1990s and increased activity in quantum technologies in the 21st century suggest notional ways wireless power might be performed. Though the focus for OAM has principally been on enhancing the capacity of communications channels, there has been some association with energy transfer [66]. In 2023, after many years of theoretical exploration, there was finally a reported instance of successful quantum energy teleportation (QET) by Ikeda [67]. Though there may be no theoretical limit on the distance over which QET might operate, the first demonstrations were "over distances roughly the size of a computer chip" [68]. Note that QET wireless power transmission does not exceed the speed of light in theory or practice.

3.10   Conclusion

This chapter reviewed fundamentals of energy, power, waves, beams, and some of the tools and models that might be used in developing a power beaming system. These included the constructs of Gaussian beams, radiance, and étendue. Beam profiles were discussed, and an example drawn from practice was presented. A range of concepts and misconceptions that sometimes arise in power beaming contexts were addressed.

3.11   Further Reading

The following texts expand further on many of the topics explored in this chapter:

B. E. A. Saleh and M. C. Teich, *Fundamentals of Photonics*, Third edition. Wiley, 2019.

N. Shinohara, *Wireless Power Transfer via Radiowaves*. Wiley, 2014.

J. L. Miller and E. Friedman, *Photonics Rules of Thumb: Optics, Electro-Optics, Fiber Optics, and Lasers*, Third edition. SPIE Press, 1996.

P. F. Goldsmith, *Quasioptical Systems: Gaussian Beam Quasioptical Propagation and Applications*. IEEE Press, 1998.

References

[1] A. Einstein, L. Infeld, and W. Isaacson, *The Evolution of Physics: From Early Concepts to Relativity and Quanta*. Simon & Schuster, New York, NY, 2007.

[2] C. Moskowitz, "Fact or Fiction?: Energy Can Neither Be Created Nor Destroyed," *Scientific American*. [Online]. Available at: https://www.scientificamerican.com/article/energy-can-neither-be-created-nor-destroyed/ (accessed December 18, 2021).

[3] "SI Base Units — BIPM." [Online]. Available at: https://www.bipm.org/en/measurement-units/si-base-units (accessed December 18, 2021).

[4] "Essentials of The SI: Base & Derived Units." [Online]. Available at: https://physics.nist.gov/cuu/Units/units.html (accessed December 18, 2021).

[5] "Resolution 2 of the 15th CGPM (1975)," *BIPM*. [Online]. Available at: https://www.bipm.org/en/committees/cg/cgpm/15-1975/resolution-2 (accessed September 27, 2023).

[6] "Electromagnetic Spectrum," *NIST*. [Online]. Available at: https://www.nist.gov/image/06phy009emspec2hrjpg (accessed September 27, 2023).

[7] D. M. Kesler, "©WiTricity Corporation, 2017," p. 13, 2017.

[8] H. A. Haus, "Waves and Fields in Optoelectronics," in *Prentice-Hall Series in Solid State Physical Electronics*. Englewood Cliffs, New Jersey: Prentice-Hall, 1984.

[9] William H Louisell, *Coupled Mode and Parametric Electronics*. Wiley, New York & London, 1960.

[10] N. Shinohara, Ed., "Wireless Power Transfer: Theory, Technology, and Applications," in *IET Energy Engineering*, no. 112. London, UK: Institution of Engineering and Technology, 2018.

[11] B. E. A. Saleh and M. C. Teich, *Fundamentals of Photonics*, Third edition., 2 vols. in Wiley series in pure and applied optics. Hoboken, New Jersey: Wiley, 2019.

[12] D. Halliday, Robert Resnick, and Jearl Walker, *Fundamentals of Physics*, Tenth edition. Wiley, 2013.

[13] D. A. Fleisch, *A Student's Guide to Maxwell's Equations*. Cambridge, UK; New York: Cambridge University Press, 2008.

[14] D. A. Fleisch and L. Kinnaman, *A Student's Guide to Waves*. Cambridge, New York: Cambridge University Press, 2015.

[15] R. McCluney, *Introduction to Radiometry and Photometry*. Boston, MA: Artech House, 1994.

[16] D. R. Paschotta, "Optical Intensity." [Online]. Available at: https://www.rp-photonics.com/optical_intensity.html (accessed October 7, 2023).

[17] B. J. Hunt, *The Maxwellians*, 1. Print. in Cornell History of Science Series. Ithaca, NY: Cornell University Press, 2005.

[18] D. J. Griffiths, *Introduction to Rlectrodynamics*, Fourth edition. Cambridge, UK; New York, NY: Cambridge University Press, 2018.

[19] A. Zangwill, *Modern Electrodynamics*. Cambridge: Cambridge University Press, 2013.

[20] P. F. Goldsmith, *Quasioptical Systems: Gaussian Beam Quasioptical Propagation and Applications*, in IEEE Press/Chapman & Hall Publishers series on Microwave Technology and RF. Piscataway, New Jersey: IEEE Press, 1998.

[21] J. D. Jackson, *Classical Electrodynamics*, Third edition. New York: Wiley, 1999.

[22] D. K. Cheng, *Field and Wave Electromagnetics.* in The Addison-Wesley Series in Electrical Engineering. Reading, MA: Addison-Wesley, 1989.

[23] Chetvorno, *Diagram of a Plane Wave Showing Electric And Magnetic Field Vectors.* 2020. [Online]. Available at: https://commons. wikimedia.org/wiki/File:Plane_electromagnetic_wave.svg (accessed December 7, 2021).

[24] E. Hecht, *Optics*, Fifth edition. Boston, MA: Pearson Education, Inc., 2017.

[25] L. A. Manning, "The Effects of Irregularity Scale on Usable Antenna Aperture," Stanford University, Stanford, CA, SU-SEL-68-051, Jul. 1968. Accessed: Oct. 05, 2023. [Online]. Available: https://apps.dtic. mil/sti/tr/pdf/AD0838819.pdf.

[26] D. L. Knepp, "Multiple Phase-Screen Propagation Analysis for Defense Satellite Communications System," Defense Technical Information Center, Fort Belvoir, VA, September 1977. doi: 10.21236/ ADA053463.

[27] W. L. Stutzman and G. A. Thiele, *Antenna Theory and Design*, Third edition. Hoboken, New Jersey: Wiley, 2013.

[28] J. D. Kraus and R. J. Marhefka, *Antennas for All Applications*, Third edition. New York: McGraw-Hill, 2002.

[29] Anthony E. Siegman, *Lasers.* University Science Books, Mill Valley, CA, 1986.

[30] X. Ji and L. Dou, "Two types of definitions for Rayleigh range," *Opt. Laser Technol.*, vol. 44, no. 1, pp. 21–25, February 2012. doi: 10.1016/j.optlastec.2011.05.006.

[31] C. S. Adams and I. Hughes, *Optics f2f: From Fourier to Fresnel*, First edition. Oxford, UK: Oxford University Press, 2019.

[32] J. W. Goodman, *Introduction to Fourier Optics*, Fourth edition. New York: W.H. Freeman, Macmillan Learning, 2017.

[33] "Gaussian Beam Propagation |Edmund Optics." [Online]. Available at: https://www.edmundoptics.com/knowledge-center/application-n otes/lasers/gaussian-beam-propagation/ (accessed August 24, 2020).

[34] R. Hansen, "A One-Parameter Circular Aperture Distribution with Narrow Beamwidth and Low Sidelobes," *IEEE Trans. Antennas Propag.*, vol. 24, no. 4, pp. 477–480, July 1976. doi: 10.1109/TAP. 1976.1141365.

[35] P. Lu, M. Wagih, G. Goussetis, N. Shinohara, and C. Song, "A Comprehensive Survey on Transmitting Antenna Systems with

Synthesized Beams for Microwave Wireless Power Transmission," *IEEE J. Microw.*, pp. 1–21, 2023. doi: 10.1109/JMW.2023.3285825.

[36] M. Duocastella and C. B. Arnold, "Bessel and Annular Beams for Materials Processing," *Laser Photon. Rev.*, vol. 6, no. 5, pp. 607–621, September 2012. doi: 10.1002/lpor.201100031.

[37] D. R. Paschotta, "Brightness." [Online]. Available at: https://www.rp-photonics.com/brightness.html (accessed October 9, 2023).

[38] "Fire From Moonlight." [Online]. Available at: https://what-if.xkcd.com/145/ (accessed March 9, 2021).

[39] R. John Koshel, "Étendue (Photon Snacks 11)." January 27, 2022. [Online]. Available at: https://wp.optics.arizona.edu/jkoshel/wp-content/uploads/sites/78/2022/01/Photon-Snacks-11.pdf (accessed October 21, 2023)

[40] N. W. Johnson *et al.*, "Space-based Solar Power: Possible Defense Applications and Opportunities for NRL Contributions," NRL/FR/7650–09-10,179, 2009.

[41] D. R. Paschotta, "M^2 factor." [Online]. Available at: https://www.rp-photonics.com/m2_factor.html (accessed July 26, 2023).

[42] J. L. Miller and E. Friedman, *Photonics Rules of Thumb: Optics, Electro-Optics, Fiber Optics, and Lasers*, Third edition. in SPIE. Bellingham, Washington: SPIE Press, 1996.

[43] "IEEE Standard for Definitions of Terms for Antennas," *IEEE Std 145-2013 Revis. IEEE Std 145-1993*, pp. 1–50, March 2014. doi: 10.1109/IEEESTD.2014.6758443.

[44] *Energy transmitted by laser in 'historic' power beaming demonstration*, October 22, 2019. [Online Video]. Available at: https://www.youtube.com/watch?v=Xb9THqrXd4I (accessed July 21, 2023).

[45] Thomas J. Nugent, Jr., David Bashford, Thomas Bashford, Thomas J. Sayles, and Alex Hay, "Long-Range, Integrated, Safe Laser Power Beaming Demonstration," in *Technical Digest OWPT 2020*, Yokohama, Japan: Optical Wireless Power Transmission Committee, The Laser Society of Japan, April 2020, pp. 12–13.

[46] P. Jacquinot and B. Roizen-Dossier, "II Apodisation," in *Progress in Optics*, vol. 3, Elsevier, 1964, pp. 29–186. doi: 10.1016/S0079-6638(08)70570-5.

[47] W. H. Hayt and J. A. Buck, *Engineering Electromagnetics*, 7. ed., Internat. ed. Boston: McGraw-Hill Higher Education, 2006.

[48] H. T. Friis, "A Note on a Simple Transmission Formula," *Proc. IRE*, vol. 34, no. 5, pp. 254–256, May 1946. doi: 10.1109/JRPROC.1946.234568.

[49] N. Shinohara, "Beam Efficiency of Wireless Power Transmission via Radio Waves from Short Range to Long Range," *J. Korean Inst. Electromagn. Eng. Sci.*, vol. 10, no. 4, pp. 224–230, December 2010. doi: 10.5515/JKIEES.2010.10.4.224.

[50] R. C. Hansen, *Microwave Scanning Antennas: Apertures.* in Microwave Scanning Antennas. Academic Press, 1964. [Online]. Available at: https://books.google.com/books?id=F0rAwgEACAAJ.

[51] Charles Capps, "Near field or Far Field?," *EDN*, August 16, 2001. [Online]. Available at: https://people.eecs.ku.edu/~callen58/501/Capps2001EDNpp95.pdf(accessedDecember28,2021).

[52] Georg Goubau, "Microwave Power Transmission from an Orbiting Solar Power Station," *J. Microw. Power*, vol. 5, no. 4, pp. 223–231, 1970.

[53] W. C. Brown, "The Technology and Application of Free-Space Power Transmission by Microwave Beam," *Proc. IEEE*, vol. 62, no. 1, pp. 11–25, January 1974. doi: 10.1109/PROC.1974.9380.

[54] G. Borgiotti, "Maximum Power Transfer between Two Planar Apertures in the Fresnel Zone," *IEEE Trans. Antennas Propag.*, vol. 14, no. 2, pp. 158–163, March 1966. doi: 10.1109/TAP.1966.1138660.

[55] W. C. Brown and E. E. Eves, "Beamed Microwave Power Transmission and Its Application to Space," *IEEE Trans. Microw. Theory Tech.*, vol. 40, no. 6, pp. 1239–1250, June 1992. doi: 10.1109/22.141357.

[56] G. Goubau and F. Schwering, "Free Space Beam Transmission," in *Microwave Power Engineering: Generation, Transmission, Rectification*, E. C. Okress, Ed., in Electrical Science Series, no. v. 1. Academic Press, 1968. [Online]. Available at: https://books.google.com/books?id=Obw3BQAAQBAJ

[57] J. Benford, John A. Swegle, and Edl Schamiloglu, *High Power Microwaves.* CRC Press, 2019.

[58] N. Shinohara, *Wireless Power Transfer via Radiowaves*, in ISTE Series. Wiley, 2014. [Online]. Available at: https://books.google.com/books?id$=$pJqOAgAAQBAJ (accessed April 21, 2020).

[59] J. V. Hutson and C. T. Rodenbeck, "Computation of Power Beaming Efficiency in the Fresnel Zone with Application to Amplitude and Phase Optimization," NRL/MR/5307–20-10,118, 2021.

[60] R. L. Forward, "Roundtrip Interstellar Travel Using Laser-Pushed Lightsails," *J. Spacecr. Rockets*, vol. 21, no. 2, pp. 187–195, March 1984. doi: 10.2514/3.8632.

[61] E. Brookner, Ed., *Practical Phased-Array Antenna Systems*, in The Artech House Antenna Library. Boston, MA: Artech House, 1991.

[62] A. Kedar, *Sparse Phased Array Antennas: Theory and Applications*, in Artech House Antennas and Propagation Library. Boston, MA: Artech House, 2022.

[63] R. C. Hansen, *Electrically Small, Superdirective, and Superconducting Antennas*. Hoboken, NJ: Wiley-Interscience, 2006.

[64] C. J. Bouwkamp and N. G. deBruijn, "The Problem of Optimum Antenna Current Distribution," *Philips Res. Rep.*, vol. 1, pp. 135–158, 1945.

[65] Z. Liu *et al.*, "Experimental Studies of Far-Field Superlens for Sub-Diffractional Optical Imaging," *Opt. Exp.*, vol. 15, no. 11, p. 6947, 2007. doi: 10.1364/OE.15.006947.

[66] V. S. Lebedev, "Theory of the Orbital Angular Momentum and Energy Transfer Processes in Collisions Involving Rydberg Atoms," *J. Phys. B At. Mol. Opt. Phys.*, vol. 31, no. 7, pp. 1579–1602, April 1998. doi: 10.1088/0953-4075/31/7/021.

[67] K. Ikeda, "Realization of Quantum Energy Teleportation on Superconducting Quantum Hardware." arXiv, May 3, 2023. [Online]. Available at: http://arxiv.org/abs/2301.02666 (accessed July 11, 2023)

[68] "First Demonstration of Energy Teleportation," *Discover Magazine*. [Online]. Available at: https://www.discovermagazine.com/the-sciences/first-demonstration-of-energy-teleportation (accessed July 11, 2023).

Chapter 4

Wavelength Selection, Beam Propagation, Safety, and Regulations

For a successful technology, reality must take precedence over public relations, for Nature cannot be fooled.

— Richard Feynman [1, p. 237]

4.1 Introduction

The selection of an operating wavelength[1] for a power beaming link is among the most critical design decisions. The wavelength of the electromagnetic wave affects the beam propagation, safety, and the applicable regulatory regimes. As such, the interplay in balancing wavelength and application-specific considerations for each of these is key in implementing a power beaming link. This chapter explores the considerations in making this selection.

Presently, the vast majority of human and robotic activities occur on the earth and within the Earth's atmosphere. This may be poised to change, as many intriguing and potentially game-changing

---

[1]Wavelength and frequency are essentially interchangeable because $c = n\lambda\nu$, where c is the speed of light in a vacuum, n is the local index of refraction, λ is the wavelength, and ν is frequency. Wavelength is principally used here because the physical effects, such as diffraction and material interactions, are more fundamentally related to the wavelength. The speed of light and the wavelength may change with the medium, but the frequency is constant in accordance with conservation of energy.

power beaming applications exist beyond the atmosphere. These applications include those between spacecraft, between spacecraft and celestial bodies, and upon celestial bodies. While this chapter principally focuses on instances within or through the Earth's atmosphere, the approaches in many cases are readily extensible to space situations. For links in space, a crucial difference is that effects on power beaming links due to the Earth's atmospheric and weather characteristics can either be neglected or replaced with considerations from the atmospheric effects of the non-Earth body concerned, assuming the body has an atmosphere. In application areas where there won't be humans, animals, or other vulnerable entities, safety, and regulatory considerations may be very different from those on Earth. Likewise, in situations where the atmospheric effects need not be considered, the design focus may instead be more on transmit and receive device efficiencies at different wavelengths.

4.2   Wavelength Selection

Wavelength, transmit aperture size, receive aperture size, and link distance constrain power beaming link efficiency as guided by the physics of diffraction, which were explored in Chapter 3. Different wavelengths propagate differently in the atmosphere and have different loss mechanisms. Different wavelengths also present a range of potential safety risks, depending on the power density of the beam and the reaction of the human body to different wavelength regimes.

If the transmit and receive aperture sizes and the link distance are held constant, making the wavelength as short as possible will maximize the beam collection efficiency in the absence of any other factors. This would not only realize minimizing diffraction but would also maximize the photon energy E as dictated by the Planck relation:

$$E = h\upsilon \qquad (4.1)$$

where h is Planck's constant and υ the photon's frequency. However, short wavelengths may not be conducive to efficient energy conversion at the transmitter and receiver, and they may not have favorable atmospheric propagation performance.

4.3   Beam Propagation

An examination of the transmissibility characteristics of the atmosphere reveals spectral regions of potential interest for power beaming. Generally, these will be the regions which provide high relative transmission, and many past demonstrations have occurred in these regions. In most cases, it is desirable to minimize the losses incurred by the atmosphere and from tropospheric weather effects. Consequently, there are two large spectral regions of primary interest for power beaming: the radio window between wavelengths of about 1 m and 1 mm (frequencies between 300 MHz and 300 GHz), and the optical window covering visible through the far infrared, with wavelengths from around 10 μm to a few hundred nanometers. These can be seen in Figure 4.1.

Because Figure 4.1 covers a very broad section of the spectrum, a tremendous amount of detail and variation within these regions is obscured. In the microwave region, there are distinct sub-windows where atmospheric transmission is much greater than at adjacent wavelengths. In the optical region, propagating energy even a few wavelengths away from a favorable region for transmission can have very poor atmospheric transmission. Figure 4.1 also considers transmission through the entire atmosphere from the surface of the Earth to space, and thus overstates atmospheric opacity for short links

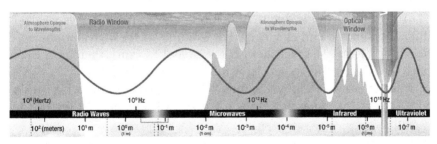

Figure 4.1.   The radio and optical atmospheric transmission windows. Adapted from [2].

that traverse less of the atmosphere. The mechanisms that define the boundaries and features within the windows vary.

4.3.1 *Absorption and Scattering*

In both the radio and optical windows, absorption arises from oxygen and water vapor in the air. The absorption mechanism for these constituents for the radio window is summarized by Goldhirsh [3, p. 4–4]. Absorption results in the conversion of the beam's energy to heat, or partial thermal conversion plus incoherent reradiation. Scattering results not from conversion but from reradiation in directions other than that which is desired.

In the optical window, Beer's Law[2] describes how light is lost during propagation of a beam due to both absorption and scattering [5, p. 12]. The atmospheric transmissivity T is given by

$$T = e^{-(\alpha(\lambda))z} = e^{-(A_a + S_a)z} \tag{4.2}$$

where A_a is the absorption coefficient per unit length in air, S_a is the scattering coefficient per the same unit length in air, $\alpha(\lambda)$ is their sum and called the attenuation or extinction coefficient, with λ as wavelength and z is the distance of propagation. Absorption occurs by both molecular and particulate mechanisms, whereas scattering occurs via Rayleigh and particulate mechanisms. "Particulates" in this context can include not only the normal definition of dust and other "dry" particles, but also water droplets from rain, fog, and other precipitation which are discussed separately. Particulates suspended in a gas-like air are also called aerosols [6].

Atmospheric losses due to absorption and scattering have a direct impact on end-to-end efficiency. For example, if the attenuation coefficient is $0.1\,\mathrm{km}^{-1}$ (sometimes expressed as 10%/km), then for a 1 km path length the end-to-end transmission is only about 90.5% ($e^{-0.1}$) of what it would be in a vacuum, and at 2 km it would be nearly 81.9% ($e^{-0.1*2}$) remaining. In some cases, dB/km is used instead of %/km.

---

[2]Also called Bougher's Law, or the Beer-Lambert Law [4, p. 112], or other permutations of these three names.

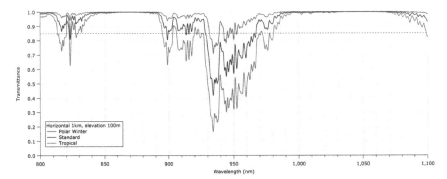

Figure 4.2. Atmospheric molecular absorption for a 1 km horizontal, near-ground, path from 800 to 1,100 nm in three atmosphere types generated using a HITRAN model [7].

4.3.1.1 Atmospheric Absorption

The transmittance in the near-IR regime over a 1 km long horizontal path, accounting only for atmospheric molecular losses is shown in Figure 4.2 for three different humidity scenarios. There are clear windows of high transmittance, particularly in the 850–880 nm and 995–1,075 nm ranges.

This molecular absorption is caused by the presence of oxygen and water in the atmosphere. Attenuation is greatest near sea level since oxygen and water levels decrease when moving away from the Earth. Other gases such as carbon dioxide can contribute attenuation, but oxygen and water dominate due to their prevalence.

The choice of operating wavelength for a microwave power beaming system is likewise partially governed by atmospheric absorption. A plot of the attenuation arising from a path directly upwards from the surface of the Earth to space from 0 to 140 GHz (about 2 mm) is shown in Figure 4.3. The relative losses as a function of wavelength substantively align with those for zenith attenuation. Precise estimated losses for a given link can be found using the MATLAB function gaspl [8], which takes as inputs the link distance, frequency, ambient temperature, ambient pressure, and atmospheric water vapor density. It is based on the International Telecommunications Union's (ITU) atmospheric gas attenuation model [9].

The industrial, scientific, and medical (ISM) bands at 2.45 GHz (about 12 cm) and 5.8 GHz (about 5 cm) have been chosen in the past

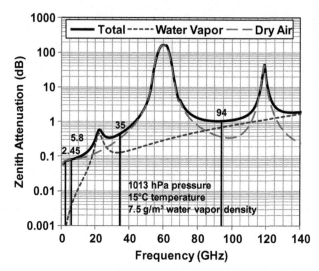

Figure 4.3.   Atmospheric attenuation from 0 to 140 GHz for a vertical path from the surface of Earth to space. Data from ITU-R P.676-10, "Attenuation by atmospheric gases," 2013. Plot created by James McSpadden and featured in [10], used with permission.

for power beaming links because their attenuation is low relative to higher frequencies, and the sizes of the transmitting and receiving antennas are of reasonable size relative to longer radio wavelengths. A microwave power beaming system designed at 22 GHz would see large amounts of attenuation due to water vapor, especially in humid climates near sea level. Similarly, oxygen would limit the range of microwave power beaming at 60 GHz. Both attenuation peaks are evident in Figure 4.3.

4.3.1.2   *Scattering*

Electromagnetic waves can scatter off particles of any size and can be defined by two regimes, often referred to as Rayleigh and Mie scattering. Both regimes largely determine the direction of the scattered light and are determined largely by the relative size of the scattering particulate compared to the wavelength. When the particle size is much smaller than the wavelength, Rayleigh scattering describes the effects. Its characteristic term is a $1/\lambda^4$ dependence, causing shorter wavelengths like blue light to scatter more than longer wavelengths like red and infrared. Rayleigh scattering is roughly uniform in all directions regardless of the direction of the incident light. Optical power beaming often uses near-IR wavelengths, which may be as much as three times as long as blue light. The Rayleigh scattering of near-IR laser light is barely more than 1% of what it is for the blue portion of sunlight.

When the particle size is of the order of the wavelength or larger, then Mie scattering describes the effects. Dust, smoke, and pollen are examples of the types of particulates that typically cause Mie scattering. The direction of Mie scattering is heavily weighted in the forward direction of the incident beam and can play a role in the spreading of the beam.

4.3.1.3 *Combined Measures*

While optical transmittance can be measured directly for a specific path with sensitive equipment (such as scintillometers for turbulence fluctuations, transmissometers for the extinction coefficient, and nephelometers for suspended particulates) it can also be estimated from more widely available measures included in weather reports, including visibility and precipitation rate. Inclement weather further complicates the loss problem by adding variable amounts of attenuation.

According to Atlas [11], the relation between visibility[3] and the extinction coefficient can be expressed as

$$V \approx \frac{2.9}{\alpha} \tag{4.3}$$

This relationship is an approximation because the value of 2.9 in the numerator is related to the threshold of contrast, a subjective quantity that can vary between observers, and has accepted values between 2.8 and 4.8. Another more nuanced approximation accounts for wavelength:

$$\alpha = \frac{3.912}{V} \left(\frac{550 \, \text{nm}}{\lambda} \right)^q \tag{4.4}$$

---

[3]Defined in aviation as the greater of (1) the greatest safe distance at which a black object of suitable dimensions, situated near the ground, can be seen and recognized when observed against a bright background; and (2) the greatest distance at which lights in the vicinity of 1000 candelas can be seen and identified against an unlit black background [12].

The exponent q depends on the visibility: it is 1.6 for V > 50 km, 1.3 for 6 km < V < 50 km, and $0.585V^{1/3}$ for V < 6 km.

Atlas similarly demonstrated that, given a measurement of the rate of rainfall R in mm/h, the extinction coefficient (in units of km^{-1}) can be estimated as

$$\alpha_{rain} \approx 0.312 R^{0.67} \tag{4.5}$$

This equation is accurate only in the case of non-orographic type rainfall. For orographic type rainfall,[4] the coefficient changes to 1.2 and the exponent to 0.33.

From a statistical perspective, for most locations on Earth most of the time, moderate or heavy precipitation does not last very long, nor does it occur frequently.

The attenuation loss in dB due to rainfall for microwaves is

$$L_{ra}(t) = \int_0^{d(t)} a[A(z,t)]^b dz,$$

$$a = \begin{cases} 4.21 \times 10^{-5} f^{2.42}, & 2.9 \le f \le 54\,\text{GHz} \\ 4.09 \times 10^{-2} f^{0.699}, & 54 \le f \le 180\,\text{GHz} \end{cases}$$

$$b = \begin{cases} 1.41 f^{-0.0779}, & 8.5 \le f \le 25\,\text{GHz} \\ 2.63 f^{-0.272}, & 25 \le f \le 164\,\text{GHz} \end{cases}$$

$$\tag{4.6}$$

where $d(t)$ is the time-dependent portion of the path between the transmitting and receiving antennas that contains the rain and $A(z, t)$ is the amount of rainfall in mm/h at time t at a distance z km measured from the ground along the path [13]. These atmospheric attenuation problems are well-known to satellite communication designers and have the potential to greatly affect a microwave power beaming link.

4.3.2 *Turbulence and Scintillation*

Turbulence and scintillation have minimal effects on power beaming in the radio window but can be consequential in the optical window. Differential temperatures in the air, along with wind, cause slight density and pressure gradients which create turbulent airflow. Changes in the density of air have a small effect on the index of refraction, and it is the accumulation of these differences along an optical

---

[4]That is, rainfall induced by mountains or other elevated land masses.

path that effectively acts like a series of randomly displaced thin lenses of random, long focal lengths, with the net effect of causing an input beam profile to be distorted in ways which can be described only statistically, and not predicted moment-by-moment in advance. To counteract some of the effects, adaptive optics measure and correct for the distortions at a high rate, typically hundreds of hertz. One important effect of these distortions is the variation in the beam intensity at the receiver. Scintillation can be seen in the form of twinkling stars and in the shimmering observed above an asphalt road on a sunny day. The scintillation index is the normalized variance of the irradiance fluctuations [5, p. 261]:

$$\sigma_I^2 = \frac{\langle I^2 \rangle - \langle I \rangle^2}{\langle I \rangle^2} \tag{4.7}$$

The refractive-index structure constant C_n^2, also called the atmospheric structure constant, is a measure of atmospheric turbulence with units of $m^{-2/3}$ [5, p. 64]. It can generally range from values as low as about $10^{-17}\,m^{-2/3}$ in calm air at altitudes well above ground level up to values as high as $10^{-12}\,m^{-2/3}$ close to the ground over hot asphalt in the middle of a sunny day. Values below about $10^{-16}\,m^{-2/3}$ are often seen at higher altitudes or in calm winds near dusk and dawn, whereas values greater than about $10^{-13}\,m^{-2/3}$ are seen closer to the ground, especially when there are strong temperature differences between the air and the ground.

The Rytov variance σ_R^2, assuming a plane wave and the Kolmogorov spectrum, is useful to determine the approximate turbulence regime propagation will be in, because it combines the atmospheric structure constant and the length of the beam path into a dimensionless number [5, p. 140]:

$$\sigma_R^2 = 1.23\,C_n^2 k^{\frac{7}{6}} L^{\frac{11}{6}} \tag{4.8}$$

Here k is the wave number $(2\pi/\lambda)$ and L is the path length. A Rytov variance much less than one means that the beam experiences weak turbulence, whereas at higher values it is in the strong turbulence regime. The scintillation index can be calculated as a function of the Rytov variance in various conditions. Equations describing the impact of turbulence have been derived for weak turbulence, and ongoing work is refining models to approximate the effects of strong turbulence. Andrews and Phillips observe that for focused beams, different conditions and models may apply [5, p. 140].

The Fried parameter, also called the atmospheric coherence width, is used to describe the size scale beyond which the atmosphere reduces the spatial coherence of parts of the beam relative to each other. For a telescope without adaptive optics, it would be the largest effective aperture for imaging. The form of the Fried parameter can vary depending on assumptions and context, but was put forth in his 1966 paper [14] as

$$r_0 \approx \left(\frac{6.88}{A} \right)^{3/5} \tag{4.9}$$

where A is a function of the propagation-path length, the wavelength, and the degree of turbulence along the path. Further explorations of the Fried parameter can be found in [[15], p. 19; [5], p. 617]. It is used in calculations of the amount of beam wander and growth due to turbulence.

These refractive index variations can broaden the width of a beam, change its intensity profile, steer it away from its intended centerline, and decrease spatial beam coherence. The mathematics and analysis techniques for these beam characteristics are beyond the scope of this book.

4.3.2.1 *Impact on Beam Size*

Turbulence within the "low" turbulence regime can increase the beam size[5] by

$$\Delta d_R = \frac{\sqrt{2}\lambda l}{\pi r_0} \tag{4.10}$$

An expression for the beamwidth d_{LT} averaged over a "long" time in turbulence relative to its expected width in a vacuum d_0 is

$$d_{\mathrm{LT}} = \left(1 + 1.33\sigma_R^2 \left(\frac{21}{kd_0^2} \right)^{\frac{5}{6}} \right)^{\frac{3}{5}} \tag{4.11}$$

These equations can be used to estimate the average beam size.

---

[5]Different derivations result in differences in the numerical constant. Here a value with more adverse impact is used.

Beam growth will modify the effective full angle divergence (for distances where the Fresnel number is greater than 1):

$$\theta_{\text{full}} = \frac{d_R}{l} = \frac{4\,\text{BPP}}{d_T} + \frac{\sqrt{2}\,\lambda l}{\pi r_0} \tag{4.12}$$

Solving for transmit diameter aperture d_T gives

$$d_T = \frac{4\,\text{BPP}}{\dfrac{d_R}{l} - \dfrac{\sqrt{2}\,\lambda l}{\pi r_0}} = \frac{l\pi r_0 4\,\text{BPP}}{d_R \pi r_0 - \sqrt{2}\lambda l^2} \tag{4.13}$$

4.3.2.2 *Impact on Beam Intensity Profile*

For any given optical beam profile, atmospheric turbulence will change that beam profile at the target via scintillation [16]. As mentioned above, the scintillation index is the normalized variance of the beam intensity. The statistics describing intensity fluctuations follow a log-normal distribution. The curves shown in Figure 4.4 are an example of the probability density function (PDF, bottom) and cumulative distribution function (CDF, top) for an instantaneous measurement of beam intensity relative to the vacuum value, averaged across a defined receive aperture width in weak turbulence conditions.

Worth noting in the probability density function is that portions of the intensity of the beam can actually be *higher* than they would be in the absence of air. This should be accounted for in the link and receiver design. Scintillation is primarily a concern for power beaming in the optical window, but it can also affect microwave power beams that traverse the ionosphere, as would likely be the case for a power beam from a solar power satellite [17].

4.3.3 *Large-Scale Refraction*

While assuming that beam propagation in the air will behave nearly the same as in a vacuum may be a reasonable approximation in many circumstances, it can fall short for power beaming in both the optical and radio windows. Differences in temperature and humidity in different air layers and regions can cause effects because of varying indexes of refraction within the regions both at small and large scales. Small-scale variations in the index of refraction discussed in previous sections affect optical power beaming more than microwave power

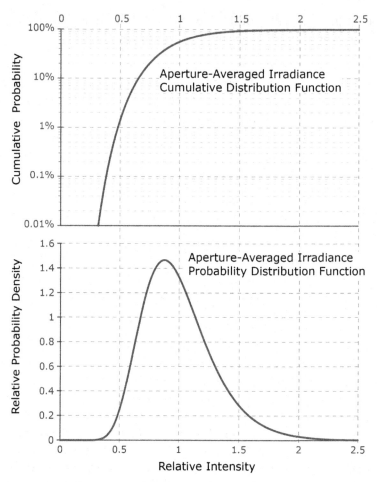

Figure 4.4.   *Probability density (top) and cumulative distribution (bottom) functions at a receiver for $\lambda = 1{,}070\,nm$, $L = 1\,km$, $C_n^2 = 5e - 15$, receive aperture $= 2.5\,cm$.*

beaming. The large-scale effects of refraction can be referenced with respect to the path the beam would take in a vacuum and the surface of the Earth. Subrefraction occurs when the path of the beam curves upward away from the Earth as compared to the path the beam would take in a vacuum. Superrefraction occurs when the path of the beam curves downward toward the Earth as compared to the path the beam would take in a vacuum. Since the atmosphere normally exhibits an index of refraction gradient with increasing altitude above the ground, a certain amount of superrefraction is typically evident. This has been termed "standard refraction" [3, pp. 10–12]. Refraction that results in the beam path being slightly farther from the Earth than standard refraction but not exceeding the boundary representing the vacuum path is considered substandard. Refraction where the beam curves further towards the Earth is considered superstandard. Ducting, in which a beam enters and is channeled through a path that has a curvature matching that of the Earth's surface, is possible but unlikely to be an important factor in power beam propagation. This is because the influences of absorption, scattering, and other atmospheric effects would likely have already had outsize impacts in circumstances where ducting might occur.

For further exploration of large-scale refractive and other atmospheric effects, readers should consult Levis *et al.* [18] and Goldhirsch [3] for the radio window and Andrews and Phillips [5] for the optical window. Rizzo *et al.* also assessed the terahertz region and compared it with the microwave and optical regions of the spectrum [19].

4.3.4 *Power Intensity Effects*

It is generally desirable to keep aperture diameters as small as practical, for reasons including mass, footprint size, and cost. For a given transmit power, smaller apertures lead to higher power intensity, which can cause problems if the intensity is high enough.

As the beam intensity increases sufficiently, even the small fraction of transmitted energy that is absorbed can cause local heating of the air, which can further drive atmospheric effects, including beam spreading. This is called thermal blooming [4, p. 294]. At high enough intensities (for 1 μm light, around 2×10^{11} W/cm^2 in clean air, and 5×10^9 W/cm^2 in non-clean air), the air can break down, meaning it changes from being principally a nonconductive medium to a partially conductive one (i.e., a partial plasma). Ali summarizes how this can happen for lasers [20]. These effects are mostly only a concern for directed energy weapons due to the need to concentrate laser light at a distant location. Because power beaming does not need to concentrate optical power nearly as much, typical power densities don't cause this problem. Additionally, beam distortions are not a great concern since the beam is already nonhomogeneous. Still, at lower intensities, the air can be heated enough to impact its density, causing a non-negligible amount of lensing and turbulence effects.

As discussed in Section 4.3.1, a small fraction of the energy per unit distance is scattered in various directions. If the source light intensity is high enough, then even this small amount of scattered energy could exceed the allowed exposure levels, presenting a safety concern.

4.4 Safety

For some applications and situations, safety and regulatory requirements will limit the maximum power intensity produced by the transmitter. Many applications need a certain amount of power to arrive at the receiver, and there may be a small, fixed amount of physical space in which the receiving aperture can reside. Thus, the transmitted power, received power, system efficiency, and the physical size available may present situations that approach or exceed regulatory limits. Regulatory requirements vary depending on the operating wavelength and other factors.

Power beaming is unlikely to be societally accepted unless it is safe. Different parts of the electromagnetic spectrum are regulated by different organizations. Safety approaches can be divided into two broad approaches:

(1) Operate the system only at power densities that are below accepted exposure limits.
(2) Operate the system with power densities that approach or exceed accepted exposure limits but ensure that they are inaccessible.

Before further examination of these approaches, the nature of potential safety and regulatory hazards will be explored.

4.4.1 *Human Health Effects*

For forms of nonionizing radiation at sufficiently high-power densities, a primary concern is the possibility of harm arising from thermal effects. At optical wavelengths shorter than 100 nm, ionizing effects may become a concern. In the visible range of 400–600 nm, photochemical effects can occur as well [21, p. 173].

The parts of the human body typically considered are the retina, the cornea, skin, and subcutaneous tissue. The penetration depths and absorption coefficients vary by wavelengths in different types of tissues, with longer wavelengths generally correlating with greater penetration depths. One notable exception is the eye. Near-IR light with wavelengths shorter than about 1,400 nm can pass through the cornea and be absorbed in the retina. It is made more hazardous because the eye can still focus this light into a higher intensity than the free space beam but does not evoke a blink response because it is invisible. Light with wavelengths from about 1,400 nm to 1 mm is absorbed by the cornea, which has similar exposure intensity limits as skin for visible through mid-IR. With the shorter wavelengths of the near-ultraviolet range from 315 to 390 nm, there is absorption in the lens of the eye, which can present a hazard [22, p. 4].

For wavelengths from the infrared to about 12 cm (approximately 2.4 GHz), the principal effect of concern is the possibility of heating. This can be a hazard at sufficiently high power densities. Parts of the body with higher water content, such as the eyes and gonads, can absorb energy at greater rates. There is no clear evidence of harmful nonthermal human health effects from microwaves [23, p. 804].

4.4.2   *Objects in the Beam*

Beyond direct exposure to the parts of the human body discussed above, objects entering the beam could pose a hazard to humans or to the objects themselves. In many cases, it may be necessary to cease or alter the beam's operation until the beam path is clear.

One case is that of a highly reflective surface being introduced into the beam path. This could reflect energy to an unintended location. The speed of a moving object will affect how fast it needs to be detected. Figure 4.5 shows regions and points for a variety of object sizes along with a range of speeds.

Depending on the beam path and local topography, people might approach a power beam path while on foot or while operating or being a passenger in a vehicle. Birds span a wide range of maximum speeds as well as sizes. Leaves, insects, and raindrops may or may not present concerns, depending on the operating context. Determining object characteristics and their corresponding cutoff thresholds to balance safety and link uptime may prove a critical part of system design.

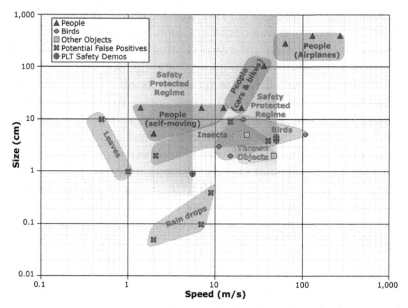

Figure 4.5. Sizes and maximum speeds of potential foreign objects for an active safety system to detect [24]. Shaded regions encompass a class of objects, while individual markers represent specific examples in a class.

4.4.3 Regulatory Agencies and Standards

4.4.3.1 Lasers

The International Electrotechnical Commission (IEC) and the United States Food and Drug Administration's (FDA) Center for Devices and Radiological Health (CDRH) are two organizations that manage laser product safety standards and/or approval. These are respectively associated with guidelines and standards maintained by the International Commission on Non-Ionizing Radiation Protection (ICNIRP) and the American National Standards Institute (ANSI). In the United States, the Occupational Health and Safety Administration (OSHA) regulates industrial usage of lasers. The portions of the US Code of Federal Regulations (CFR) that regulate lasers are collectively referred to as the Federal Laser Product Performance Standard (FLPPS). These include CFR Title 21 Sections 1040.10 and 1040.11 [25]. These regulations are largely based on the ANSI Z136.1-2022 and related standards [21]. The comparable IEC standard is IEC 60825-1:2014 [26].

Lasers are categorized by most regulatory bodies into four broad groups, ranging from the safest, Class 1, to the potentially most hazardous, Class 4. Most lasers for power beaming will incorporate a Class 4 laser but will use protection and mitigation methods to permit the system's operation as Class 1 or Class 1M. Class 1M systems are considered safe unless the beam is viewed with an optical aid, such as a telescope. The laser classes are determined by an assessment of the maximum accessible emission limit (AEL). For more information on laser classifications, developers should consult the IEC 60825-1 or ANSI Z136.1 standards.

For laser beams outside pointed above the horizon, in the United States the Federal Aviation Administration (FAA) should be engaged to provide a written indication of "objection" or "non-objection" concerning laser operations per Part 6, Chapter 29 of [27]. As of 2012, it is a federal crime in the United States to aim a laser at an aircraft [28]. For implementers associated with the US Department of Defense, the Laser Clearinghouse (LCH) is required to be engaged per DODI 3100.11 [29, p. 74]. Other implementers may engage voluntarily. For example, astronomical observatories using guide star lasers frequently do so.

For optical power beaming, some benefits can be had by utilizing guidance and resources created for the optical communications community, such as this overview [30] by Lafon *et al.*

4.4.3.2 *Microwaves*

Radiofrequency and microwave safety are governed by a range of bodies worldwide. Whether recommended exposure limits are legally binding depends on the specific country. A survey of many nations' approach to RF safety and whether exposure limits are legally binding can be found in a report for the Dutch National Institute for Public Health and the Environment [31]. In the United States, bodies involved with the regulation of RF exposure include the Federal Communications Commission (FCC), OSHA, and the FDA. For most countries, exposure guidelines derive either from the IEEE C95. 1-2019 standard, which covers from 0 Hz to 300 GHz (1 mm wavelength), or the ICNIRP Guidelines for Limiting Exposure to Electromagnetic Fields (100 kHz to 300 GHz).

Both IEEE and ICNIRP have supporting documents that address nuances and adjunct matters to the exposure guidelines. The FCC offers Bulletin 65 which compliance and measurement matters [32]. Most of the standards and associated documents can be downloaded free of charge. The FCC's Electromagnetic Compatibility Division has a comprehensive "Frequently Asked Questions" page on its website concerning RF Safety that presents an excellent place to start for those new to this topic [33].

4.4.4 *Power Density and Allowed Exposure*

As with ionizing radiation, the three principal means to minimize exposure to nonionizing hazards are reducing the time of exposure from the source, increasing the distance from the source, and utilizing shielding from the source. The latter two of these have the effect of reducing the power density presented by the source, while the former can limit the energy transferred. Together, they influence the maximum permitted power density and allowed exposure.

Since many power beaming systems may operate nearly continuously, perhaps pausing only for outages due to weather or safety deactivations, the levels considered here will assume essentially uninterrupted exposure. In some cases, this will result in estimates that could be as much as several orders of magnitude below the actual safe thresholds for power density, because the real exposure may be very brief. For the ANSI Z136.1 standard, "continuous" exposure is considered as 10–30,000 s (more than 8 h). The maximum permissible exposure (MPE) is shown as a function of wavelength in Figure 4.6 for a wavelength range likely to be of interest for optical power beaming. Below 1,400 nm, the MPE is for ocular exposure is more restrictive than the skin exposure limit. For wavelengths from 1,500 to 1 mm, the skin MPE is 1,000 W/m^2 for exposure areas smaller than 100 cm^2. The skin MPE decreases linearly for areas of up to 1,000 cm^2, at which point it becomes 100 W/m^2 [21, Sec. 8.4.2]. The peak in MPE that appears at 1,300 nm is a result of changes in laser safety standards that took effect around 2013 [34].

In cases with brief exposures, pulsed beams, extended sources, or large exposure areas, the MPE can be very different than shown in Figure 4.6. A rigorous treatment of laser safety analyses, practices, and implementations is beyond the scope of this book. System developers should utilize laser safety resources and have close engagement with a trained Laser Safety Officer (LSO). Determining the MPE, Nominal Hazard Zone (NHZ), and other key quantities is essential

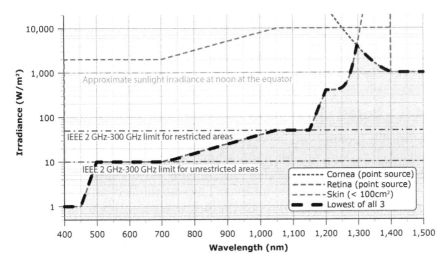

Figure 4.6. MPE for long durations (10–30,000 s) taking the most restrictive result of ocular (for point sources) and skin limit calculations using ANSI Z136.1-2022 and assuming skin exposure area <100 cm².

in the safe development, testing, and operation of any optical power beaming system and requires careful attention.

Per IEEE Std C95.1^{TM}-2019, the microwave power intensity exposure reference level (ERL) over the range of 2 GHz to 300 GHz (wavelengths of 15 cm to 1 mm) for unrestricted environments is $10\,\text{W/m}^2$, and the ERL for restricted environments is $50\,\text{W/m}^2$. These correspond to the same irradiance levels as shown in Figure 4.6 at 600 and 1,100 nm, respectively. These ERLs assume a 30-min average time for whole-body exposure [35, Tbls. 7 and 8].

As a point of reference, the maximum power density generally available on Earth under ideal seasonal conditions near the middle latitudes at noon is 1,000 W/m^2 [36][p. 6]. This corresponds to the irradiance MPE for 1,400 nm in Figure 4.6.

While the exposure limits above include margin to avoid being too close to harm thresholds,[6] they should be heeded by power beaming system implementers.

---

[6]A study by Wasserman *et al.* in which birds were exposed to $250\,\text{W/m}^2$ of continuous 2.45 GHz microwave energy for 18 weeks found no discernible harm, despite the exposure level being 25 times higher than that for an unrestricted environment [37].

To be useful, power beaming systems may need to operate at intensities that are significantly higher than the allowed irradiance exposure limits. Thus, safety measures may be required to ensure safe operation.

4.4.5   *Safety Measures*

Historically, power beaming experiments and tests relied on observers designated to look out for approaching vehicles, people, animals, or other potential situations that had bearing on the link's operation to ensure safe operating conditions. They would alert test operators, perhaps via handheld radio, for the power transmitter to be turned off.

Microwave power beaming pioneer Dickinson described some of the safety and other preparations that were made prior to the notable 1975 demonstration at the NASA Goldstone facility [38]:

(1) Posting people as lookouts for and listening for airborne objects such as birds, planes, hang gliders, gliders, etc.
(2) People with binoculars and surveillance TVs scouring the area around the receiver tower looking for prospectors, gold hunters, tame and wild burros.
(3) Notice to Air Missions (NOTAMs) issuance.
(4) Alerts to people with pacemakers.
(5) 24-h previous alert to Southern California Edison for additional MW of 60 Hz power.
(6) Explosive munitions warnings, etc.

While perhaps suitable for experiments and demonstrations, for systems to be deployed for operation, such manually intensive approaches to safety are likely to be impractical.

In some cases, it may be possible to limit access to irradiance levels higher than the MPE or ERL limits by operating either at low power densities or by using barriers, distancing, or other physical means to make the power beam and any reflections or spillover inaccessible. This approach is considered "passive" in that a system that senses the operation environment and responds to changing conditions is not required. For other situations, an "active" safety approach may be needed. The general approach for active safety has similarities to the Observe, Orient, Decide, Act (OODA) loop [39].

The process of reacting to a potential beam intrusion is referred to as the D^3 process, according to these elements:

- **Detect:** Sense an object approaching the beam, or another potentially unsafe or undesirable condition.
- **Decide:** Interpret the sensor data to determine the required course of action, which could be one of the following desist options.
- **Desist:** Take action to ensure safe operation:
  - ○ **Divert:** Redirect the beam to an alternate location that is consistent with safe operation.
  - ○ **Dim:** Reduce the beam output power by some amount that has been determined to meet safe operation requirements.
  - ○ **Defocus:** Spread the transmitted energy by altering the amplitude and/or phase to lower the power density to meet safe operation requirements.
  - ○ **Deactivate:** Turn off the beam quickly enough to meet safe operation requirements.

The flow chart associated with these elements and how they are connected is shown in Figure 4.7.

Often, the simplest and fastest action in response to the detection of other than normal operating conditions will be deactivation. Critically, not all of the desist options may be practical depending on the modality employed and the system implementation. For instance, an optical power beaming system might not be able to divert the beam without sweeping it unsafely across an area with unknown risks. Similarly, it may not be possible to defocus a beam to ensure safety, or defocusing might introduce other undesirable effects, like interference with communication systems.

A major factor affecting the safety process time requirement is the distance separating the foreign object (FO) detection zone from the power beam, as depicted in Figure 4.8.

One concept for detecting objects approaching the power beam is to use a "beam break" or "light curtain" system that places emitters and detectors around the power beam, and to turn off the power transmission if one of these units loses its signal, much in the same way as the safety system for a residential garage door. Challenges associated with using multiple 1-to-1 emitter-detector pairs, especially at moderate distances outdoors, led the company PowerLight

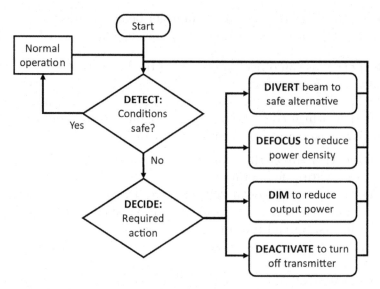

Figure 4.7.   A process for ensuring safe operation of a power beaming link.

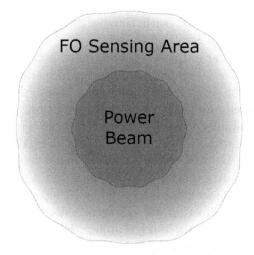

Figure 4.8.   Schematic representation of a safety sensor's FO detection zone around a power beam.

Technologies to create its "enhanced light curtain" which enables a dense web of beam paths by coupling the emitters and detectors in a many-to-many configuration. A prototype tested by the Navy's Laser Safety Review Board confirmed its effectiveness [40]. It has been shown to have a D^3 time as short as 1 ms. If the distance between the beam break "ring" and the power beam is 5 cm, it implies that a detectable object traveling as fast as 50 m/s (112 mph) would not be exposed to hazardous light.

Other means of power beaming path monitoring and FO detection have been explored by different groups, including using cameras, lidar, radar, and more. It is possible that advances in the development of similar systems for autonomous vehicle systems will be applicable to power beaming safety systems as well.

In addition to monitoring the beam path and external conditions, it is good practice to monitor the overall system health status. This might include whether the ratio of output power to input power is within an expected range on either end of the beam, and at subsystem boundaries. It can also include quantities such as the transmitter and receiver subsystem temperatures, voltages, and currents.

4.5   Other Regulations

In addition to the regulations that directly and indirectly concern safety discussed in the preceding section, power beaming system implementers must also consider those pertaining to spectrum usage. These regulations are most likely for wavelengths in the radio atmospheric transmission window. While spectrum allocation for the optical region by bodies has been considered in the context of the ITU and Federal Communications Commission, there is currently no regulatory framework in place [41,42]. Thus, this discussion focuses on spectrum allocation challenges for the microwave portion of the spectrum.

Because of its utility for communications, radar, navigation, and other applications, the microwave spectrum is heavily used and may be considered to be effectively fully occupied. The wavelengths that are attractive for power beaming because of their favorable atmospheric transmission qualities are attractive for many other uses for the same reasons. Though there exist Industrial, Scientific, and Medical (ISM) frequency bands that are comparatively less regulated, they do not present a free-for-all bonanza for microwave power beaming. Indeed, many of the most popular applications with the most users reside in the ISM bands, such as Wi-Fi, and any disruption would almost certainly provoke an intense, adverse reaction.

In the United States, the FCC and the National Telecommunications and Information Administration (NTIA) manage spectrum allocations. The current allocations can be found online at the FCC's website as the FCC ONLINE TABLE OF FREQUENCY ALLOCATIONS, 47 C.F.R. §2.106 [43]. A graphical depiction of the allocations from January 2016 is available from the NTIA. The allocations cover from 9 kHz to 300 GHz (about 33 m to 1 mm). A representative section of the graphical depiction showing the allocations between 2 and 3 GHz appears in Figure 4.9.

The scale of the challenge for spectrum allocation for microwave power beaming has been explored by McSpadden [45], and the ITU has released several documents concerning the topic, the most recent being a recommendation from September 2022 [46]. Changes in spectrum allocation typically occur very slowly through a multi-year process undertaken by the ITU and other players who are part of the spectrum management establishment. The division of the ITU responsible for international frequency management is the Radiocommunication Sector (ITU-R), which administers the Radio Regulations (RR). The RR is a binding international treaty that governs the use of the radio spectrum [47].

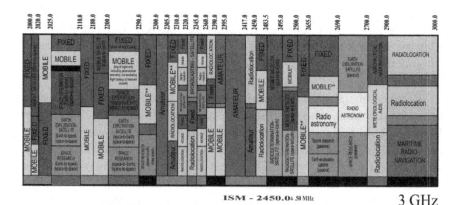

Figure 4.9. Representative section of the 2016 United States Frequency Allocations chart published by the NTIA [4]. US government work not subject to copyright.

The RR is updated through a process involving the World Radio-communication Conference (WRC), a meeting that occurs every 4 years, and lasts for four weeks. In 2023, WRC-23 is slated to be held in Dubai, United Arab Emirates [48]. The WRC-23 will establish the draft agenda for WRC-27, and the preliminary agenda for WRC-31. Topics that are not included in the 4-year study cycles will typically not receive consideration during the WRC meeting itself. It is evident that securing spectrum allocation for microwave power beaming is likely to require considerable advance preparation and sustained engagement with the ITU-R, its study groups, and other stakeholders.

4.6   Conclusion

Wavelength selection for a power beaming system has important implications for diffraction-limited performance, atmospheric propagation, safety, spectrum allocation concerns, and available device efficiencies. The effects of absorption, scattering, turbulence, scintillation, and refraction arising from the atmosphere were reviewed. Safety systems are of paramount importance for power beaming, and they were likewise discussed, with consideration given to health effects, detection of beam incursions, regulatory agencies and standards, power density matters, and safety measures. Spectrum allocation, a prime driver for microwave power beaming frequency selection, was introduced. Each of these areas will likely need attention for any power beaming system development.

4.7   Further Reading

The following texts expand further on many of the topics explored in this chapter:

Atmospheric Propagation

F. G. Smith, *The Infrared & Electro-Optical Systems Handbook — Volume 2 — Atmospheric Propagation of Radiation*, vol. 2. in The Infrared and Electro-Optical Systems Handbook, vol. 2, 1993.

L. C. Andrews and R. L. Phillips, *Laser Beam Propagation through Random Media*, Second edition. S.l.: International Society of Optical Engineering, 2005.

C. A. Levis, J. T. Johnson, and F. L. Teixeira, *Radiowave Propagation: Physics and Applications*. Hoboken, New Jersey: Wiley, 2010.

Safety

T. E. Johnson and H. Cember, *Introduction to Health Physics*, Fifth edition. New York: McGraw-Hill Education, 2017.

A. R. Henderson and K. Shulmeister, *Laser Safety*. Bristol; Philadelphia, PA: Institute of Physics, 2004.

S. M. Michaelson and J. C. Lin, *Biological Effects and Health Implications of Radiofrequency Radiation*. New York: Plenum Press, 1987.

R. Kitchen, *RF and Microwave Radiation Safety Handbook*, Second edition. Oxford; Boston, MA: Newnes, 2001.

References

[1] R. P. Feynman, *What Do You Care What Other People Think? Further Adventures of a Curious Character*. New York, NY: W. W. Norton, 2018.

[2] "Introduction to the Electromagnetic Spectrum | Science Mission Directorate." [Online]. Available at: https://science.nasa.gov/ems/01_intro (accessed October 28, 2021).

[3] J. Goldhirsh, *Propagation of Radio Waves in the Atmosphere*, vol. 1. Johns Hopkins University, 2004.

[4] F. G. Smith, The Infrared & Electro-Optical Systems Handbook — Volume 2 — Atmospheric Propagation of Radiation, in *The Infrared and Electro-Optical Systems Handbook*, vol. 2. Ann Arbor, MI: Environmental Research Institute of Michigan, 1993.

[5] Larry C. Andrews and Ronald L. Phillips, *Laser Beam Propagation Through Random Media*, Second edition. S.l.: International Society of Optical Engineering, 2005.

[6] "Introduction to Aerosols," CAICE. [Online]. Available at: https://caice.ucsd.edu/introduction-to-aerosols/ (accessed October 14, 2023).

[7] "Spectral Calculator-Hi-Resolution Gas Spectra." Available at: https://spectralcalc.com/info/about (accessed October 21, 2023).

[8] "RF Signal Attenuation due to Atmospheric Gases — MATLAB Gaspl." [Online]. Available at: https://www.mathworks.com/help/phased/ref/gaspl.html (accessed January 15, 2023).

[9] "RECOMMENDATION ITU-R P.676-13 — Attenuation by atmospheric gases and related effects," International Telecommunications Union, Geneva, Switzerland, Aug. 2022.

[10] P. Jaffe and J. McSpadden, "Energy Conversion and Transmission Modules for Space Solar Power," *Proceedings of the IEEE*, vol. 101, no. 6, pp. 1424–1437, June 2013. doi: 10.1109/JPROC.2013.2252591.

[11] D. Atlas, "Optical Extinction by Rainfall," *J. Meteor.*, vol. 10, no. 6, pp. 486–488, December 1953. doi: 10.1175/1520-0469(1953)010 <0486:OEBR>2.0.CO;2.

[12] "Visibility | SKYbrary Aviation Safety." [Online]. Available at: https://skybrary.aero/articles/visibility (accessed October 14, 2023).

[13] R. Olsen, D. Rogers, and D. Hodge, "The aRb Relation in the Calculation of Rain Attenuation," *IEEE Trans. Antennas Propagat.*, vol. 26, no. 2, pp. 318–329, March 1978. doi: 10.1109/TAP.1978.1141845.

[14] D. L. Fried, "Optical Resolution through a Randomly Inhomogeneous Medium for Very Long and Very Short Exposures," *J. Opt. Soc. Am.*, vol. 56, no. 10, p. 1372, Oct. 1966. doi: 10.1364/JOSA.56.001372.

[15] J. L. Miller and E. Friedman, *Photonics Rules of Thumb: Optics, Electro-Optics, Fiber Optics, and Lasers*, Third edition, in SPIE. Bellingham, WA: SPIE Press, 1996.

[16] L. C. Andrews, R. L. Phillips, and C. Y. Young, *Laser Beam Scintillation with Applications*. SPIE, 2001. doi: 10.1117/3.412858.

[17] "Ionospheric Scintillation | NOAA / NWS Space Weather Prediction Center." [Online]. Available at: https://www.swpc.noaa.gov/phenomena/ionospheric-scintillation (accessed August 26, 2023).

[18] C. A. Levis, J. T. Johnson, and F. L. Teixeira, *Radiowave Propagation: Physics and Applications*. Hoboken, New Jersey: Wiley, 2010.

[19] L. Rizzo, J. F. Federici, S. Gatley, I. Gatley, J. L. Zunino, and K. J. Duncan, "Comparison of Terahertz, Microwave, and Laser Power Beaming under Clear and Adverse Weather Conditions," *J. Infrared Milli. Terahz. Waves*, vol. 41, no. 8, pp. 979–996, August 2020. doi: 10.1007/s10762-020-00719-w.

[20] A. W. Ali, "On Laser Air Breakdown, Threshold Power and Laser Generated Channel Length," Naval Research Laboratory, Washington, DC, NRL Memorandum Report 5187, September 1983. [Online]. Available at: https://apps.dtic.mil/sti/pdfs/ADA133211.pdf (accessed December 7, 2022).

[21] "ANSI Z136.1-2022 — American National Standard for the Safe Use of Lasers." Laser Institute of America, Orlando, FL, August 3, 2022.

[22] B. Rockwell, Ed., *Laser Safety Guide*, Twelfth edition. Orlando, FL: Laser Institute of America, 2015.

[23] T. E. Johnson and H. Cember, *Introduction to Health Physics*, Fifth edition. New York: McGraw-Hill Education, 2017.

[24] T. Nugent, "Improving Performance Metrics for Power Beaming," presented at the *3rd Optical Wireless Power Transfer Conference*, Yokohama, Japan, Apr. 2021.

[25] "CFR — Code of Federal Regulations Title 21." [Online]. Available at: https://www.accessdata.fda.gov/scripts/cdrh/cfdocs/cfcfr/ CFRSearch.cfm?FR=1040.10 (accessed September 2, 2023).

[26] "IEC 60825-1:2014 — Safety of Laser Products – Part 1: Equipment Classification and Requirements." International Electrotechnical Commission, May 15, 2014. [Online]. Available at: https://webstore.iec.ch/ publication/3587 (accessed September 2, 2023).

[27] "Order JO 7400.2P – Procedures for Handling Airspace Matters." April 20, 2023. [Online]. Available at: https://www.faa.gov/document Library/media/Order/7400.2P_Basic_dtd_4-20-23–COPY_FINAL.pdf (accessed September 2, 2023).

[28] "Laser Laws & Enforcement | Federal Aviation Administration." [Online]. Available at: https://www.faa.gov/about/initiatives/lasers/ law_enforcement_guidance (accessed September 2, 2023).

[29] "ANSI Z136.6-2015 — American National Standard for Safe Use of Lasers Outdoors." Laser Institute of America, Orlando, FL, USA, October 05, 2015.

[30] R. Lafon, J. Wu, and B. Edwards, "Regulatory Considerations: Laser Safety and the Emerging Technology of Laser Communication," presented at the *Commercial Laser Communications Interoperability and Regulatory Workshop*, Washington, DC, June 12, 2017.

[31] Rianne Stam, "Comparison of International Policies on Electromagnetic Fields," Netherlands National Institute for Public Health and the Environment, 2017. [Online]. Available at: https://www.rivm. nl/sites/default/files/2018-11/Comparison%20of%20international%20 policies%20on%20electromagnetic%20fields%202018.pdf (accessed September 2, 2023).

[32] "Evaluating Compliance with FCC Guidelines for Human Exposure to Radiofrequency Electromagnetic Fields," Federal Communications Commission Office of Engineering & Technology, OET Bulletin 65, August 1997. [Online]. Available at: https://transition.fcc.gov/oet/ info/documents/bulletins/oet65/oet65.pdf (accessed September 2, 2023).

[33] "RF Safety FAQ." [Online]. Available at: https://www.fcc.gov/engineering-technology/electromagnetic-compatibility-division/radio-frequency-safety/faq/rf-safety (accessed September 2, 2023).

[34] K. Schulmeister, "The Upcoming New Editions of IEC 60 825-1 And ANSI 21 36.1 — Examples on Impact for Classification and Exposure LimITS," *Conference Proceedings*, 2013.

[35] "IEEE Standard for Safety Levels with respect to Human Exposure to Electric, Magnetic, and Electromagnetic Fields, 0 Hz to 300 GHz," *IEEE Std C95.1^{TM}-2019 (Revision of IEEE Std C95.1-2005/Incorporates IEEE Std C95.1-2019/Cor 1-2019)*, pp. 1–312, October 2019. doi: 10.1109/IEEESTD.2019.8859679.

[36] M. A. Green, *Solar Cells: Operating Principles, Technology and System Applications*, Repr. [der Ausg.] Englewood Cliffs, NJ, 1982. Kensington, NSW: Univ. of New South Wales, 1998.

[37] F. E. Wasserman, T. Lloyd-Evans, S. P. Battista, D. Byman, and T. H. Kunz, "The Effect of Microwave Radiation (2.45 GHz CW) on the Molt of House Finches (Carpodacus Mexicanus)," *Space Solar Power Rev.*, vol. 5, pp. 261–270, 1985.

[38] Richard M. Dickinson, "FW: 4 Ds - maybe 2 more," July 31, 2020.

[39] D. Ford, *A Vision So Noble: John Boyd, the OODA Loop, and America's War on Terror*. CreateSpace, 2010.

[40] T. J. Nugent, Jr., D. Bashford, T. Bashford, T. J. Sayles, and A. Hay, "Long-Range, Integrated, Safe Laser Power Beaming Demonstration," in *Technical Digest OWPT 2020*, Yokohama, Japan: Optical Wireless Power Transmission Committee, The Laser Society of Japan, April 2020, pp. 12–13.

[41] "Optical Spectrum (>3 THz) — WRC-12 to Consider Procedures For Free-Space Optical Links." [Online]. Available at: https://www.itu.int/net/ITU-R/information/promotion/e-flash/4/article3.html (accessed September 4, 2023).

[42] J. Thayer, "Lasering in on the Federal Communications Commission: Can the FCC Regulate Laser Communications?," *Intellectual Property Brief*, vol. 6, no. 2, p. 30, 2015.

[43] "FCC Online Table of Frequency Allocations 47 C.F.R. §2.106." Federal Communications Commission, July 1, 2022. [Online]. Available at: https://transition.fcc.gov/oet/spectrum/table/fcctable.pdf (accessed September 4, 2023).

[44] "United States Frequency Allocations." U.S. Department of Commerce National Telecommunications and Information Administration Office of Spectrum Management, January 2016. [Online]. Available at: https://www.ntia.gov/sites/default/files/publications/january_2016_spectrum_wall_chart_0.pdf (accessed September 4, 2023).

[45] J. McSpadden, "Solar Power Satellite Frequency Selection," presented at the *2014 IEEE 25th Annual International Symposium on Personal, Indoor, and Mobile Radio Communication (PIMRC)*, Washington, DC, September 2, 2014.

[46] "RECOMMENDATION ITU-R SM.2151-0 — Guidance on frequency ranges for operation of wireless power transmission via radio frequency beam for mobile/portable devices and sensor networks," International Telecommunications Union, Geneva, Switzerland, September 2022.

[47] "ITU-R: Managing the Radio-Frequency Spectrum For The World," ITU. [Online]. Available at: https://www.itu.int:443/en/media centre/backgrounders/Pages/itu-r-managing-the-radio-frequency-spectrum-for-the-world.aspx (accessed September 4, 2023).

[48] "WRC-23 — World Radiocommunication Conferences (WRC)." [Online]. Available at: https://www.itu.int/wrc-23/ (accessed September 4, 2023).

Chapter 5

Components and Subsystems for Microwave Power Beaming

*Divide each of the difficulties ... into as many parts as possible
and as might be required in order to resolve them better.*

— René Descartes [1]

5.1 Introduction

The microwave region of the electromagnetic spectrum has been a
principal area of activity for power beaming development. Extending from 1,000 to 300,000 MHz (or 30 cm to 1 mm wavelength) [2],
there are subregions that have very favorable transmission through
the atmosphere and its weather. This chapter addresses component, subsystem, system design, and implementation considerations
for these microwave power beaming systems. Specific operating frequency selection was explored in Chapter 4.

5.2 System Overview

Microwave-based power beaming systems have been substantially
developed and demonstrated since the 1960s. The transmission of
microwave or millimeter wave power for power beaming can involve
either single radiating elements or phased arrays, on both the transmitter and receiver. Precise phase and magnitude variations applied
across the elements of a phased array can achieve greater directivity,

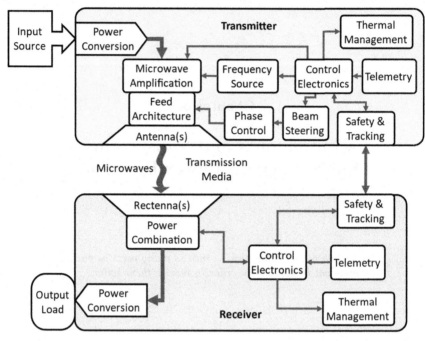

Figure 5.1. Functional block diagram for a microwave-based power beaming system.

beam steering, and sidelobe minimization compared to fixed elements. Although long-range microwave power beaming systems may seem complex, most of the subsystems required are relatively straightforward. A microwave or millimeter wave power beaming system can be formulated or analyzed by looking at segments for DC-RF conversion, RF feeding architecture, transmitting aperture, receiving aperture, and RF-DC conversion and management.

A functional block diagram for a microwave-based power beaming system is shown in Figure 5.1.

Note that a given system may not have all the elements indicated or may be configured differently. In particular, amplification may occur downstream of phase control. Interconnects depicted between subsystems may represent power, data, or both. The principal power transmission path is along the left side of the diagram and is shown by thicker lines.

5.2.1 *System Efficiencies*

There will be inefficiencies in each segment of the system. Every inefficiency detracts from the ultimate purpose of delivering energy. Many major sources of inefficiency arise from the power conversions needed within the subsystems.

For the transmitter, a prime source of inefficiency is typically in the conversion of DC power to RF power. DC-to-RF conversion and amplification can happen in different stages and may require separate DC-DC conversion prior to DC-to-RF conversion by a solid state or vacuum electronics device. A carefully selected or specifically tailored DC-DC converter can be in excess of 95% efficient [3, p. 671]. Low-power DC-RF devices, such as voltage-controlled oscillators (VCOs), may not be especially efficient but their loss tends to be negligible compared to the RF amplifier employed. Depending on the operating frequency, solid-state RF amplifiers can have power-added efficiency (PAE) exceeding 80% [4], and vacuum electronics have been reported at 60% [5].

Receiver elements have been demonstrated at a range of frequencies, with the lower frequency and higher power density generally correlating with higher efficiency [6]. The best of these rectifying antenna elements ("rectennas") have exceeded 90% efficiency [7, p. 72]. A plot showing rectenna element efficiencies at different frequencies appears in Figure 5.19.

Losses in efficiency also occur with both transmit and receive apertures, in the transmission media, and elsewhere. These are explored in both Chapters 4 and 7. Arraying conversion elements incurs additional losses arising from power division (typically at the transmitter) and recombination (typically at the receiver). These losses are sustained whether the power is at RF or DC.

5.3 Transmitters

The transmitter consists of one or more sources that convert electricity into microwaves or millimeter waves, subsystems needed to support this conversion, subsystems that provide for beam shaping and steering, and safety and control subsystems. This section reviews options for how some of these might be implemented.

5.3.1   *RF Power Sources*

Both the operating frequency of the system and the required power levels will have a large influence on which RF source is appropriate. There are two prevalent categories: the generation of a small signal followed by amplification, and the generation of high-power signals directly. The choice of low or high-power RF oscillators for power beaming is dependent upon the DC prime power available and the RF power levels needed at an antenna's input port. If an active array is used, the input RF source power will need to be sufficient for distribution among a network of amplifiers located prior to the radiating elements.

RF oscillators can be divided into three categories: harmonic oscillators, relaxation oscillators, and VCOs. Harmonic oscillators include both feedback and negative resistance oscillators. In the microwave range, negative resistance oscillators are generally the better option since the feedback oscillators exhibit more phase shifts in the feedback path at these higher frequencies. The active oscillating device is placed into an RC, LC, or RLC circuit to provide continuous, spontaneous resonant oscillations at the design frequency. Relaxation oscillators use an energy-storing element and a nonlinear switching device connected within a feedback loop to provide non-sinusoidal waveforms such as sawtooth or square waves. VCOs use a varactor diode or diodes in a tuned circuit to output frequency-selective RF energy. Different biasing of the varactor changes the resonant frequency of the tuned circuit.

Many different active devices can provide RF energy. Some of the popular options are presented in Table 5.1. Choosing the proper active device is a function of its frequency range of operation and the RF output power the device and its support circuit can generate. A system approach is necessary to identify the power levels needed at different locations within an antenna array for instance, especially if that array is an active phased array.

Numerous architectures supporting space-based solar power beaming have focused on high-power generation from such apparatus as traveling-wave tubes (TWTs), klystrons, and magnetrons. Examples are shown in Figure 5.2.

TWTs are linear vacuum tubes that amplify input RF energy by allowing this energy to absorb power from an electron beam as it travels down the tube. Two types of TWTs exist: helix and coupled

Table 5.1. Active devices used in oscillator circuits with their approximate highest operating frequencies [8, pp. 180–182].

Active device	Frequency (GHz)
Triode vacuum tube	1
Bipolar transistor (BJT)	20
Heterojunction bipolar transistor (HBT)	50
Metal-semiconductor field-effect transistor (MESFET)	100
Gunn diode, fundamental mode	100
Magnetron tube	100
High electron mobility transistor (HEMT)	200
Klystron tube	200
Gunn diode, harmonic mode	200
IMPATT diode	300
Gyrotron tube	600

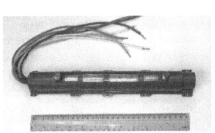  

Figure 5.2. Microwave power generation, from left to right: traveling-wave tube [9], *klystron* [10], *and continuous wave magnetron* [11].

cavity. Helix TWTs exhibit wideband performance up to two octaves and are thus attractive for radar applications. Coupled-cavity TWTs have a series of cavities along the path of the RF energy which makes them narrowband with fractional bandwidths of 10%–20%. TWTs operate from 400 MHz to 50 GHz with gains of 40–70 dB. They output RF from a few watts to megawatts. Klystrons were first successfully demonstrated in 1937 [12]. They are essentially narrowband coupled-cavity TWTs that are designed to have gains of up to around 60 dB with output powers in the tens of megawatts.

Magnetrons were developed for radar applications during the Second World War [13]. They have been particularly popular for power beaming due to their mature high-power performance, availability, and their ability to convert DC power to RF directly. Magnetrons take in DC power and convert that DC into RF energy, commonly at levels in the tens of kilowatts. Magnetrons are divided into three categories: negative resistance, cyclotron frequency, and traveling wave cavity. The negative resistance magnetrons have low efficiencies at higher frequencies. They are useful at frequencies of less than 500 MHz which makes them a poor fit for power beaming in most cases because of the large apertures that would likely be required. The cyclotron varieties are useful for frequencies greater than 100 MHz, but the cavity magnetrons are more efficient at microwave frequencies and have very high peak power levels.

In the case of space solar, DC power might be provided by photovoltaic solar arrays spread over a wide area that is matched with the area of the transmit aperture. The amount of DC power would depend upon the efficiency and arrangement of these solar arrays. This DC energy would then be appropriately distributed to drive TWTs or cavity magnetrons. Each TWT or magnetron could be associated with a passive subarray consisting of antenna elements. If an active phased array is used as the subarray within a much larger array to facilitate beam steering, each TWT or magnetron would need to provide adequate power to each antenna element.

Thermal management concerns for DC-RF conversion elements will be strongly influenced by the device efficiency and operating environment. Obviously, the less efficient conversion will result in greater generation of waste heat and thus greater attention to thermal management. For systems operating in air, the ambient temperature and rate of airflow will drive whether it is feasible to use heatsinks or passive thermal approaches, or whether fluid-based cooling methods are needed. In all situations, the three principal methods of heat transfer (conduction, convection, and radiation) will apply [14, p. 201].

5.3.2   *Transmitter Antenna Architecture*

Primarily due to cost and complexity, many past microwave beaming tests have used a single parabolic dish or other monolithic aperture as the transmitter. Such apertures can represent simple, reliable, and high-directivity options. While these may suffice for some links with stationary transmitters and receivers, they do not intrinsically allow for beam steering or a high degree of dynamic beam shaping. Applications with single apertures that require pointing or focusing might need mechanisms to accomplish this. These might be slow, imprecise, and failure prone. For these reasons, there are instances of power beaming when a phased array transmitter may be the preferred choice.

The transmission of microwave/millimeter wave power can benefit greatly from the use of a phased array, as this can permit the possibility of control over both the amplitude and phase from each radiating element. A single antenna element is limited by its own beam pattern and generally has no native reconfigurability. A phased array, consisting of an organized grouping of many individual radiating elements, can circumvent these weaknesses. An array can achieve higher gain, lower sidelobe levels, beam focusing, and beam steering.

While comprehensive coverage of phased array design considerations is beyond the scope of this book, an overview of some of the factors for phased arrays for power beaming is presented. Readers are encouraged to consult texts solely devoted to phased arrays [15–17] and other resources [18] for additional information.

The design of transmit antenna arrays for power beaming involves the following considerations:

(1) Array element selection:

  (i) Design simplicity
  (ii) Directivity
  (iii) Radiation efficiency
  (iv) Suitability to array geometry

(2) Array geometry selection:

  (i) Physical limitations (size, weight)
  (ii) Directivity
  (iii) Sidelobe/grating lobe reduction

(3) Amplitude taper selection:

  (i) Width of the main beam vs. sidelobe level
  (ii) Dynamic steerability requirements

(4) Feeding architecture selection:

   (i) Physical limitations
   (ii) Beam steering accuracy requirements
   (iii) Power handling

These considerations are in many cases interrelated and subject to manufacturability and cost factors, which in turn will be shaped by the frequency selected and the output power level.

5.3.3 *Array Element Selection*

Power beaming applications typically require a narrow beam that contains as much of the total transmitted power as possible. Although the incorporation of many omnidirectional elements into an array can result in a directive beam, the results are generally better if each individual element is relatively directive. Antenna element gain is another important metric. The overall gain is the product of the antenna's directivity and its radiation efficiency. Radiation efficiency measures what percentage of the power entering the transmitting antenna is successfully converted from input signal to electromagnetic radiation. If an antenna has very high directivity, but low radiation efficiency, that antenna will likely be ineffective for power beaming since much of the power entering the antenna is dissipated within the antenna before transmission. Prospective array elements are shown in Figure 5.3.

Dipole antennas can generally be designed and fabricated without complex computations or expensive manufacturing techniques. Typical dipole antennas efficiently radiate a donut-shaped pattern having a maximum directivity of approximately 1.64 dBi. This means dipoles are usually poor choices in a typical power beaming system. This is true even when a multitude of dipoles are placed into an array format without a ground plane present.

Parabolic dish antennas are passive radiators that usually perform better in terms of pattern quality and gain when compared to passive antenna arrays of similar size. Parabolic dishes can have aperture efficiencies up to 75%, with typical values from 55% to 65% [24, p. 415]. Passive dishes can be large and still radiate efficiently, assuming spillover and feed blockage are minimized. Parabolic dishes are generally difficult to put into array configurations due to their

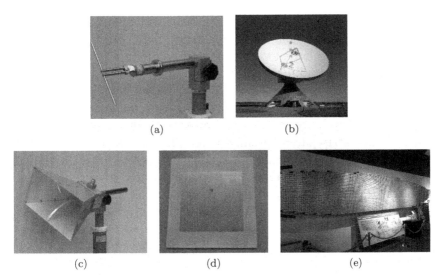

Figure 5.3. Candidate elements: (a) half-wave dipole antenna [19], *(b) parabolic dish antenna* [20], *(c) horn antenna* [21], *(d) patch antenna* [22], *and slotted waveguide array* [23].

cumbersome volumetric geometries. They are too heavy for many airborne applications, and their large swept volumes are often undesirable, even Cassegrain variants. However, their excellent passive directivity and gain make them good stand-alone transmitters for statically configured power beaming or where the slow, gimballed movement of the dish is adequate for beam realignment.

Horn antennas tend to have reasonably high gains, but they are large, heavy, and expensive. Their awkward rigid flared shape makes them difficult to put into arrays. Like the parabolic dish, a horn can be used as a stationary single-element transmitter. For gimballed power beaming, a parabolic dish could be superior primarily due to lower swept volumes. Several horn configurations have aperture efficiencies exceeding 90% [25, Figure 5.4].

Microstrip antennas can generally be designed and fabricated without complex computations or expensive manufacturing techniques. A very popular type of microstrip antenna is a planar patch antenna which consists simply of a copper trace attached to a rectangular copper area, both etched on a circuit board. Individual patches have directivities on the order of 5–7 dB, and they focus their energy upwards in a direction normal to the patch antenna's surface.

Patch antenna radiation efficiency can vary widely and depends primarily on the loss tangent of the dielectric material supporting the etched copper. Higher dielectric constants will reduce the patch's surface area, thereby reducing gain. The relatively low gain and broad 3-dB beamwidth on the order of 60° make single-patch antennas bad choices for power beaming. However, arraying numerous patch antennas together can be effective. The accessibility, low cost, and scalability of microstrip patch antennas can make them a desirable option for integration into transmitting arrays. But unless both the design and the dielectric selection are performed with considerable rigor, passive patch arrays will usually have lower gains when compared to parabolic dishes and horns with similarly sized radiation apertures. This is particularly true as the aperture size grows since the passive patch array requires ever-increasing lengths of lossy feed lines to distribute power to each of its patch radiators. In fact, very large passive patch arrays become resistive elements due to these feeding losses. If a large patch array is desired to provide a narrow beam for a given power beaming application, such as ones requiring beam steering, RF amplification must be placed prior to the radiating elements. This will allow for the RF power densities coming out of the radiating aperture to be higher, overcoming prior losses in the patch array's feed network. This amplification is often accomplished, along with phase shifting, with modules placed prior to each patch radiator.

Cavity-backed slotted waveguides are very efficient series-fed radiators emitting power normal to their slotted top surfaces with reduced back lobe radiation. A single-slotted waveguide is essentially a linear array of slots that act collectively and efficiently to produce high directivity and high gains [26]. These linear slotted waveguide arrays can be placed adjacent to one another in both orthogonal directions. One disadvantage is that the long length of a single slotted waveguide limits phase-tuning resolution in that direction, resulting in substantially quantized phase sidelobes. Slotted waveguides can be challenging to fully characterize using 3D electromagnetic simulation tools due to the number of unknowns necessary to fully model the slots and their uncertainties. These include the surface roughness, thickness, presence of coating material, and resolution of the slot cutout. Slotted waveguides can also be somewhat expensive to construct due to the challenges of forming the waveguide and cutting

Table 5.2.   Comparison of relative attributes for different elements.

Element	Directivity	Efficiency	Complexity	Cost
Dipole	Low	Very High	Low	Low
Microstrip	Medium	Medium	Low	Low
Dish	Very High	Medium	High	Very High
Horn	High	Medium	High	Very High
Slotted Waveguide	Very High	Very High	High	Medium

accurate slots. However, recent advances in both design resources and manufacturing make them more accessible than in the past [27,28].

A comparison of attributes of the element options reviewed above is shown in Table 5.2.

Further information on antenna element design can be found in textbooks [24,29,30] and design guides [31].

5.3.4 *Array Geometry Selection*

Phased arrays may be sized to a limited aperture area, typically defined by their application. Array geometry, operational frequency, and transmission efficiency are used to formulate the array's architecture. The directivity of an array is proportional to its number of elements. Arrays with more elements yield superior directivity/gain, which can be roughly estimated by correlating each increase in the order of magnitude of elements with a 10 dB increase in directivity [32]. Higher numbers of elements also enable reduced sidelobe levels. Higher operational frequencies allow for more elements to be placed within a given surface.

A planar array of N elements will possess the same directivity regardless of its operating frequency. A 10 × 10 array with 1 wavelength spacing at 2.4 GHz will be a little over a meter long on each side, but its directivity will be the same as a 10 × 10 array with 1 wavelength spacing at 24 GHz, which will measure only 10 cm on a side. A 100 × 100 array with 1 wavelength spacing at 24 GHz will be the same size as the 10 × 10, 1 wavelength, 2.4 GHz array, but will have a higher directivity.

A generalized equation for planar array directivity from Mailloux [33, p. 86] is

$$D = N d_E(\theta_0, \phi_0) \epsilon_T \tag{5.1}$$

where N is the number of elements, d_E is the average element directivity, and ϵ_T is the taper efficiency. Taper efficiency is defined by

$$\epsilon_T = \frac{|\sum a_n|^2}{N \sum |a_n|^2} \tag{5.2}$$

where the coefficients a_n represent voltage, currents, or incident wave amplitude of each array element [33, p. 71]. As a_n increases, so too will ϵ_T. Note that selecting elements with high directivity will limit the range of angles the array will be able to scan across effectively.

It has been known since at least 1945 that there is no theoretical limit on array directivity [34]. However, this requires "close element spacings and with extreme amplitude and phase changes across the array" [24, p. 303]. A balance of directivity and grating lobe reduction can be realized with a uniformly excited array with element spacing of around 0.7 wavelengths. Element spacings larger than this increase the undesirable appearance of grating lobes, which may have equal or even greater magnitude than the main power beam. This results in losses in system efficiency and increases in potential safety risks. A depiction of grating lobes is shown in Figure 5.4.

5.3.5   *Amplitude Taper Selection*

Reduction of grating lobe and side lobe magnitude can be achieved through judicious amplitude taper selection. Amplitude tapering involves applying different amplitudes to the transmit elements of an array. There are various weighting schemes for accomplishing this, including Chebyshev, Taylor, Gaussian, and more [33,36]. A few general guidelines apply to each:

(1) Different amplitude tapering schemes increase or decrease the directivity. Generally, greater directivity correlates to a narrower beam.
(2) Different amplitude tapering schemes increase or decrease the magnitude of the sidelobes. Generally, the larger the quantity of side lobes, the smaller the resulting magnitude of each.

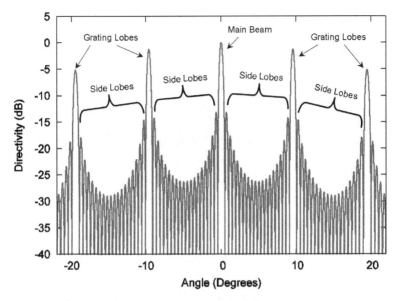

Figure 5.4. *An example of how large grating lobes can appear at unintended angles away from the main beam* [35]. *Image used under CC BY-SA 3.0.*

In power beaming, maximizing the total power illuminating the receiver usually requires that the total energy contained within the sidelobes be minimized. When this is achieved, the peak gain of the main lobe is less important, provided the receiver is designed to be large enough to intercept the entire beam at the desired distance of transmission, and that it can operate effectively at the resulting power density.

Methods for optimizing the amplitude taper for power beaming, as well as phase focusing, are given by Hutson [36]. Retrodirective methods can also be used to aid in this optimization, as outlined by Wang and Lu [37] and others [38, p. 1428].

One consideration for amplitude tampering is thermal management. Since power beaming transmitters might operate in a continuous rather than pulsed mode, the amount of waste heat that could build up may be significantly larger, which could reduce the performance or lifetimes of electronic components. This could be particularly true in the center of the transmit array if the tapering scheme that is used results in greater amplitudes being concentrated there. An alternative is to alter the scheme to reduce the amplitude

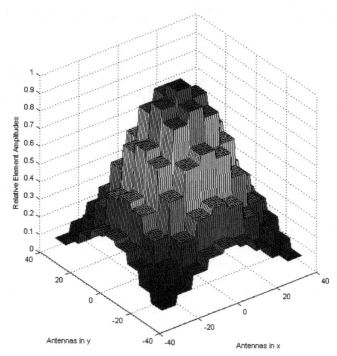

Figure 5.5.　A 72 × 72 element array with 81 8 × 8 subarrays implementing an approximate 30 db Taylor-weighted taper.

in the center of the array, but this will have consequences for the resulting antenna pattern that is synthesized.

Consider an example in which sources are applied to subarrays to approximate a 30 dB Taylor-weighted taper. Figure 5.5 shows such an array aperture composed of 9 × 9 subarrays. Each subarray has 8 × 8 elements with all 64 subarray elements driven by the same level of RF power. Thus, the total transmitting aperture consists of 72 × 72 antenna elements, for a total of 5,184.

If the 72 × 72 antenna elements are spaced $\lambda_0/2$ in both x and y directions with no progressive phase shift, the resultant array pattern shown in Figure 5.6 is generated. This pattern represents 90 different elevation cuts with each rotationally stepped to bisect the xy-plane every $2°$. These 2D elevation pattern cuts convey sidelobe and pattern shape information for the entire upper radiation hemisphere. This array has worst-case sidelobes of -27 dB at $\pm13.5°$ along the principal plane cuts, as seen in Figure 5.6. Single-element roll-off

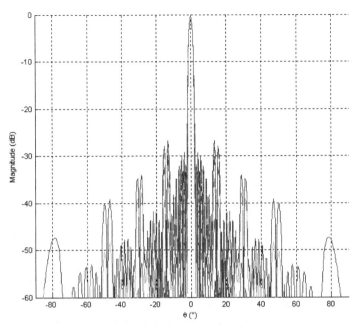

Figure 5.6. *The array pattern that results from the amplitude taper applied in Figure 5.5.*

is accounted for using a typical raised-cosine pattern. The double-hump sidelobes, typical of subarray amplitude quantization, are readily seen. The sidelobe suppression can be improved by reducing the subarray size or by increasing the overall size of the array.

The source placements are determined by the power level of the various subarrays. If all sources were of equal magnitude, placing one at each subarray would result in the entire array being uniform, producing 13 dB sidelobes. To accomplish the taper, sources need to be placed in a non-uniform manner. For instance, the centermost subarray could be tied to a single source, but the subarrays moving away from the center need to be grouped in various ways to different sources to accomplish the power distribution across the aperture. The overall size of the transmitting aperture is based on the desired transmitting gain and the beamwidths necessary to avoid unnecessary spillover losses. Feed network implementation would need to account for the avoidance of undesirable phase shifts between elements and subarrays.

5.3.6   *Transmitter Feeding Architecture*

The previous section focused on how to produce desirable radiation patterns using an arrangement of amplitude-quantized power weights appropriately delivered to each radiating element. Delivering these weights to the proper locations on the aperture can require expensive and complex RF circuitry. This circuitry also can be used to provide different phases to the radiating elements for electronic focusing and steering. Different approaches can be used to attain varying levels of performance at different efficiency and cost constraints.

5.3.6.1   *Static Feeding Architectures: Equal Taper, Equal Phase*

The simplest and most inexpensive feed networks contain purely passive structures. Two examples are presented in Figure 5.7. The layout seen in Figure 5.7(a) is often referred to as a "corporate-H" feed network. This corporate feeding allows for equal phase distribution since the electrical lengths from the input to all the outputs are similar. Amplitude tapering can also be easily applied to this type of network by simply replacing the 50/50 T-shaped splits with unbalanced splitters.

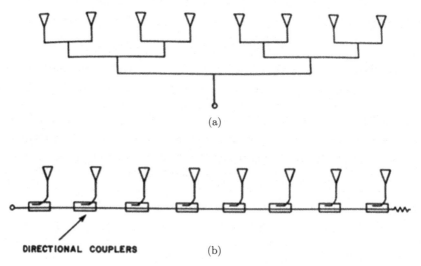

(a)

DIRECTIONAL COUPLERS   (b)

Figure 5.7.   Passive (a) corporate feed network and (b) linear feed network [33].

Another commonly used splitting network is the "series-fed" linear feeding of Figure 5.7(b). This feed type requires that the elements tap their power from the main line every half wavelength to deliver the same phase to each element. This is true at only one frequency, so the feeding is inherently narrowband. Maintaining equal power to each radiator requires different unbalanced splitters at each junction, adding complexity. A major advantage of the series architecture is the overall reduced length of the feeding compared with the corporate layout. This shortened feeding can increase the radiating efficiency of an aperture. Since power beaming is typically focused at a single frequency, a narrowband highly efficient series feeding architecture may be appropriate, especially if sophisticated beam control is not required.

5.3.6.2   *Dedicated Phase Shifter and Amplifier Per Unit*

The corporate and series feed networks of Figure 5.7 support a static beam pattern. When the receiver is moving relative to the transmitter, the array's pattern needs to steer. Phase profiles at the various antennas need to change to steer the beam. Direct control of the amplitudes and phases provided to each antenna element can be accomplished by splitting the input source signal evenly using microwave circuit components such as Wilkinson power dividers [39] and hybrid couplers. The power provided to each antenna can be individually conditioned with a dedicated variable gain amplifier and phase shifter positioned prior to the radiating element as seen in Figure 5.8. The downside is that this architecture can be expensive. An array consisting of 10,000 antennas would require 10,000 amplifiers, 10,000 phase shifters, and the associated control circuitry to generate the required output beam. Amplifiers are costly, and phase shifters, especially at high frequencies, can be more expensive still. These items could dominate the total system material cost.

5.3.6.3   *Ehyaie Method*

One beam steering method which reduces the number of costly phase shifters is the Ehyaie method [22] pictured in Figure 5.9. By using a collection of hybrid 3-dB couplers connected in series with a hindmost phase shifter terminated to the ground, both the amplitudes and phases of the signals diverted toward each antenna element can be

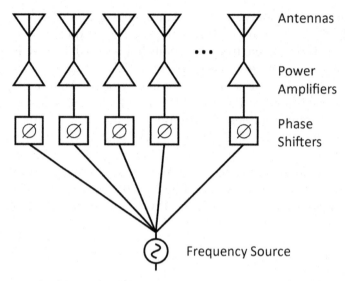

Figure 5.8. A simple block diagram showing a classical feeding architecture. Adapted from [40].

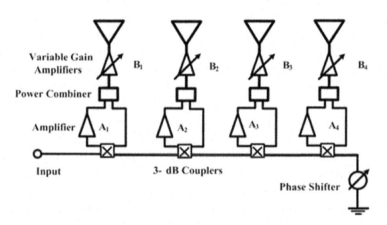

Figure 5.9. The Ehyaie method allows the use of a single-phase shifter for the entire array, but requires two amplifiers per antenna [41].

controlled. The advantage is that only one phase shifter is required, reducing costs. The input signal's phase is dependent on the length of the wires in the system. The reflected signal's phase is determined by the setting of the phase shifter. The input and reflected signals appear on separate ports of the hybrid coupler.

By amplifying the reflected signal an appropriate amount, and then combining the input signal and the reflected signal, a phase shift can be accurately controlled. This works on the principle of vector addition, but there are some downsides to this methodology. Although it reduces the required number of phase shifters, the method necessitates that there be two amplifiers per antenna element. In some high-power applications, this could undo the phase shifter cost savings. Additionally, the phase shift variability is limited to a maximum of 90° due to the destructive interference rules of vector addition. Combining signals of 0°- and 90°-phases results in destructive interference and power loss. To make matters worse, if signals of 0° and 180° are summed together, there is complete destructive interference. This results in a much smaller signal with no phase shift being achieved. There are some extension techniques that incrementally improve performance, but fundamental limitations remain.

5.3.6.4 *Four-Bus Feeding Architecture*

Another method for reducing the phase shifter count is the "four-bus" feeding architecture proposed in [40] and shown in Figure 5.10.

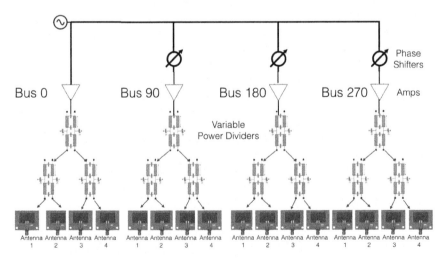

Figure 5.10. Four-bus feeding network topology [40].

This four-bus approach significantly reduces the number of phase shifters required although insertion losses cause the resultant efficiency to be less than desirable for many applications. The 4-bus method divides the original source signal into four equal parts. Appropriate phase shifts are then applied to each so that the relative phases between the four signals are 0°, 90°, 180°, and 270°. These signals are then separately amplified to the appropriate levels.

After amplification, the bussed signals then enter corresponding variable power dividing networks as shown in Figure 5.11. These networks divert the power disproportionally to the various antenna elements. If a 45° phase shift is needed at a particular antenna, equal amounts of power need to come from the 0° and 90° buses with negligible amounts emanating from the 180° and 270° buses. These appropriate powers are then combined before entering the antenna. For a 60° shift, more energy needs to be provided from the 90° bus. When a 240° phase shift is required, power will need to come from the

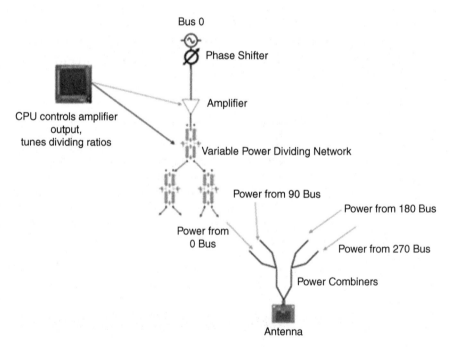

Figure 5.11.   An illustration of how the power from each bus is directed through the variable power-dividing network to reach its destinations [40].

180° and 270° buses. This variable power dividing strategy employs the same vector addition principles used in the Ehyaie Method to cover the entire 360° spectrum.

Variable power dividers are available commercially, and they have been customized for various frequency bands over the decades [42–46]. Depending on the topology, power may be divided equally or unequally. For example, the coupled port of a directional coupler may be −20 dB compared to the original signal. A variable power divider uses a DC voltage control signal to change how much power flows from its two output ports. One of the most popular ways of accomplishing variable power division is an active methodology that implements varactor diodes, as shown in Figure 5.12.

The attraction of the methods without dedicated phase shifters and amplifiers is their probable lower costs. However, managing the complex variable power dividing networks that could result may produce a heavier computational burden on the processing system, and may take up more physical space than the phase shifters and amplifiers it replaced. The vector addition process also causes destructive interference that can result in power loss, which is almost always undesirable for power beaming. Additionally, the insertion losses from the cabling through many stages could be large.

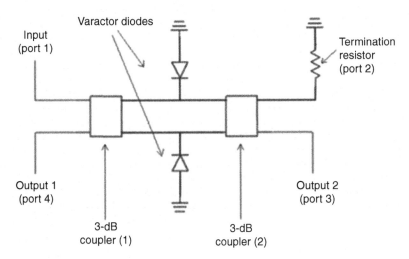

Figure 5.12.   Varactor-diode-based variable power divider [42].

5.3.7　*Transmitter System Examples*

In many past demonstrations, such as [47], power beaming transmitters have been implemented by repurposing transmitters originally developed for another purpose. The examples in this section were specifically developed for microwave power beaming.

The transmitter used for the March 21, 2023 Virtus Solis microwave power beaming demonstration in Indianapolis, IN is shown in Figure 5.13.

Figure 5.13. The transmitter from Virtus Solis' March 2023 10 GHz, 100 m microwave power beaming demonstration. Image courtesy and © Virtus Solis. Used with permission.

Figure 5.14. The transmitter from Emrod's September 2022 5.8 GHz, 34 m microwave power beaming demonstration. Image courtesy and © Emrod. Used with permission.

The transmitter has a 1.92-m² aperture and 6,400 transmitting antenna elements [48]. The power output was reported as 640 W [49].

Figure 5.14 shows the transmitter from Emrod's September 27, 2022 demonstration at Airbus' Munich Area Site as part of a 36-m 5.8 GHz microwave power beaming demonstration.

Observe that both the Virtus Solis and Emrod transmitters are constructed of individually manufactured subarrays. The array in Virtus Solis' case is 10 by 10 subarrays, and in Emrod's case is 8 by 8 subarrays minus those that would be present at each corner. The use of subarrays permits the construction of arrays of arbitrarily large sizes and means that design and production efforts can be largely focused on the subarray. Mechanically and electrically integrating the subarrays into the full transmitter remains a critically important task, as does the implementation of the power distribution architecture to feed them.

5.4　Receivers

The receiver segment collects and converts the energy in the transmitted microwave or millimeter beam. It also includes subsystems to support these functions and those needed for interfacing with the consumers of the received power. This section reviews options for these elements.

5.4.1　*Rectenna Element*

The functional element that converts radiated RF energy back to DC at the receiver in a power beaming system is the rectenna. This word is a portmanteau of the words "rectifier" and "antenna" as it performs these functions together. In this section, the considerations for rectenna elements will be explored, and subsequent sections will examine the combination of these elements into arrays.

5.4.1.1　*Rectifying Subelement*

The rectifying subelement of a rectenna can be either a diode or transistor [50]. Most rectennas have used diodes for this function, but using transistors offers the possibility of two-way power transfer, wherein an aperture could operate as a transceiver in either receive or transmit mode. For this section, the focus will be principally on considerations for diode-based rectennas.

A schematic of the basic diode-based rectenna is illustrated in Figure 5.15(a). A photo of an actual LP rectenna is shown in Figure 5.15(b) to show the corresponding rectenna components etched on a printed substrate.

This LP rectenna is designed in coplanar stripline (CPS) which provides advantages for component placement. An antenna is used to capture the incident RF energy at frequency f. This energy is then passed through the harmonic rejection filter with minimal loss and on to the Schottky diode where the RF energy is rectified to form DC power. The DC energy is then passed through the DC bypass filter to appear as the voltage V_D across the load resistor R_L.

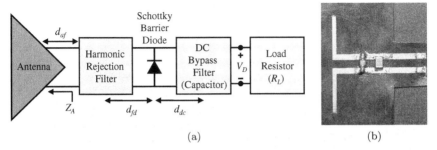

(a)   (b)

Figure 5.15. (a) Rectenna schematic with appropriate component distances. (b) Photograph of an LP dipole rectenna. Copper strips (not seen) are etched on the substrate's backside between the dipole and the diode to form the harmonic rejection filter. A chip capacitor forms the DC bypass filter. The load resistor is not shown. Adapted from [51].

The remaining energies at the various harmonic frequencies $2f, 3f, \ldots$ created from the Schottky diode's nonlinear process are reflected back to the diode by both the harmonic rejection filter and the DC bypass filter. This "trapping effect" results in additional mixing of the harmonic frequencies and ultimately in the generation of more DC power. The harmonic rejection filter also keeps unwanted harmonic energy from re-radiating into free space via the antenna. If allowed to re-radiate, this harmonic energy could interfere with various electronic devices near the rectenna that are operating in the same frequency band. Minimal fundamental f energy passed by the harmonic rejection filter is lost to the antenna for re-radiation since Schottky diodes have shown RF-to-DC conversion efficiencies on the order of 80% [51]. In past designs, low-pass filters have been used to suppress the harmonic energy with very low loss. However, in more recent designs, band stop filters have been used to provide much greater harmonic suppression while maintaining the low loss in the passband. In addition, the harmonic rejection filter is designed to match the real part of the diode's impedance to the antenna's input impedance Z_A. The DC bypass filter also serves two additional purposes. First, by acting as a short-circuited tuning stub, it tunes out the reactance of the Schottky diode based on the DC bypass filter's position in the rectenna circuit. Secondly, it blocks any RF signals ($f, 2f, 3f, \ldots$) from reaching the resistive load. This allows the DC voltage across the load resistor to be level with minimal amplitude versus time variation.

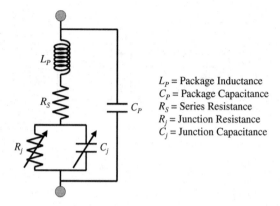

Figure 5.16.   Schottky diode equivalent circuit.

When designing an efficient rectenna, consideration must be given to choosing the proper diode, the type of transmission line on which to distribute the power between the rectenna's various components, the type of antenna, the spacing between the diode and the DC bypass filter (capacitor), and the resistance value of the load. The first component to consider is the nonlinear Schottky diode since the design of the other rectenna parts depends directly upon the diode's performance. The diode conversion efficiency (η_D) is key in determining the rectenna's performance and is based upon the diode's equivalent circuit, diagramed in Figure 5.16.

The diode efficiency is defined as the following ratio:

$$\eta_D = \frac{\text{DC output power}}{\text{RF power incident on diode}} \tag{5.3}$$

The diode's input impedance at the fundamental frequency f is defined as [36,37]

$$Z_D = \frac{\pi R_S}{\cos\theta_{\text{on}}\left(\frac{\theta_{\text{on}}}{\cos\theta_{\text{on}}} - \sin\theta_{\text{on}}\right) + j\omega R_S C_j \left(\frac{\pi - \theta_{\text{on}}}{\cos\theta_{\text{on}}} + \sin\theta_{\text{on}}\right)} \tag{5.4}$$

where θ_{on} refers to the diode's forward-bias turn-on angle. The real part of the diode's impedance from Equation (5.4) is generally on the order of a couple hundred ohms. This resistance value determines the type of transmission line to use for the diode to achieve high RF-to-DC conversion efficiency. By choosing a transmission line with a characteristic impedance equal to the diode's real impedance,

RF mismatch at the diode can be eliminated. If a mismatch is present, the RF-to-DC conversion efficiency suffers. A commonly used transmission line that can be designed to have characteristic impedances on the order of several hundred ohms is CPS. CPS is composed of two parallel conducting strips which can propagate energy from one location to another with low loss.

The imaginary part of the diode impedance that results from Equation (5.4) is tuned out using the length of the transmission line between the diode and the DC bypass filter (typically a chip capacitor). The capacitor acts electrically as a short for RF energy. Therefore, the combination of the capacitor and the length of the CPS transmission line between the diode and capacitor form a short-circuited tuning stub. The CPS has a topology that is readily suited to construct the short-circuited tuning stub. This tuning stub's length is adjusted by changing the capacitor's placement on the CPS. With the tuning stub counteracting the diode's reactance and the characteristic impedance of the CPS matching the real impedance of the diode, the RF energy that is incident upon the diode will ideally see no reflection or mismatch across the terminals of the diode. This allows for maximum capture of the received energy.

An experimental technique described by Strassner and Chang [52,53] and shown in Figure 5.17 utilizes a microstrip to CPS balun to determine the distance between the diode and the capacitor in conjunction with various source powers and load resistances. This circuit provides a direct approach for experimentally determining the diode's RF-to-DC conversion efficiency versus the microwave power incident upon the diode. In essence, it allows for a way to design the rectifier portion of the rectenna. Microwave energy is inputted into the microstrip line at a predetermined level based upon the CP antenna

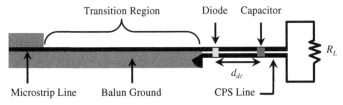

Figure 5.17. Circuit designed for the direct determination of η_D and d_{dc}. The black and dark gray traces are etched on the top and bottom of the substrate, respectively [38,39].

gain and the assumed power density levels at the rectenna array's surface. The electromagnetic fields are then rotated in the shaded transition (balun) region to match the microstrip line to the CPS. Once the energy strikes the diode, DC power is formed and passed through the capacitor to the load R_L. Various values for R_L are tried for maximum efficiency determination. The distance between the capacitor and R_L can be largely arbitrary since only DC energy is propagated between the two. This direct measurement approach has the advantage of including the effects of the higher order harmonics created from the diode mixing.

Several other types of rectifying circuits are possible. They have differing theoretical maximum efficiencies, as shown in Figure 5.18.

The circuit selected for a given system will be guided by cost and performance considerations.

A further consideration is the expected incident power density, as it can significantly affect the resulting conversion efficiency. In general, higher power densities will result in greater efficiency, provided the components of the rectenna element are not damaged. A plot of demonstrated results for rectennas at different frequencies and power densities adapted from Kazemi [55] is shown in Figure 5.19.

The plot illustrates how efficiency generally begins to flatten as the operating frequency increases. Predicting the conversion efficiency of higher frequencies from the extensive measured data gathered at 2.45 GHz can sometimes be used to approximate the likely performance of elements at higher frequencies.

5.4.1.2 *Antenna Subelement*

The other rectenna subcomponent is the antenna. The impedance Z_A looking into the terminals of the antenna must be purely real and must match the characteristic impedance of the transmission line that connects the antenna to the harmonic rejection filter. The antenna is chosen based on the necessary polarization, gain, and bandwidth as well as how the antenna is to be connected to the rest of the rectenna circuitry. As far as printed rectennas are concerned, the ones that have achieved the highest efficiencies have used dipole antennas etched on thin substrates over ground planes. These types

Rectifying Circuit	Maximum Efficiency	R_G and R_L Relationship for Maximum Efficiency
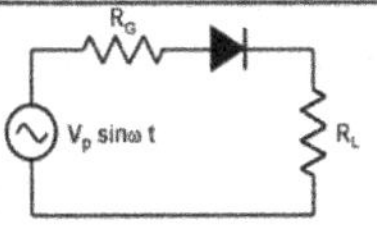	20.3%	$R_L = R_G$
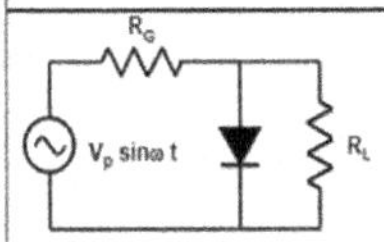	20.3%	$R_L = R_G$
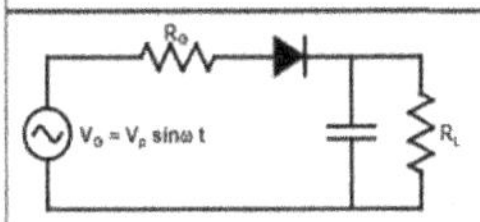	46.1%	$R_L = 2.695\,R_G$
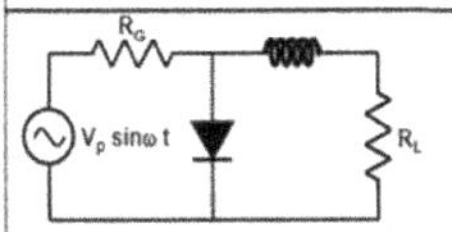	46.1%	$R_L = R_G/2.695$
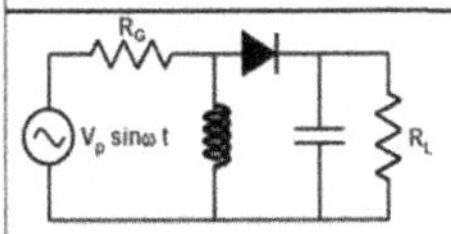	81.1%	$R_L = R_G$
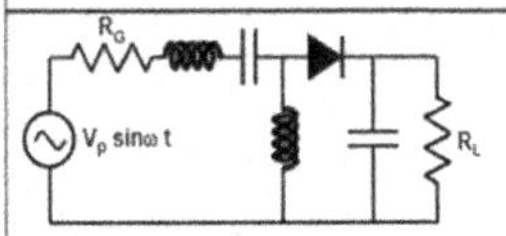	100%	$R_L = R_G/2$
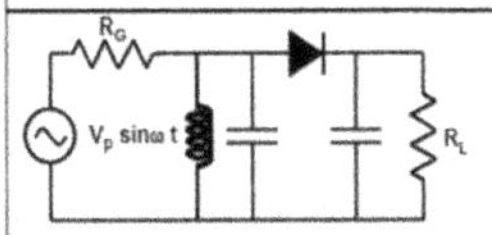	100%	$R_L = 2\,R_G$
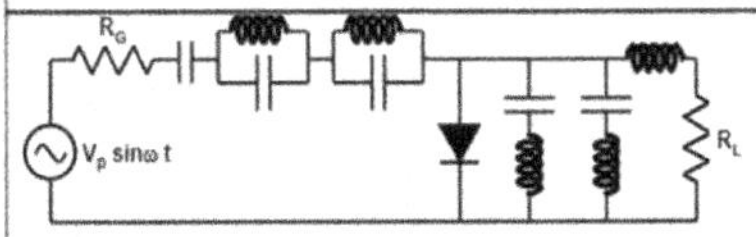	100%	$R_L = \dfrac{\pi^2}{8}\,R_G$
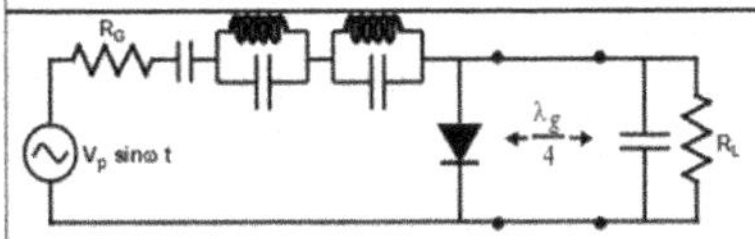	100%	$R_L = \dfrac{\pi^2}{8}\,R_G$

Figure 5.18.   Maximum theoretical efficiencies for different rectenna circuits [54]. *Image courtesy and © James McSpadden. Used with permission.*

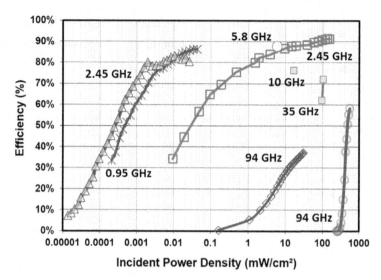

Figure 5.19. Rectenna element efficiency as a function of incident power density for different frequencies and rectennas. Adapted from Kazemi [55], used under CC BY 4.0. Data source references in [55].

of antennas are easily fed using CPS. A reflecting plane is normally placed approximately a $\lambda_0/4$ behind the dipole antenna to focus the antenna's energy in one direction. Since air acts as the dielectric between the dipole and the reflecting ground plane, the antenna can achieve close to 100% radiating efficiency. Having such a high radiation efficiency is key to achieving the best possible rectenna efficiency.

Rectennas have been developed to address a wide range of specialized applications and circumstances. A selection of these are presented in Appendix B.

5.4.2 *Rectenna Array*

Most of the principles concerning the transmitting array can be applied to the receiving array. In some applications, the transmitting and receiving arrays may even be identical. The rectenna array serves as both the absorber of the microwave energy from the transmitter and the rectifier of the microwave energy to DC power [56]. A diagram of a typical rectenna array is shown in Figure 5.20.

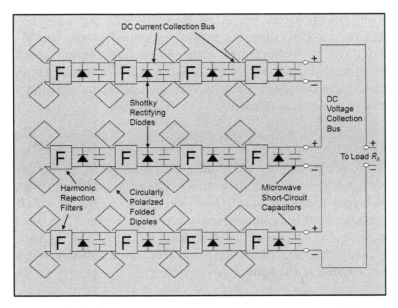

Figure 5.20.   Array consisting of 12 individual rectennas [57].

This example will be used to explore different considerations in the design of rectenna arrays. Other arrays will have similar functional elements and will need to address factors like element spacing and effective power combining.

The antennas in Figure 5.20 are CP folded dipoles that send captured microwave energy at frequency f through the harmonic rejection filters (Fs) to the Schottky rectifying semiconductor diodes. The DC power that then arises from the diode rectifiers is passed through the capacitors along the current collection bus to the peripheral voltage collection bus. Each of the CP folded dipoles is essentially two antennas positioned at the same spot. This reduces the number of rectennas needed to cover an area by one-half when compared to an LP system. Thus, the number of capacitors and diodes is also cut in half, resulting in potential cost savings. CP also has a depolarization advantage, in that a misalignment of the antenna with the incoming beam will not result in decreased received power.

In designs where dipoles and CPS transmission lines are used, columns of parallel-cascaded rectennas are joined in series to produce large DC powers at the output of the array. The rectennas on each column produce DC currents that are summed at the end of that column. The DC voltages of each column are summed resulting in the voltage V_A across a load resistor R_A. The rectenna array's load resistance necessary for achieving the maximum rectenna array RF-to-DC conversion efficiency η_A is defined by

$$R_A = R_L \frac{N_x}{N_y} \tag{5.5}$$

where N_x is the number of columns in the array, N_y is the number of rectennas in each column, and R_L is the optimal load resistance for each individual rectenna. The diodes are connected in parallel in each column, and the columns are connected in series. Equation (5.5) assumes that the incident power striking the surface of the rectenna array is arriving as a plane wave and that the number of rectennas in each column is the same.

5.4.3 *Antenna-to-Capacitor Distance (d_{ac}) Determination*

Each CP rectenna in the array is isolated RF-wise from the next adjacent rectenna by the capacitors which appear as short circuits to the incident microwave energy that strikes them. Energy accepted by a particular CP antenna travels to both the nearest capacitor as well as the nearest filter F. The antenna-to-filter distance d_{af} and the filter-to-diode distance d_{fd} can be arbitrary, but the antenna-to-capacitor d_{ac} distance, illustrated in Figure 5.21, must be properly determined to allow for efficient rectenna array operation. The test circuit for determining the proper d_{ac} is shown in Figure 5.21(a). This circuit includes the antenna, filter, and a microstrip-to-CPS balun. Figure 5.21(a) components are simulated in an electromagnetic simulator without the capacitor and tweaked for a desired input match at 5.8 GHz. This circuit is then etched. Placing the capacitor at various d_{ac} locations on the fabricated circuit results in the experimental input port S_{11} matches shown in Figure 5.21(b). For this specific example, the distance $d_{ac} = 8.5$ mm gives the same measured input match at 5.8 GHz as the measured case when no capacitor is used [28]. As expected, this distance corresponds to a quarter wavelength within the CPS environment.

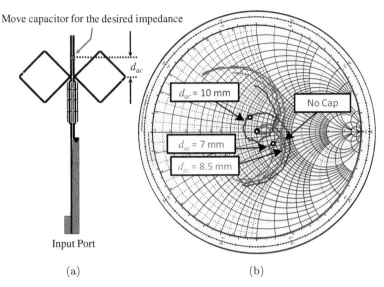

Figure 5.21. (a) Circuit for determining antenna-to-capacitor spacing. (b) Resulting input match [57].

5.4.4   Free-Space Measurement

The setup for determining the rectenna array's η_A is depicted in Figure 5.22.

The RF-to-DC conversion efficiency η_A is defined in terms of the rectenna array's aperture area A_A^{eff} as

$$\eta_A = \frac{P_{dc}}{P_r} = \frac{4\pi d^2 \left(\frac{V_A^2}{R_A}\right)}{P_t G_t \left(\theta_t, \phi_t\right) A_A^{\text{eff}} \left|\hat{\rho}_t \cdot \hat{\rho}_r^*\right|^2 \left(10^{\frac{L_a(z)}{10}}\right) \left(10^{\frac{L_{ra}(t)}{10}}\right)} \qquad (5.6)$$

where $A_A^{\text{eff}} = 4ab$. It is important to make sure the rectenna's aperture is positioned such that each of its antennas point towards the transmitter, i.e., $\theta_r = 0°$. The rectenna array is composed of numerous rectenna elements, each receiving power according to the transmit gain distribution $G_{xy}(x, y, d)$. The average transmit gain seen across the rectenna array's aperture is [53]

$$G_{\text{avg}}(a, b, d) = \frac{1}{4ab} \int_{-b}^{b} \int_{-a}^{a} G_{xy}(x, y, d) \, dx \, dy \qquad (5.7)$$

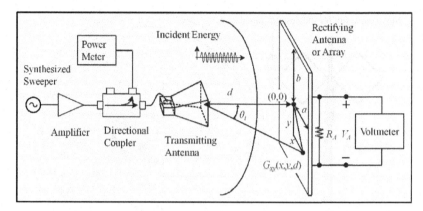

Figure 5.22. Laboratory setup for measuring η_A. The rectenna array output power is defined by the square of the voltage V_A divided by the load resistance R_A [57].

This considers the fact that the power striking the rectenna array's aperture is not exactly a plane wave. Often, the transmit power density is greatest at $(0,0)$ and decreases towards the rectenna array's edges. Per [53], the RF-to-DC conversion efficiency can now be expressed as

$$\eta_A = \frac{\pi d^2 \left(\frac{V_A^2}{R_A} \right)}{ab P_t G_{\text{avg}} (a, b, d) |\hat{\rho}_t \cdot \hat{\rho}_r^*|^2 \left(10^{\frac{L_a(z)}{10}} \right) \left(10^{\frac{L_{ra}(t)}{10}} \right)} \tag{5.8}$$

Some past designs compensate for the tapered power density present at the rectenna array's surface to increase η_A. One way this compensation is accomplished is by tuning the load resistance or by changing the lengths between various components within each rectenna individually.

5.4.5   *Retrodirectivity*

An additional consideration in microwave power beaming-link design is retrodirectivity. Retrodirectivity utilizes a signal, often called a pilot signal, sent from the power beaming receiver to the power beaming transmitter with a retrodirective antenna (RDA) array. The RDA array can use the phase differences in the received signal to adjust the phase of the transmitted power at the transmit elements. This permits the beam to be more effectively steered.

Phase differences can arise from displacements of transmitter elements across the transmitting aperture. For example, this might occur for a solar power satellite power beaming transmitter comprised of elements forming a kilometer-scale aperture. Because it is very difficult to maintain the rigidity of the structure over such a large distance, the resulting small deviations from the transmitter plane could degrade the directivity of the transmitted beam without retrodirective beam steering. A feedback loop is established by a pilot signal sent from the rectenna array to the transmitting phased array to determine the proper angle θ_t and set the proper aperture phase taper to keep the beam on the rectenna array [58].

Using a pilot signal and retrodirective scheme can also apply to sending power to multiple receivers. Figure 5.23 shows the general concept of how this could be implemented at room-scale. The approach is extensible to other scenarios as well.

A comprehensive presentation of retrodirective techniques is beyond the scope of this book. For additional information, readers

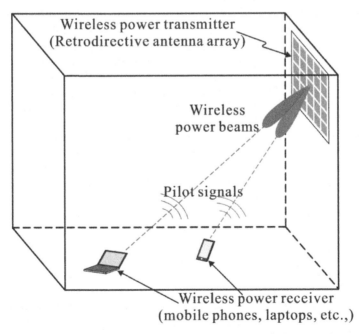

Figure 5.23.   Retrodirective approach to room-scale power beaming. Adapted from Trinh-Van et al. [59]. Used under CC BY 4.0.

are encouraged to consult some of the many resources on this topic, including [60,61, Sec. 13.2, 62].

5.4.6　*Resistive Loading Optimization*

Power density variability on the surface of a rectenna array degrades the overall RF-to-DC conversion efficiency if element loading is uniform. This was recognized and compensated for in Dickinson and Brown's record efficiency demonstration in 1975, in which 22 different sets of radial groupings were matched to four different loads to maximize conversion efficiency for the incident power density [63, p. 35].

In 2013, researchers at the Sichuan University in Chengdu, China revisited this issue by looking at a way to simplify variable loading for increased efficiency. To do so, they built the connectorized rectenna array shown in Figures 5.24(a) and 5.24(b). The 50 cm × 50 cm array had 72 rectenna elements arranged along a $\lambda_0/2$ equilateral triangular lattice. The connectorized components could be individually tested to verify functionality. Two loading schemes were used for testing when the 16-dB transmitting horn antenna was 12.2 cm away from the rectenna array. In Figure 5.24(c), R_1 was 10 Ω. In Figure 5.24(d), R_2 and R_3 were 15 Ω and 25 Ω, respectively. The move to the two-resistor topology resulted in RF-to-DC conversion efficiencies that were 5% higher [64].

To optimize the receiver efficiency, each element should ideally be able to adapt its loading to a peak power point. This approach is like the practice of maximum power point tracking (MPPT) that has been widespread in the solar industry for many years. Many approaches to accomplish this have been implemented. Additional information can be found in [65,66]. MPPT techniques can also offset the effects of temperature and other factors affecting rectenna output.

5.4.7　*Temperature Considerations*

While MPPT can also compensate for rectenna changes that arise because of temperature, it is possible to reduce temperature effects through different design choices. Increasing the incident power density on a rectenna increases the energy available for conversion, and often increases the efficiency of the conversion, up to some limit.

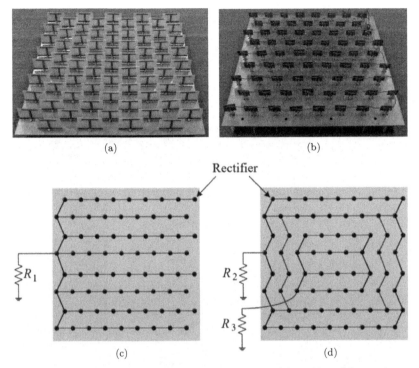

Figure 5.24. Connectorized dipole rectenna array: (a) dipoles, (b) rectifier circuits, (c) loading for plane wave, and (d) loading for "short distance." Figures reproduced from [64] courtesy of The Electromagnetics Academy.

The considerations of utility and safety in determining the power density to be created at the receiver are important as well and are discussed in Chapters 4 and 8.

Higher incident power densities will cause higher temperatures to occur at the sites where rectification occurs, likely degrading performance. One approach to address this is to spread out rectifier sites for heat distribution purposes, as was taken by Zhang *et al.* [64]. Zhang's team used 16 diodes in an array to disperse the heat for each RF input, rather than a single diode.

5.4.8 Receiver System Examples

Though no large-scale, commercially produced, high-power microwave power beaming receiver examples were on the market as

Figure 5.25.   The rectenna receiver array from Virtus Solis' March 2023 10 GHz, 100 m microwave power beaming demonstration. Image courtesy and © Virtus Solis. Used with permission.

of Fall 2023, several intriguing prototypes had been developed. Figure 5.25 shows a receiver that was used for the Virtus Solis demonstration conducted on March 21, 2023, in Indianapolis, IN.

This rectenna receiver measures 1.32 m × 1.62 m and consists of 1,944 rectenna elements [48]. The link length for the demonstration was 100 m, and nearly 70 W was output from the receiver [49].

Figure 5.26 shows the receiver used for Emrod's September 27, 2022 demonstration at Airbus' Munich Area Site as part of a 36-m 5.8 GHz microwave power beaming demonstration [67].

Power received as part of the demonstration was used to power a model city, a hydrogen electrolyzer, and a beer fridge [68]. Data

Figure 5.26. The rectenna receiver from Emrod's September 2022 5.8 GHz, 36 m microwave power beaming demonstration at Airbus in Munich. Image courtesy and © Emrod. Used with permission.

shown on the right side of Figure 5.14 indicates that more than 500 W of power was received.

Strassner and Chang developed the printed CP rectenna receiver array shown in Figure 5.27(a) [53].

The array used an MACOM MA4E1317 Schottky diode and a chip capacitor to achieve 78% RF-to-DC conversion for an incident power density of $8 \, \text{mW/cm}^2$ at 5.61 GHz. The rectenna uses folded dipole antennas called dual rhombic loops which have the advantages of being circularly polarized, giving them higher gains than traditional

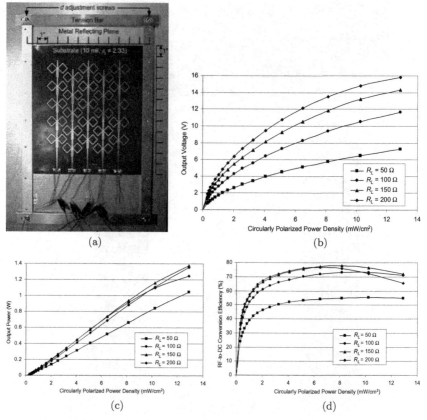

Figure 5.27. (a) CP Rectenna Array. Rectenna array CP measured performance: (b) rectified voltage, (c) output power, and (d) η_A for various resistive loading $(R_L = R_A)$.

dipole antennas, and being easily implemented into CPS circuits. The antennas are placed $\lambda_0/4$ above a ground plane for a 3 dB radiation enhancement in the desired direction. A 3×3 inner section of the array was connected, and the resulting measured data is shown in Figures 5.27(b)–5.27(d). Figure 5.27(d) shows that there is a plateau in the efficiency curves. The array's load resistance was optimized to obtain high efficiencies. These curves are typical of rectennas operating in the microwave frequency range.

Many other recent microwave power beaming receivers appear in the literature, such as [47].

5.5 Notional Link Examples

These examples outline technical considerations and approaches for meeting notional requirements in instances where microwave power beaming might be used. For the development of real-world links, additional consideration of factors such as regulatory requirements, mass producibility, marketability, and economics may need consideration. System developers should consult systems engineering [69] and product design [70] resources as appropriate.

5.5.1 *Consumer Electronics Charging*

Several examples of consumer electronics microwave power beaming have already been deployed to market or are available as development kits for companies interested in incorporating power beaming into their products. Accordingly, this subsection refers to those in aggregate rather than formulating a single notional link. A further discussion of the consumer electronics application case is found in Chapter 8 — Applications. Though specific requirements will vary with the vendor and application instance, in general, each offering adheres to something along these lines:

1. The system shall comply with all applicable safety and regulatory requirements.
2. The transmitter input shall plug into a standard electrical outlet.
3. The receiver output shall provide enough power to [satisfy a specific device charging; power requirement that is at least several mW] at [a distance that is typically at least 10 cm].

In addition to consumer electronics charging, links of this type can be used for powering sensors and Internet of Things (IoT) devices.

Companies producing hardware that create power beaming links for these cases include Reach [71], Ossia [72], Energous [73], GuRu Wireless [74], Powercast [75], and others. These companies use various frequencies in the industrial, scientific, and medical (ISM) band ranges, including those near 915 MHz, 2.45 GHz, 5.8 GHz, and 24 GHz. Powercast has a number of implementation examples in a video series featuring their technology [76], and they offer a product for wireless charging of a game controller as seen in Figure 5.28.

Figure 5.28. A wirelessly charged game controller by Powercast is available on the consumer market [77]. *Image courtesy and* © *Powercast. Used with permission.*

Some of these companies offer developer kits that permit prospective clients to determine if the technology is a good fit for their application. Powercast's development kit is shown in Figure 5.29.

Though closer to power harvesting than power beaming, there have also been two excellent papers by Talla *et al.* that delve into the challenges of powering sensors and other devices with Wi-Fi frequencies, and which include remarkable hardware demonstrations and results [78,79].

5.5.2 *Outdoor Horizontal Point-to-Point*

In this example, the following top-level requirements are assumed:

(1) The system shall comply with all applicable safety and regulatory requirements.
(2) The transmitter input shall plug into a standard electrical outlet.

Figure 5.29. The P1110-EVAL-01 development kit by Powercast for battery recharging. Image courtesy and © Powercast. Used with permission.

(3) The receiver output shall provide at least 300 W continuous electrical output.

(4) The link shall operate over distances of at least 300 m.

One use case for a system meeting these parameters is a temporary electrical power link. Electrical extension cords are generally limited to lengths of about 61 m (200 ft), and rated for delivering less than 2 kW [80]. A power beaming link could exceed this length limitation and could avoid constraints that might be imposed along the path between the transmitter and receiver, such as bodies of water, chasms, or inaccessible property. This is one of the use cases of interest by the New Zealand company Emrod [81].

As discussed in Chapter 4, the selection of the operating frequency will drive many other aspects of the system design. For this application, an ISM band will be chosen to decrease the chances of regulatory challenges. Three bands of interest have center frequencies at 2.45, 5.8, and 24.125 GHz. Since higher frequencies permit smaller apertures and less divergence, 24.125 GHz will be provisionally selected. Atmospheric attenuation is higher at this frequency than the others, but since the link is over a relatively short distance, this is not an overriding consideration. Device efficiencies are also lower at this frequency, but they have been demonstrated to be high enough to make this selection tenable. Several authors have reported peak transmit power added efficiencies of 40% or higher for custom RF amplifier devices [82,83], and even as high as 48% [84]. Attention should be given to the gain of the amplifiers, with reported results ranging from 8 to 25 dB in the three papers cited. On the receive side, several authors have reported rectenna element efficiencies for 24 GHz and above in excess of 50% [6], and even as high as 70% [85]. For purpose-built narrowband devices, it is likely that both transmit and receive element efficiencies can be increased. Losses in frequency sources and conversion in the systems supporting the transmit and receive elements are inevitable but can be minimized through judicious power system design.

Aperture sizing may be driven by transportability considerations. In many countries, consumer utility vehicles like pickup trucks and minivans can carry 125 cm by 250 cm or 4-foot by 8-foot sheets of construction materials, such as plywood. Constraining the transmit and receive apertures so that each consists of two such sheets promotes their likely transportability. Using the smaller of these two options, a square aperture with 2.43 m sides results when the sheets are placed adjacent to each other in the same plane along one of their long edges.

Using the Goubau relationship [86] to find the beam collection efficiency results in approximately 78% at a distance of 300 m. Using this figure and estimates for device efficiencies from above, the link efficiency contributions are shown in Table 5.3.

The link efficiency suggests that meeting the requirement of providing 300 W at the receiver will require at least 1,218 W at the

Table 5.3. Notional link efficiency for point-to-point microwave power beaming link operating at 24 GHz with 2.43 m square apertures over 300 m.

		Input (W)
Transmitter power and thermal management	97%	1,218
Transmitter device power added efficiency (PAE)	48%	1,181
Beam collection efficiency	78%	567
Rectenna element efficiency	70%	442
Receiver power management	97%	310
	Power to load	**300**
End-to-end efficiency	**25%**	

transmitter. This could be easily sourced by a 120 V/15 A circuit in North America, or a comparable circuit elsewhere.

The mean transmitted power per unit area assuming a uniform amplitude distribution is about 200 W/m^2. This exceeds the IEEE whole-body 30-min averaged exposure limit for this frequency for both unrestricted and restricted areas, by factors of twenty and four respectively [87, p. 51]. Very near the transmitter, fields will be even higher. These conditions likely necessitate an active safety system both near the transmitter and having coverage along the entire link, perhaps something similar to what was used in 2019 for a laser power beaming demonstration [88]. Elevating the link above a level easily accessible to people can add a measure of safety.

The system is likely to exhibit some degradation in link efficiency in adverse weather, which will vary depending on where it is used.

Each aperture will need structural elements to maintain flatness and orientation. A deployable tower or fold-out stand could be used for pointing. Implementing one or more of a retrodirective beam control or automated closed-loop tracking and pointing system could assist users in ensuring the apertures are properly aligned.

5.5.3 *Space to Earth*

In this example, the following top-level requirements are assumed:

(1) The system shall comply with all applicable safety and regulatory requirements.

(2) The transmitter input shall come from sunlight collected outside Earth's atmosphere.

(3) The receiver on the Earth shall provide electrical output.

Link examples for power beaming for space solar have been examined exhaustively by many authors and organizations [86,89,90]. Rather than start afresh, a selection of these links will be reviewed and discussed.

Frequency selection for microwave power beaming for space solar is influenced by atmospheric opacity and consideration of weather effects. Unlike a short, terrestrial point-to-point link, in most microwave-based concepts the power beam will traverse the entire atmosphere from top to bottom. This may occur at an oblique incident angle, increasing the amount of air mass and the possibility of sources of attenuation. As a result, atmospheric transmission windows have typically been selected for the frequency.

Goubau explored power beaming link considerations for space solar in 1970 [86], several years before the extensive NASA and US Department of Energy studies that addressed the same topic [91,92]. In the wake of these studies, Woodcock graphically outlined the constraints on microwave power beaming links, as seen in Figure 5.30.

Many later studies developed additional prospective power beaming links for space solar. Table 5.4 shows a collection of links developed by the Japan Aerospace Exploration Agency (JAXA) compared with one of the NASA links [89].

The land area demands for each of the schemes presented thus far are relatively large. There have also been explorations of smaller receive areas for military and remote installation applications. One collection of these is shown in Table 5.5, where microwave and millimeter wave links are contrasted with laser-based links.

In all of these cases, power densities at the transmitter, receiver, and points along the link path are of prime interest. Power densities at the transmitter may be constrained by thermal performance [93]. At the receiver, higher power densities may present safety and interference hazards, as discussed in Chapter 4. Along the beam's path, consideration must be given to birds, aircraft, and spacecraft.

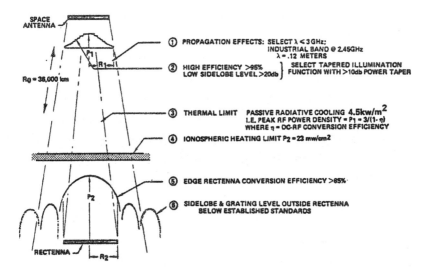

Figure 5.30. Woodcock's depiction of six design constraints on a microwave power beaming link for space solar from geosynchronous orbit [92].

Table 5.4. Link parameters for space solar presented in a 2007 URSI report [89, p. 13].

Model	Old JAXA Model	JAXA1 Model	JAXA2 Model	NASA-DOE Model
Frequency	5.8 GHz	5.8 GHz	5.8 GHz	2.45 GHz
Diameter of transmitting antenna	2.6 kmϕ	1 kmϕ	1.93 kmϕ	1 kmϕ
Amplitude taper	10 dB Gaussian	10 dB Gaussian	10 dB Gaussian	10 dB Gaussian
Output power (beamed to earth)	1.3 GW	1.3 GW	1.3 GW	6.72 GW
Maximum power density at center	63 mW/cm^2	420 mW/cm^2	114 mW/cm^2	2.2 W/cm^2
Minimum power density at edge	6.3 mW/cm^2	42 mW/cm^2	11.4 mW/cm^2	0.22 W/cm^2
Antenna spacing	0.75 λ	0.75 λ	0.75 λ	0.75 λ
Power per one antenna (Number of elements)	Max. 0.95 W (3.54 billion)	Max. 6.1 W (540 million)	Max. 1.7 W (1,950 million)	Max. 185 W (97 million)
Rectenna Diameter	2.0 kmϕ	3.4 kmϕ	2.45 kmϕ	10 kmϕ
Maximum Power Density	180 mW/cm^2	26 mW/cm^2	100 mW/cm^2	23 mW/cm^2
Collection Efficiency	96.5%	86%	87%	89%

Table 5.5.   Prospective power beaming links for situations with limited available receiver area [90].

Factors	Parameter	μ wave MEO	mm-wave MEO	Optical MEO	Optical GEO
Geometric	Frequency (GHz)	5.8	35	194,000	194,000
	Wavelength (μm)	5.17E + 04	8.57E + 03	1.55	1.55
	Transmit aperture diameter (m)	500	350	2.5	2.5
	Link distance (km)	20,000	20,000	20,000	35,786
	Receive aperture diameter (m)	1,000	500	11	50
	Beam collection efficiency (%)	13.4%	47.4%	38.6%	95.7%
Implementation	Intercepted sunlight power (kW)	268,606	131,617	269	5,368
	Space segment conversion efficiency (%)	18.3%	7.4%	11.6%	11.6%
	Transmit power (kW)	49,042	9,734	31	620
	Receiver segment conversion efficiency (%)	73%	47%	62%	62%
Atmospheric Effects	Minumum clear sky losses at sea level (%)	1%	6%	5%	5%
	Cloud/weather losses at sea level (%)	minimal	varies	varies	varies
Receiver	Power density at receiver center (W/m^2)	9	32	160	994
Peak	Average power density across receiver (W/m^2)	8	22	107	257
Output	Receiver average output power density (W/m^2)	6	10	66	158
Parameters	Receiver output power (kW)	4,742	2,024	6	311
		Input	Calculation		

Previous work suggests that power densities of 250 W/m^2 are acceptable for birds [94], and levels up to 230 W/m^2 will not cause adverse ionospheric effects [7, p. 44].

Beam steering may necessitate the control of the signal's phase at each antenna element. Keeping grating lobes and sidelobes to a minimum can improve transmission efficiency, reduce interference, and mitigate safety issues. Phase control may be needed for each array element or subarray, possibly levying requirements on amplifiers and feed networks.

The transmitter will be subject to the challenges of surviving in the space environment, including temperature extremes, radiation, micrometeorites, space debris, and solar activity. The orbit selection will affect the risks posed by these. Each of these environmental challenges is a field unto its own, though a summary treatment can be found in Wertz *et al.* [95, Ch. 7].

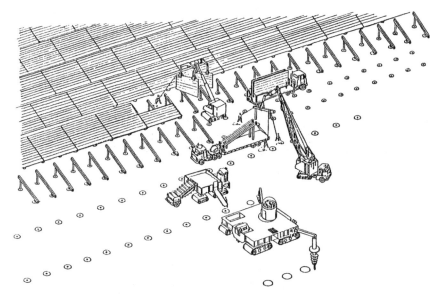

Figure 5.31.   Depiction of rectenna receiver construction [96, p. 32].

The rectenna receiver elements might resemble those seen in Figure 5.24 or Figure 5.27. They would be combined into larger assemblies and might be emplaced in the manner shown in Figure 5.31.

The overall receiver size and shape would depend on the particulars of the system design. For a system with satellites in equatorial geostationary orbits, receiver sites away from the Earth's equator or east or west of the satellite's orbital position might be elliptical. A notional layout of a microwave receiver site is shown in Figure 5.32.

The receiver will be subject to the siting constraints that face much power infrastructure, as well as many of the challenges of implementing power plants of other varieties. General guidance on the considerations for the development of power plants, many aspects of which will apply to the rectenna receiver, can be found in [97–99]. These include space requirements, transmission costs, land costs, taxes, and zoning. For the rectenna receiver specifically, prior examinations have considered the impacts of lightning [100], and the potential effects on bird migration [96, App. C], regional energy factors [96, Sec. VI], and more.

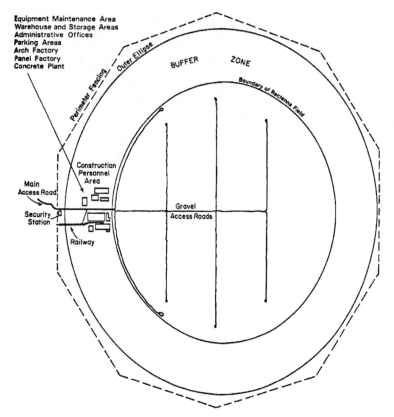

Figure 5.32. Possible layout of a rectenna receiver site [96, p. 31]. *Sites might be elliptically shaped if away from the area beneath the satellite's orbital position.*

Further reviews of system, subsystem, and component hardware considerations for microwave-based power beaming for space solar can be found in many of the resources listed in Table 8.2, as well as in [101].

5.6   Conclusion

The selection of components and subsystems in the design and implementation of power beaming systems will be driven by the modality utilized. This chapter reviewed options and considerations for systems operating in microwave and millimeter wave regimes and explored notional link examples.

5.7 Further Reading

For more on microwave-based power beaming systems, see Chapters 7–9 of *Wireless Power Transfer: Theory, Technology, and Applications* edited by Naoki Shinohara, as well as other texts written and edited by Naoki Shinohara. Works by William C. Brown and Richard Dickinson, such as the 1974 Proceedings of the IEEE paper "The Technology and Application of Free-Space Power Transmission by Microwave Beam" and the 2003 Acta Astronautica paper "Wireless Power Transmission Technology State of the Art" provide valuable context as well.

For more on phased array design, see:

R. J. Mailloux, *Phased Array Antenna Handbook*, 3rd edition. Norwood, MA: Artech House, 2018.

E. Brookner, Ed., *Practical Phased-Array Antenna Systems*. Boston, MA: Artech House, 1991.

A. J. Fenn, *Adaptive Antennas and Phased Arrays for Radar and Communications*. Boston, MA; London: Artech House, 2008.

References

[1] R. Descartes, *Discourse on the Method of Rightly Conducting One's Reason and Seeking Truth in the Sciences*. 1637. [Online]. Available at: https://earlymoderntexts.com/assets/pdfs/descartes1637.pdf (accessed June 16, 2021).

[2] "Electromagnetic Radiation — Microwaves |Britannica." [Online]. Available at: https://www.britannica.com/science/electromagnetic-radiation/Microwaves (accessed October 28, 2022).

[3] P. Jaffe *et al.*, "Sandwich Module Prototype Progress for Space Solar Power," *Acta Astronaut.*, vol. 94, no. 2, pp. 662–671, February 2014. doi: 10.1016/j.actaastro.2013.08.012.

[4] D. Schmelzer and S. I. Long, "A GaN HEMT Class F Amplifier at 2 GHz with >80% PAE," in *2006 IEEE Compound Semiconductor Integrated Circuit Symposium*, November 2006, pp. 96–99. doi: 10.1109/CSICS.2006.319923.

[5] R. N. Simons, J. D. Wilson, and D. A. Force, "High Power and Efficiency Space Traveling-Wave Tube Amplifiers with Reduced Size and Mass for NASA Missions," in *2008 IEEE MTT-S International Microwave Symposium Digest*, Atlanta, GA, USA: IEEE, June 2008, pp. 319–322. doi: 10.1109/MWSYM.2008.4633167.

[6] C. R. Valenta and G. D. Durgin, "Harvesting Wireless Power: Survey of Energy-Harvester Conversion Efficiency in Far-Field, Wireless Power Transfer Systems," *IEEE Microw. Mag.*, vol. 15, no. 4, pp. 108–120, June 2014. doi: 10.1109/MMM.2014.2309499.

[7] P. E. Glaser, F. P. Davidson, and K. I. Csigi, Eds., *Solar Power Satellites: A Space Energy System for Earth*, in Wiley-Praxis Series in space Science and Technology. Chichester, New York: Wiley published in association with Praxis Publishing, Chichester, 1998.

[8] A. V. Räisänen and A. Lehto, *Radio Engineering for Wireless Communication and Sensor Applications*, in Artech House mobile Communications Series. Boston, MA: Artech House, 2003.

[9] Fry-09, *English: The Traveling-Wave Tube "Shtormovka." Desined in town Fryazino Russia in the 1980's for the First in the Series Communications Satellites "Gorizont." Chief designer of TWT "Shtormovka" is Myakinkov Yu. P.* 2009. [Online]. Available at: https://commons.wikimedia.org/wiki/File:TWT_Shtormovka.jpg (accessed October 14, 2023).

[10] M. rf, *English: K41 = Klystron Reflective with Heated Cathode*, 2023. [Online]. Available at: https://commons.wikimedia.org/wiki/File: K41_Klystron_reflective_vacuum_tube_01.jpg (accessed October 14, 2023).

[11] H. H. L. Page, *Magnetron in its Box*, 2005. [Online]. Available at: https://commons.wikimedia.org/wiki/File:Magnetron1.jpg (accessed October 14, 2023).

[12] N. H. Pond, Ed., *The Tube Guys*. West Plains, Missouri: Russ Cochran, Publisher, 2008.

[13] A. S. Gilmour, *Microwave and Millimeter-Wave Vacuum Electron Devices: Inductive Output Tubes, Klystrons, Traveling-Wave Tubes, Magnetrons, Crossed-Field Amplifiers, and Gyrotrons*. Boston, MA: Artech House, 2020.

[14] J. R. Welty, Ed., *Fundamentals of Momentum, Heat, and Mass Transfer*, Fifth edition. Danver, MA: Wiley, 2008.

[15] R. J. Mailloux, *Phased Array Antenna Handbook*, Third edition, in Artech House Antennas and Propagation Library. Norwood, MA: Artech House, 2018.

[16] E. Brookner, Ed., *Practical Phased-Array Antenna Systems*, in The Artech House Antenna Library. Boston, MA: Artech House, 1991.

[17] A. J. Fenn, *Adaptive Antennas And Phased Arrays for Radar and Communications*, in Artech House Radar Library. Boston, MA; London: Artech House, 2008.

[18] "Microwaves101 | Phased Array Antennas." [Online]. Available at: https://www.microwaves101.com/encyclopedias/phased-array-antennas (accessed November 10, 2022).

[19] S. Mess-Elektronik, *English: UHA 9125 D, Half-Wave Dipole Antenna, 1.0 − 4 GHz.* 2007. [Online]. Available at: https://commons.wikimedia.org/wiki/File:Half_%E2%80%93_Wave_Dipole.jpg (accessed October 14, 2023).

[20] R. B. Freak Munich aka Makro, *Deutsch: Die größte Erdfunkstelle der Welt bei Raisting (in der Nähe des Ammersees) ist eine Bodenstation für die Kommunikation mit Nachrichtensatelliten und ist mit ihren großen Parabolantennen weithin sichtbar.* 2008. [Online]. Available at: https://commons.wikimedia.org/wiki/File:Erdfunkstelle_Raisting_2.jpg (accessed October 14, 2023).

[21] S. Mess-Elektronik, *English: Broadband Microwave Horn Antenna. Its Bandwidth Is 0.8–18 GHz. The Coaxial Cable Feedline Is Attached to the Connector Visible at Top. This Type Is Called A Ridged Horn; The Curving Metal Ridges or Fins Visible Inside The Mouth of the Horn Reduce Its Cutoff Frequency, Increasing The Bandwidth.* 2007. [Online]. Available at: https://commons.wikimedia.org/wiki/File:Schwarzbeck_BBHA_9120_D.jpg (accessed October 14, 2023).

[22] AB2013, *Français: Une Antenne Patch (Ou Antenne Planaire) Avec Alimentation Coaxial.English: A Coaxial Feeded Patch Antenna.* 2013. [Online]. Available at: https://commons.wikimedia.org/wiki/File:Antenne_patch.jpg (accessed October 14, 2023).

[23] Daderot, *English: Exhibit in the National Electronics Museum, 1745 West Nursery Road, Linthicum, Maryland, USA. All Items in this Museum Are Unclassified. The Museum Permitted Photography without Restriction.* 2014. [Online]. Available at: https://commons.wikimedia.org/wiki/File:AWACS_antenna,_Airborne_Warning_and_Control_System_-_National_Electronics_Museum_-_DSC00416.JPG (accessed October 14, 2023).

[24] W. L. Stutzman and G. A. Thiele, *Antenna Theory and Design*, Third edition. Hoboken, New Jersey: Wiley, 2013.

[25] E. Lier and J. Matthews, "Performance Comparison of High-Gain Horn Antennas," in *2005 IEEE Antennas and Propagation Society International Symposium*, vol. 3A, July 2005, pp. 753–756. doi: 10.1109/APS.2005.1552365.

[26] L. Ripoll-Solano, L. Torres-Herrera, and M. Sierra-Perez, "Design, Simulation and Optimization of a Slotted Waveguide Array with Central Feed and Low Sidelobes," in *2018 IEEE-APS Topical Conference on Antennas and Propagation in Wireless Communications (APWC)*, Cartagena des Indias: IEEE, September 2018, pp. 886–889. doi: 10.1109/APWC.2018.8503799.

[27] M. I. M. Ghazali, K. Y. Park, V. Gjokaj, A. Kaur, and P. Chahal, "3D Printed Metalized Plastic Waveguides for Microwave Components," *Int. Symp. Microelectron.*, vol. 2017, no. 1, pp. 000078–000082, October 2017. doi: 10.4071/isom-2017-TP33_096.

[28] L. Josefsson and S. Rengarajan, Eds., *Slotted Waveguide Array Antennas: Theory, Analysis and Design.* Institution of Engineering and Technology, 2018.

[29] C. A. Balanis, *Antenna Theory: Analysis and Design*, Third edition. Hoboken, NJ: John Wiley, 2005.

[30] J. D. Kraus and R. J. Marhefka, *Antennas for All Applications*, Third edition. New York: McGraw-Hill, 2002.

[31] Shiv Prasad Tripathy, *The 100-Page Book on Antenna Design Parameters.* Independently published, 2020.

[32] "Microwaves101 | Microwave Rules of Thumb." [Online]. Available at: https://www.microwaves101.com/encyclopedias/microwave-rules-of-thumb (accessed November 9, 2022).

[33] R. J. Mailloux, *Phased Array Antenna Handbook*, in Artech House Antenna Library. Boston: Artech House, 1994.

[34] C. J. Bouwkamp and N. G. deBruijn, "The Problem of Optimum Antenna Current Distribution," *Philips Res. Rep.*, vol. 1, pp. 135–158, 1945.

[35] Mr. PIM at the English Wikipedia, "Typical_antenna_pattern_with_grating_lobes.jpg (968 × 699)." [Online]. Available at: https://upload.wikimedia.org/wikipedia/commons/a/aa/Typical_antenna_pattern_with_grating_lobes.jpg (accessed November 1, 2022).

[36] Judith V. Hutson and Christopher T. Rodenbeck, "Computation of Power Beaming Efficiency in the Fresnel Zone with Application to Amplitude and Phase Optimization," NRL/MR/5307--20-10,118, 2021.

[37] X. Wang and M. Lu, "Microwave Power Transmission Based on Retro-Reflective Beamforming," in *Wireless Power Transfer — Fundamentals and Technologies*, E. Coca, Ed., InTech, 2016. doi: 10.5772/62855.

[38] P. Jaffe and J. McSpadden, "Energy Conversion and Transmission Modules for Space Solar Power," *Proc. IEEE*, vol. 101, no. 6, pp. 1424–1437, Jun. 2013. doi: 10.1109/JPROC.2013.2252591.

[39] "Wilkinson Power Splitters." [Online]. Available at: https://www.microwaves101.com/encyclopedias/wilkinson-power-splitters (accessed November 13, 2022).

[40] M. H. Szazynski, "Wireless Power Transfer: A Reconfigurable Phased Array with Novel Feeding Architecture," Purdue University, 2018. [Online]. Available at: https://docs.lib.purdue.edu/dissertations/AA I10808942 (accessed November 13, 2022).

[41] Danial Ehyaie, "Novel Approaches to the Design of Phased Array Antennas," University of Michigan, Ann Arbor, MI, 2011. [Online]. Available at: https://www.eecs.umich.edu/radlab/html/NEWDISS/ Ehyaie.pdf (accessed November 13, 2022).

[42] S. Bulja and A. Grebennikov, "A Novel Variable Power Divider with Continuous Power Division," *Microw. Opt. Technol. Lett.*, vol. 55, no. 7, pp. 1684–1686, Jul. 2013. doi: 10.1002/mop.27603.

[43] M.-C. J. Chik, W. Li, and K.-K. M. Cheng, "A 5 GHz, Integrated Transformer Based, Variable Power Divider Design in CMOS Process," in *2013 Asia-Pacific Microwave Conference Proceedings (APMC)*, November 2013, pp. 366–368. doi: 10.1109/ APMC.2013.6695148.

[44] S. Oh *et al.*, "An Unequal Wilkinson Power Divider with Variable Dividing Ratio," in *2007 IEEE/MTT-S International Microwave Symposium*, June 2007, pp. 411–414. doi: 10.1109/MWSYM.2007. 380475.

[45] Y. Peng, H.-L. Zhang, and Y.-Z. Hu, "Design of An Orthogonal Power Divider with Reconfigurable Power Division Ratio," in *2016 IEEE International Conference on Microwave and Millimeter Wave Technology (ICMMT)*, June 2016, pp. 342–344. doi: 10.1109/ ICMMT.2016.7761769.

[46] P.-W. Li and K.-K. M. Cheng, "A Novel Power-Divider Design with Variable Dividing Ratio," in *2009 Asia Pacific Microwave Conference*, December 2009, pp. 1020–1023. doi: 10.1109/APMC.2009. 5384355.

[47] C. T. Rodenbeck *et al.*, "Terrestrial Microwave Power Beaming," *IEEE J. Microw.*, vol. 2, no. 1, pp. 28–43, January 2022. doi: 10.1109/JMW.2021.3130765.

[48] "Beaming Day 2023: Live 100m Wireless Power Transfer Demonstration," Virtus Solis. [Online]. Available at: https://virtussolis.space/ blog/beaming-day-2023-live-100m-wireless-power-transfer-demonstration (accessed August 13, 2023).

[49] John Bucknell, "Virtus Solis May 2023 Newsletter," June 4, 2023.

[50] J. Ye, C. Yang, and Y. Zhang, "Design and Experiment of a Rectenna Array Base on Gaas Transistor for Microwave Power Transmission," in *2016 IEEE International Conference on Microwave and Millimeter Wave Technology (ICMMT)*, June 2016, pp. 323–326. doi: 10.1109/ICMMT.2016.7761763.

[51] J. O. McSpadden, L. Fan, and K. Chang, "Design and Experiments of a High-Conversion-Efficiency 5.8-GHz Rectenna," *IEEE Trans. Microw. Theory Tech.*, vol. 46, no. 12, pp. 2053–2060, December 1998. doi: 10.1109/22.739282.

[52] B. Strassner and K. Chang, "5.8 GHz Circular Polarized Rectifying Antenna for Microwave Power Transmission," in *2001 IEEE MTT-S International Microwave Sympsoium Digest (Cat. No.01CH37157)*, vol. 3, May 2001, pp. 1859–1862. doi: 10.1109/MWSYM.2001.967270.

[53] B. Strassner and K. Chang, "Highly Efficient C-Band Circularly Polarized Rectifying Antenna Array For Wireless Microwave Power Transmission," *IEEE Trans. Antennas Propag.*, vol. 51, no. 6, pp. 1347–1356, June 2003. doi: 10.1109/TAP.2003.812252.

[54] J. McSpadden, "Rectenna Technologies for SSP Applications," presented at the IEEE Wisee 2020 Workshop on Space Solar Power 8th Annual IEEE International Conference on Wireless for Space and Extreme Environments, Venice, Italy, October 13, 2020. [Online]. Available at: https://docplayer.net/142502734-Rectenna-technologies-for-ssp-applications.html (accessed October 14, 2020).

[55] H. Kazemi, "61.5% Efficiency and 3.6 kW/m^2 Power Handling Rectenna Circuit Demonstration for Radiative Millimeter Wave Wireless Power Transmission," *IEEE Trans. Microw. Theory Tech.*, pp. 1–1, 2021. doi: 10.1109/TMTT.2021.3110966.

[56] W. C. Brown and E. E. Eves, "Beamed Microwave Power Transmission and Its Application to Space," *IEEE Trans. Microw. Theory Tech.*, vol. 40, no. 6, pp. 1239–1250, June 1992. doi: 10.1109/22.141357.

[57] Bernd Herbert Strassner, "Nonlinear Harmonic Modeling of Phemt Devices for Increased Power Amplifier Efficiencies," Master's thesis, Texas A&M University, 1997. [Online]. Available at: https://hdl.handle.net/1969.1/ETD-TAMU-1997-THESIS-S765 (accessed January 22, 2023).

[58] L. H. Hsieh *et al.*, "Development of a Retrodirective Wireless Microwave Power Transmission System," in *IEEE Antennas and Propagation Society International Symposium. Digest. Held in Conjunction with: USNC/CNC/URSI North American Radio Sci. Meeting (Cat. No.03CH37450)*, vol. 2, June 2003, pp. 393–396. doi: 10.1109/APS.2003.1219259.

[59] S. Trinh-Van, J. Lee, Y. Yang, K.-Y. Lee, and K. Hwang, "Improvement of RF Wireless Power Transmission Using a Circularly Polarized Retrodirective Antenna Array with EBG Structures," *Appl. Sci.*, vol. 8, no. 3, p. 324, February 2018. doi: 10.3390/app8030324.

[60] V. Fusco and N. Buchanan, "Developments in Retrodirective Array Technology," *IET Microw. Antennas Propag.*, vol. 7, no. 2, pp. 131–140, 2013. doi: 10.1049/iet-map.2012.0565.

[61] R. C. Hansen, *Phased Array Antennas*, Second edition, in Wiley Series in Microwave and Optical Engineering. Hoboken, New Jersey: Wiley, 2009.

[62] Y. Li and V. Jandhyala, "Design of Retrodirective Antenna Arrays for Short-Range Wireless Power Transmission," *IEEE Trans. Antennas Propag.*, vol. 60, no. 1, pp. 206–211, January 2012. doi: 10.1109/TAP.2011.2167897.

[63] R. M. Dickinson and W. C. Brown, "Radiated Microwave Power Transmission System Efficiency Measurements," Jet Propulsion Lab., California Inst. of Tech., Pasadena, CA, United States, Technical Report NASA-CR-142986, JPL-TM-33-727, May 1975. [Online]. Available at: https://ntrs.nasa.gov/search.jsp?print=yes&R=19750 018422 (accessed April 25, 2020).

[64] W. Huang, B. Zhang, X. Chen, K.-M. Huang, and C.-J. Liu, "Study on an S-Band Rectenna Array for Wireless Microwave Power Transmission," *Prog. Electromagn. Res.*, vol. 135, pp. 747–758, 2013. doi: 10.2528/PIER12120314.

[65] "MPPT Algorithm." [Online]. Available at: https://www.mathworks.com/solutions/power-electronics-control/mppt-algorithm.html (accessed November 24, 2022).

[66] A. O. Baba, G. Liu, and X. Chen, "Classification and Evaluation Review of Maximum Power Point Tracking Methods," *Sustain. Futur.*, vol. 2, p. 100020, 2020. doi: 10.1016/j.sftr.2020.100020.

[67] "Press Release: Emrod Successfully Demonstrates Power Beaming Technology to Unlock Space-Based Solar Power," Emrod Energy. [Online]. Available at: https://emrod.energy/press-release-emrod-successfully-demonstrates-power-beaming-technology-to-unlock-space-based-solar-power/ (accessed September 29, 2022).

[68] "Satellite Power Grid Would Beam Energy around the Globe Just Like Data," *New Atlas.* [Online]. Available at: https://newatlas.com/energy/emrod-space-solar-wireless-energy/ (accessed October 10, 2022).

[69] *NASA Systems Engineering Handbook.* NASA Center for AeroSpace Information, 2007. [Online]. Available at: https://www.nasa.gov/sites/default/files/atoms/files/nasa_systems_engineering_handbook.pdf (accessed October 25, 2022).

[70] K. T. Ulrich and S. D. Eppinger, *Product Design and Development*, Fifth edition. New York: McGraw-Hill/Irwin, 2012.

[71] "High-Performance, Scalable Wireless Power-at-a-Distance," Reach. [Online]. Available at: https://reachpower.com/ (accessed January 10, 2023).

[72] O. Inc, "Next Generation 5.8GHz Cota Developer Kit." [Online]. Available at: https://info.ossia.com/products/cota-58ghz-quick-receiver (accessed January 4, 2023).

[73] "5.5W WattUp® PowerBridge Developer Kit." [Online]. Available at: https://energous.com/products/developer-kits/wattup-5.5w-active-energy-harvesting (accessed January 4, 2023).

[74] "Technology," GuRu. [Online]. Available at: https://guru.inc/techn ology-2/ (accessed January 4, 2023).

[75] "Development Kits," Powercast Co. [Online]. Available at: https:// www.powercastco.com/products/development-kits/ (accessed January 4, 2023).

[76] "Powered by Powercast," Powercast Co. [Online]. Available at: https://www.powercastco.com/products/powered-by-powercast/ (accessed January 5, 2023).

[77] "Amazon.com: Wireless Charging Grip + PowerSpot Wireless Power Transmitter Bundle Pack: Video Games." [Online]. Available at: https://www.amazon.com/Wireless-Charging-PowerS pot-Transmitter-Bundle/dp/B08FTJPDBS/ (accessed January 18, 2023).

[78] V. Talla, B. Kellogg, B. Ransford, S. Naderiparizi, S. Gollakota, and J. R. Smith, "Powering the Next Billion Devices with Wi-Fi," in *Proceedings of the 11th ACM Conference on Emerging Networking Experiments and Technologies*, Heidelberg Germany: ACM, December 2015, pp. 1–13. doi: 10.1145/2716281.2836089.

[79] V. Talla, B. Kellogg, S. Gollakota, and J. R. Smith, "Battery-Free Cellphone," *Proc. ACM Interact. Mob. Wearable Ubiquitous Technol.*, vol. 1, no. 2, pp. 1–20, June 2017. doi: 10.1145/3090090.

[80] "LifeSupplyUSA 200 ft. Orange 16/3 SJTW Indoor/Outdoor Heavy-Duty Extra Durability 10 Amp 125V 1250-Watt w/Lighted end Extension Cord 163200FTOR," The Home Depot. [Online]. Available at: https://www.homedepot.com/p/200-ft-Orange-16-3-SJTW-Indoor-Outdoor-Heavy-Duty-Extra-Durability-10-Amp-125V-1250-Watt-w-Lighted-end-Extension-Cord-163200FTOR/317876899 (accessed January 13, 2023).

[81] "Wireless Power Use Cases | Emrod," Emrod Energy. [Online]. Available at: https://emrod.energy/use-cases/ (accessed January 13, 2023).

[82] M. R. Duffy, G. Lasser, M. Roberg, and Z. Popovic, "A 4-W K-Band 40% PAE Three-Stage MMIC Power Amplifier," in *2018 IEEE BiC-MOS and Compound Semiconductor Integrated Circuits and Technology Symposium (BCICTS)*, San Diego, CA: IEEE, Oct. 2018, pp. 144–147. doi: 10.1109/BCICTS.2018.8550981.

[83] M. A. Reece, S. Contee, and C. W. Waiyaki, "K-Band GaN Power Amplifier Design with a Harmonic Suppression Power Combiner," in *2017 IEEE Topical Conference on RF/Microwave Power Amplifiers*

for Radio and Wireless Applications (PAWR), January 2017, pp. 92–95. doi: 10.1109/PAWR.2017.7875582.

[84] C. F. Campbell, K. Tran, M.-Y. Kao, and S. Nayak, "A K-Band 5W Doherty Amplifier MMIC Utilizing 0.15μm GaN on SiC HEMT Technology," in *2012 IEEE Compound Semiconductor Integrated Circuit Symposium (CSICS)*, October 2012, pp. 1–4. doi: 10.1109/CSICS.2012.6340057.

[85] P. Koert and J. T. Cha, "Millimeter wave technology for space power beaming," *IEEE Trans. Microw. Theory Tech.*, vol. 40, no. 6, pp. 1251–1258, June 1992. doi: 10.1109/22.141358.

[86] Georg Goubau, "Microwave Power Transmission from an Orbiting Solar Power Station," *J. Microw. Power*, vol. 5, no. 4, pp. 223–231, 1970.

[87] "IEEE Standard for Safety Levels with Respect to Human Exposure to Electric, Magnetic, and Electromagnetic Fields, 0 Hz to 300 GHz," *IEEE Std C951^{TM}-2019 Revis. IEEE Std C951-2005 Inc. IEEE Std C951-2019Cor 1-2019*, pp. 1–312, October 2019. doi: 10.1109/IEEESTD.2019.8859679.

[88] Thomas J. Nugent, Jr., David Bashford, Thomas Bashford, Thomas J. Sayles, and Alex Hay, "Long-Range, Integrated, Safe Laser Power Beaming Demonstration," in *Technical Digest OWPT 2020*, Yokohama, Japan: Optical Wireless Power Transmission Committee, The Laser Society of Japan, April 2020, pp. 12–13.

[89] "Report of the URSI Inter-Commission Working Group on SPS," 2007. [Online]. Available at: https://www.ursi.org/files/ICWGReport070611.pdf.

[90] P. Jaffe *et al.*, "Opportunities and Challenges for Space Solar for Remote Installations," U.S. Naval Research Laboratory, Washington, DC. Memo Report NRL/MR/8243--19-9813, October 2019. [Online]. Available at: https://apps.dtic.mil/sti/pdfs/AD1082903.pdf (accessed April 21, 2020).

[91] "Solar Power Satellite System Definition Study Part II — Volume IV — Microwave Power Transmission Systems," Boeing Aerospace Company, D180-22876–4, December 1977. [Online]. Available at: https://space.nss.org/wp-content/uploads/SSP-Boeing-CR151668-1977-Part2Vol4-Microwave-Power-Transmission.pdf.

[92] "Solar Power Satellite Microwave Power Transmission and Reception," NASA Lyndon B. Johnson Space Center, Houston, TX, NASA Conference Publication 2141, January 1980. [Online]. Available at: https://space.nss.org/wp-content/uploads/NASA-CP2141-Satellite-Microwave-Power-Transmission-And-Reception.pdf (accessed January 26, 2023).

[93] S. Spencer, B. Nguyen, and P. Jaffe, "Thermal Analysis of Space-Based Solar Power System Study Photovoltaic DC to RF Antenna Module (PRAM)," in *10th International Energy Conversion Engineering Conference*, Atlanta, GA: American Institute of Aeronautics and Astronautics, July 2012. doi: 10.2514/6.2012-4050.

[94] F. E. Wasserman, T. Lloyd-Evans, S. P. Battista, D. Byman, and T. H. Kunz, "The Effect of Microwave Radiation (2.45 GHz CW) on the Molt of House Finches (Carpodacus Mexicanus)," *Space Sol. Power Rev.*, vol. 5, pp. 261–270, 1985.

[95] J. R. Wertz, D. F. Everett, and J. J. Puschell, Eds., *Space Mission Engineering: The New SMAD*, in Space Technology Library, no. v. 28. Hawthorne, CA: Microcosm Press: Sold and Distributed Worldwide by Microcosm Astronautics Books, 2011.

[96] Rice University, Houston, TX (USA), "Satellite Power System (SPS). Rectenna Siting: Availability and Distribution of Nominally Eligible Sites," DOE/ER/10041-T10, 6740362, November 1980. doi: 10.2172/6740362.

[97] K. Gayathri, "Selection and Location of Power Plants: 14 Considerations," Engineering Notes India. [Online]. Available at: https://www.engineeringnotes.com/power-plants-2/selection/selection-and-location-of-power-plants-14-considerations/29588 (accessed January 27, 2023).

[98] P. G. Hessler, *Power Plant Construction Management: A Survival Guide*, Second edition. Tulsa, OK: PennWell Corporation, 2015.

[99] R. K. Hegde, *Power Plant Engineering*. Chennai, India: Pearson India, 2015.

[100] "Electrostatic Protection of the Solar Power Satellite and Rectenna — Part II — Lightning Protection of the Rectenna," Rice Unversity, Houston, TX. NASA Contractor Report 3345, November 1980. [Online]. Available at: https://space.nss.org/wp-content/uploads/NASA-CR3345-Lightning-Protection.pdf (accessed January 27, 2023).

[101] J. O. McSpadden and J. C. Mankins, "Space Solar Power Programs and Microwave Wireless Power Transmission Technology," *IEEE Microw. Mag.*, vol. 3, no. 4, pp. 46–57, December 2002. doi: 10.1109/MMW.2002.1145675.

Chapter 6

Components and Subsystems
for Optical Power Beaming

*Scientists didn't go out to design a CD machine: they designed
a laser. But we got all sorts of things from a laser which we
never remotely imagined, and we're still finding things for a
laser to do.*

— Robert Winston [1]

6.1 Introduction

Occupying the wavelength region of the electromagnetic spectrum
from approximately 100 nm to 1 mm, optical-based power beaming
systems have seen compelling demonstration and development, primarily in the visible and near-infrared regime from 400 to 1,600 nm.
Their attractiveness stems from their ability to employ dramatically
smaller apertures than those needed for systems operating at longer
wavelengths. This chapter walks through component, subsystem, and
system design and implementation considerations for laser power
beaming systems. A more in-depth look at the ramifications of the
selection of a particular wavelength is found in Chapter 4.

6.2   System Overview

As with any power beaming system, the elements are grouped
into transmitter and receiver categories. The transmitter category
includes the light source, as well as beam shaping, receiver tracking,
beam pointing, and safety subsystems. For the purposes of this discussion, the light source will be a laser, though other sources are also
possible such as light-emitting diodes (LEDs). The receiver category
includes optical power converters, optics, power management and
distribution (PMAD) electronics, thermal management, and safety-
related hardware. Herein, the optical power converters will be photovoltaic (PV) cells, though other options exist. A schematic for a
typical system is shown in Figure 6.1.

Depending on the sophistication of the system, not all of these
subsystems might be present. Although not discussed in detail in
this book, power beaming systems also generally include some form
of controls, communications, and a user interface.

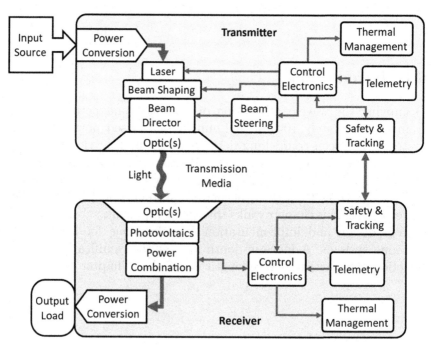

Figure 6.1.   An optical power beaming system schematic.

6.2.1 *System Efficiencies*

As the point of power beaming systems is to deliver energy, inefficiencies work against this goal. Some inefficiencies arise from the conversion of electrical energy to optical energy and back from optical energy to electrical energy, and others from subsystems that support these conversions.

For laser transmitters, the theoretical maximum wall-plug efficiency[1] varies with the materials used, the emitted wavelength, and the operating conditions. Diode lasers have been reported with wall-plug efficiencies of 73% [3] when cooled to room temperature and up to 85% at −50°C [4]. Diode-pumped fiber lasers, which have higher beam quality, have been manufactured with up to 50% wall-plug efficiency [5].

On the receiver side, maximum theoretical efficiency is also dependent on materials, wavelength, and operating conditions. PV laser power converter cells have been demonstrated at room temperature approaching 70% efficiency [6], with efficiencies reported near 80% for temperatures at 100°K [7]. Efficiencies tend to increase with increasing power densities or decreasing temperatures. The graph in Figure 6.2 shows the theoretical limit of monochromatic PV conversion efficiency, based on a detailed balance model.

Poor thermal management at both the transmitter and receiver can be a key source of inefficiency. For the laser, a chiller is often used to manage the laser temperature. This can be a critical subsystem since laser lifetime degrades exponentially at elevated temperatures. The chiller coefficient of performance (CoP) measures the ratio of heat rejected to its own power consumption, so maximizing system efficiency requires choosing a chiller with a high CoP. This is best accomplished by choosing components in parallel to maximize the combined performance. For the receiver, forced air cooling or other cooling means may be needed, resulting in a loss in system efficiency. Electrical power conversion on both ends and other ancillary subsystems may further lower system efficiency. Losses along the transmission path, such as atmospheric losses, were treated

---

[1]Wall-plug efficiency, the "total electrical-to-optical power efficiency of a laser system" [2] is generally the measure of greatest interest for a power beaming transmitter, rather than slope efficiency or quantum efficiency.

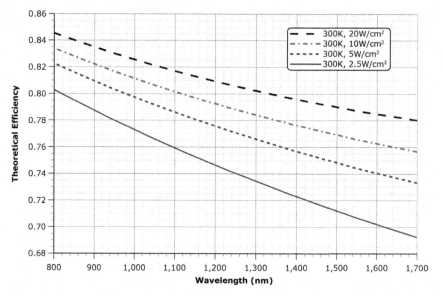

Figure 6.2.   Maximum theoretical efficiency of laser PVs for varying light intensities and wavelengths, based on the detailed balance model.

in Chapter 4. See Section 7.5.5 for a definition of end-to-end link efficiency.

6.3   Transmitters

The transmit end consists of one or more light sources that convert electricity into light, optics to shape the light beam (including collimation and focus), receiver tracking, and pointing the light beam toward the receiver. By establishing the emanated wavelength, phase, and amplitude distribution over time and spatial dimensions, the transmitter is key in affecting how much energy will be sent toward the receiver.

Any device that emits light can, in theory, be used for transmitting optical power. The most commonly used type of device is the laser, although LEDs have been used as well [8]. The reason to use lasers is because of their radiance (also known qualitatively as brightness) relative to other sources and their narrower wavelength spectrum, as radiance affects how easily a source can be focused into a narrowly directional beam and wavelength affects the efficiency of

Figure 6.3. (a) 2 kW, 976 nm diode laser. Image courtesy of Coherent Corp. (b) 6 kW, 1,070 nm fiber laser. Image courtesy of IPG Photonics.

absorption and conversion by PVs. See Chapter 3 for more details on the radiance of emitters.

Other books delve into the theory and history of the various types of lasers [9–11] from gas lasers (such as HeNe and COIL) to liquid (dye) lasers, and up through solid state (including Nd:YAG, diode, and fiber) lasers. There is a rich history of efforts to increase laser power, radiance, and efficiency at various wavelengths. For power beaming, diode and fiber lasers are of principal interest because of their relatively high efficiency and beam quality, and comparatively lower costs and complexity. A step-by-step approach for diode laser selection can be found in [12], though its principles are extensible in large part to fiber or other laser selection. An example of a diode laser appears in Figure 6.3(a), and an example of a fiber laser appears in Figure 6.3(b).

Diode or fiber lasers can also be employed in arrays for transmitters, most commonly to achieve higher optical power output than is available from a single laser. Arrays can use coherent [13] or incoherent beam combining [14], or spectral beam combining [15]. Coherent beam combination (CBC) and spectral beam combination (SBC, sometimes also called wavelength beam combination, WBC) can increase the effective radiance of a light source, thereby extending the usable transmission range for a given set of optical apertures. Certain types of beam combining can also be useful for overcoming propagation challenges such as turbulence [16].

Note that different vendors may measure and report beam quality differently, and there is no substitute for a careful review of the manufacturer's product documentation. Engineers responsible for designing power beaming systems should pay particular attention to what might be missing from stated specifications. The beamwidth measurement, such as Full Width at Half Maximum or the $1/e^2$ point, deserves special scrutiny. Many parameters may need to be measured independently, either to acquire or confirm performance data, to gain assurance that a given product will be suitable for a power beaming application.

6.3.1   *Transmitter Thermal Management*

Diode lasers and fiber lasers pumped by diode lasers are more efficient at lower temperatures. Both also undergo rapid degradation of lifetime at higher temperatures. Low-power systems (those below several hundred watts output, depending on wavelength, type, and operating conditions) may be able to manage temperatures with passive cooling, in the form of passive heat sinks, or low-power active cooling via forced-air convection or thermoelectric coolers (TECs). Higher power levels (above approximately 500 W optical) usually require phase-change cooling methods, such as vapor-compression refrigeration. Though TECs can be used for waste heat up to about 100 W, they are much less efficient than compressor-based chillers (with a CoP of less than 1). Most chillers use water as the medium between the chiller and the laser, or a mixture of chemicals such as glycol to expand the storage temperature range. Diode lasers, especially high-power stacks, used to be cooled by flowing deionized water directly over the chip mounts. This put the water in contact with the anode and cathode, driving the need for it to be deionized. More recently, there has been growing use of small bars and single emitters all mounted on a cold plate, enabling regular water to flow through channels inside the cold plate. The ability to use plain, clean water simplifies logistics.

Vendor guides [17] and comprehensive books [18,19] on thermal control for electronics and systems may be consulted for demanding power beaming transmitter designs.

6.3.2   *Beam Shaping*

An essentially uniform "top hat" profile, where the intensity is a constant value within the beam area, and zero outside, is often desirable for power beaming to permit the receiver to be designed to accept an even irradiance over its area. In practice, a top hat beam is not perfectly flat, and real beams can often be well-approximated by a super-Gaussian function.[2]

Designs can use non-imaging optics or imaging optics that don't optimize image quality, since the relevant measure of performance is how much of the transmitted energy makes it successfully to the receiver. This measure is called "power in the bucket" in the laser industry [20]. The simplest type of design images the light source, frequently the exit aperture of an optical fiber, at the receiver. The intensity profile at a fiber tip for multimode high-power lasers is usually relatively flat (i.e., close to top hat), and therefore a properly transmitted beam will be a roughly top-hat profile, limited by diffraction and atmospheric effects.

Some optical designs of beam expanders have an internal focus point between optics. One example is a Keplerian design. These types of designs should be avoided for high-power and higher beam-quality systems, otherwise, the beam intensity at the focal point can exceed the atmospheric breakdown point, as described in Section 4.3.4. Surpassing the breakdown threshold will produce a plasma, which will distort the beam and absorb some of its power. A Galilean beam expander design has no internal focus and thus avoids this pitfall. Figure 6.4 shows the path of light rays through the two options.

Optical beam shaping can be done with refractive, reflective, or diffractive optics, or combinations of these. Some of their respective qualities, including benefits and limitations of each, are summarized in Table 6.1.

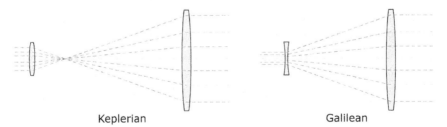

Keplerian   Galilean

Figure 6.4.   Ray paths through beam expanders with and without an internal focus.

---

[2]Instead of $e^{-(r/\sigma)^2}$ it becomes $e^{-((r/\sigma)^2)^N}$ where N is the super-Gaussian factor.

Table 6.1. Qualities of refractive, reflective, and diffractive optics.

Factor/effect	Refractive	Reflective	Diffractive
Cost	Expensive and harder to source beyond diameters of 10–15 cm.	Beyond 10–15 cm, the price grows approximately 2.8 power of diameter.	High non-recurring engineering (NRE) cost for design. Reproduction costs can be low in volume.
Chromatic aberration	Yes, but can be offset with multiple optics.	None.	Depends on implementation.
Optical efficiency and absorption	Some bulk absorption, but materials can be chosen to minimize it. Also thickness dependent.	Only what gets through the reflective coating, and then depends on the substrate. Reflective coatings can be as high as 99.9% efficient.	Depends on the substrate for transmissive optics. Can be very high. A full discussion of factors in [21].
Flexibility in choice of materials	Low. Materials must be largely transparent at the wavelength of interest.	High. The material must be reflective at the wavelength of interest or be able to hold the desired reflective coating.	Medium-High. Material (reflective or transmissive) must be etchable or machinable to the desired feature size, and able to hold coating.
On- vs. off-axis	Almost always on-axis. Allows for zoom/focus via linear stage motion.	Either. Off-axis can eliminate partial beam blockage due to support structures.	Either, but off-axis may increase manufacturing complexity.
Weight	High. Thickness grows with lower $f/\#$ (for a given aperture size). The entire thickness is solid.	Low. Only reflective surfaces and structures to maintain shape are needed.	Low. Only grating surface or layer with structure to keep shape is needed.
Coatings	Anti-reflection coating is often needed.	The reflective coating is needed if the material is not reflective.	Depends on implementation.

In some cases (frequently at longer ranges through the atmosphere), finer and faster beam shaping is required in the form of adaptive optics (AO). Often implemented with deformable mirrors, AO is used to reduce the impact of distortions on an optical wavefront [22]. In astronomy and microscopy, it's used with detectors to improve image quality. Directed Energy Weapons Systems (DEWS) are the primary current user for very high-power, large aperture AO for shaping an outgoing laser beam to counteract the profile- and beam size-degrading effects of scintillation from the atmosphere, although industrial materials processing systems are starting to adopt AO [23]. Because power beaming does not seek to create the smallest spot size, unlike DEWS, AO is generally not required for distances less than approximately 2 km close to the ground (depending on atmospheric conditions), much longer distances at high elevations, and not at all in space. One potential exception to this is for tip/tilt, which can be considered as zeroth-order AO.

While the maximum range to a receiver will determine the aperture size and influence the optical design, the minimum expected range will also impact the design. It is relatively straightforward to design a set of beam shaping optics that can adjust the beam zoom and focus within a max/min distance ratio of 2×, possibly 3×. But for a wider range, the optical design becomes more complicated, with the potential cost of additional surfaces, each of which will lose some light. This may require larger, more expensive optics and more design and control complexity.

In addition to Dickey's comprehensive texts on beam shaping [24,25], optics manufacturers offer useful overviews of beam shaping that may be helpful for power beaming implementers [26,27], as does Paschotta's photonics encyclopedia [28].

6.3.3 *Beam Tracking*

Tracking a power beaming receiver is a much easier technical problem than target tracking for directed energy weapons because the power beaming receiver can be cooperative. It may accomplish this by providing a signal or beacon to help locate and keep the beam on it. Some tracking systems can use cameras with only passive features on the receiver, such as retroreflectors or patterned colors.

This approach is often less costly but is subject to challenges under changing lighting conditions. For this reason, beacons or active feedback to the transmitter from the receiver is generally preferred. The beacon signal can be isolated via wavelength filtering, pattern recognition, pulse signature identification, or a combination of these.

Maintaining safety requires that light with hazardous power densities does not spill past the receiver. Optimal use of the PV converters also drives designs toward minimal beam wander. Minimizing the amount of potentially under-illuminated cells around the perimeter means an acceptable beam wander is often preferred to be less than 10% of the diameter of the beam at the receiver, though the exact amount will depend on the receiver size and other design constraints. The tracking system effectively operates in angle space; therefore longer ranges directly translate into smaller allowed angular error. For example, a tracking accuracy of ±5 cm at a range of 200 m would demand ±250 microradians (μrad) accuracy, whereas at 1 km it would mean that the accuracy could only be ±50 μrad.

As mentioned in Chapter 4, low-order atmospheric distortions can steer the beam off its nominal direction, sometimes at rates of hundreds of hertz. Control systems require the input signal to be sampled at a faster rate than the output signal. Treatment of control system design is beyond the scope of this book. Readers should consult one or more of the numerous guides [29,30] and texts on this subject [31,32].

Sources of error for tracking include:

- Alignment error (both fixed and temperature dependent) between the power beam's and tracking system's optical axes.
- Electronic noise in the tracking sensor.
- Resolution limits of the sensor.
- Atmospheric scintillation, resulting in distortion of the beacons and a lagging apparent position (due to speed of light limits) of the receiver.

The design of the beam tracking subsystem will need to account for error sources and the cumulative effects of their combination.

6.3.4 *Beam Steering*

Once the system has determined where the beam needs to be pointed through the tracking methods described in the previous section, the beam must actually be pointed. This is accomplished with the beam steering subsystem.

The possible position of the receiver relative to the transmitter defines the transmitter's required field of regard (FOR),[3] which in turn defines the types of beam steering hardware needed. For a stationary transmitter and receiver, a narrow-angle ("fine") steering system can suffice to overcome minor changes in relative position and the angle of the two ends due to winds, thermal changes, and other factors. This fine and fast steering is often accomplished with a fast-steering mirror (FSM). The angular range of available FSMs is usually less than a couple of degrees, often less than a few milliradians. But they can have resolution of microradians or even nanoradians and response bandwidth of hundreds of hertz or higher.

For a moving target, such as a UAV, or if a single transmitter is shared between multiple disparate receivers, then a relatively much wider FOR is accomplished with a "coarse" wide-angle steering mechanism. In this situation, a fine/fast steering system typically accompanies the wide-angle steering mechanism. The coarse/wide-angle hardware is usually the final optical assembly along the beam path in a system, except for a protective window, and is referred to as the beam director. Gimbaled mirrors of various optical path designs such as coudé path, coelostat, or others can address an FOR of an entire hemisphere or more. While directed energy weapons systems may require the coarse/wide-angle steering system to move very fast with extreme accuracy, laser power beaming can usually accommodate a system that can move slower. It might take tens of seconds up to minutes to traverse its entire range, and its resolution only needs to be moderately finer than the FOR of the fine/fast steering hardware (i.e., multiple steps within the fine/fast FOR), depending on the stability of the coarse system when commanded to a specific point. For example, if the fine/fast system has an FOR of $\pm 1°$, then

---

[3]FOR is the range of angles (e.g., in azimuth and elevation) that a steerable optical system can be pointed to.

Figure 6.5.   Fast Steering Mirror, 2" diameter. Image courtesy Optics In Motion LLC.

the coarse/slow system should have a resolution of less than roughly 0.5° (providing five steps within the fine system's 2°-wide FOR), or smaller if the coarse system wanders/bounces within one or more steps around its set point. Figure 6.5 shows an FSM with a 2-inch mirror; smaller diameters are more common, but the larger diameter is better suited to high power intensity.

Sources of noise and error in the beam steering system include:

- Noise in steering motor drive electronics.
- Noise in steering motor position sensor.
- Nonlinear response of the post-FSM optics to the shifting beam position and angle from the FSM.
- Changes in optical surfaces due to heating.
- Inherited errors from the tracking system.
- Vibration impacting FSM, camera, optics, and/or housing.

As with the beam tracking subsystem, the design of the beam steering subsystem will need to account for error sources and the cumulative effects of their combination.

6.3.5   *Transmitter System Examples*

Integrating the laser, thermal control, beam shaping, beam tracking, and beam steering is guided by the end application and system

Figure 6.6. Laser power beaming transmitter example showing components and subsystems in context. Image courtesy of PowerLight Technologies.

requirements. Figure 6.6 shows the optical power beaming transmitter developed by PowerLight Technologies (formerly LaserMotive) for the 2009 NASA Centennial Challenge for Power Beaming [33] and modified for a demonstration in 2012 with a Lockheed Martin Stalker uncrewed aircraft [34]. This system employed two diode lasers, shown in the figure on the bottom right. Beam shaping (variable focus and collimation) was accomplished with optics along the beam path, and tracking and steering were done with the gimbaled mirror that appear in the middle of the figure on the right. The primary mirrors (on the left) were f/10, resulting in the need to fold the beam multiple times, as seen in Figure 6.6.

Another example of an integrated laser power beaming transmitter is shown in Figure 6.7. This transmitter was used in 2019 for testing and demonstrations in Seattle, WA, and Bethesda, MD, delivering over 400 W of power from the receiver output across a distance of more than 300 m [35].

The lower housing of the system near the bottom of the figure contains the 976 nm diode laser, chiller, beam shaping, tracking, and steering optics, as well as controls and user interface. The top of the mast holds the coarse steering mirror that directs the beam to the receiver, and safety sensors.

Figure 6.7.   Laser power beaming transmitter in operation in Bethesda, MD [35]. *Image from US Government work not protected by copyright.*

6.4   Receivers

The critical second segment of an optical power beaming system is the receiver. This examination focuses on receivers comprised of bandgap-tuned PV cells, but it is possible to employ a receiver that uses a heat engine or other method for power recovery. Full treatments of the details of PV cells [36] and thermophotovoltaic cells [37] can be found elsewhere, and are beyond the scope of this text.

6.4.1   *Photovoltaic Arrays*

A single laser-PV cell usually outputs power from milliwatts up to 20–30 W. The power output is limited by the cells' sizes, which have ranged from 2 mm up to about 70 mm in width, the balancing of ohmic losses with grid line shadowing, and thermal degradation of

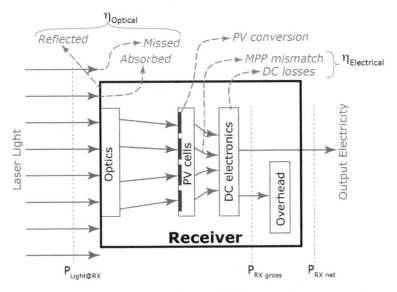

Figure 6.8. Factors affecting the overall efficiency of a power receiver.

efficiency. Because many applications need hundreds or thousands of watts of electrical power, a receiver array is built using many PV cells to match the required power output.

A PV array will necessarily have an efficiency that is limited by and functionally less than its constituent cells, even if they closely approach ideal performance. This is the case because of a variety of factors, under the general categories of optical and electrical loss factors. Figure 6.8 depicts these factors in context and they are discussed in more detail below. A useful metric to determine the design efficiency of a receiver is the overall receiver efficiency (gross or net — see Figure 6.7 for definitions) divided by the efficiency of a single PV cell. Using capital eta to represent the ratio of lowercase eta efficiencies, this is

$$H = \frac{\eta_{\mathrm{RX}}}{\eta_{\mathrm{PV}}} \tag{6.1}$$

An ideal receiver would have $H = 1$. Some experimental results have shown H values between 0.35–0.67, but the highest reported value to date is $H = 0.86$ [38, p. 14] by PowerLight with expectations of achieving $H > 0.95$ in the near future.

Many losses will manifest as heat requiring dissipation, which will be addressed in Section 6.4.5. Other losses, notably reflections, may create laser safety hazards. These are addressed in Section 4.4.

The power output of arrays of cells can also degrade over time, depending on the operating conditions. This can arise from oxidation of grid lines on PV cells, outgassing from adhesives, and other factors. Accounting for these and other typical losses that are not specific to power beaming receiver arrays is discussed in [39, p. 414].

6.4.2 *Optical Loss Factors*

6.4.2.1 *Missed Light*

Some light might miss the array either due to a beam shaping design which overfills the receiver or due to beam pointing errors that allow the beam to partially shift off the receiver. In some array designs, light might pass through the receiver if there are gaps. Atmospheric scattering will also cause some light to miss the array, as described in Section 4.3.1.

6.4.2.2 *Absorbed Light*

Whenever light impinges on anything other than active PV absorber material, some of it will be absorbed. It's important to choose materials to either transmit or reflect as much of the light as possible and to consider the following mitigations and factors:

- Anti-reflection coatings for lenses and windows.
- Reflectivity of metal and dielectric coatings for mirrors.
- Absorption in glass.
- Absorption in other materials.

Any absorbed light can present a potential source of heat that requires management.

6.4.2.3 *Reflected Light*

Transmissive optical surfaces will reflect some amount of light, similar to the impedance mismatch for an antenna. This can be minimized with anti-reflection coatings, which can reduce the loss per surface from about 4% down to as little as <0.1%. Even reflective surfaces will have some imperfections that lead to light being scattered instead of specularly reflected. This can be minimized by polishing.

6.4.3 *Electrical Loss Factors*

The efficiency of PV conversion is discussed in depth in the references at the beginning of this section. Presented here are summaries of some of the principal electrical loss factors for power beaming receiver arrays.

6.4.3.1 *Temperature Effects*

PV cells lose on the order of 1–2 mV/°C per junction, which translates to roughly 0.15%–0.18%/°C loss in III–V materials such as GaAs and InGaAs. High power densities and ineffective thermal control can both contribute to higher cell temperatures, resulting in losses that may further compound thermal problems.

6.4.3.2 *Intra-array Mismatch*

Series-connected cells have matching currents, regardless of the relative illumination of each cell, due to the conservation of charge [40]. Similarly, parallel-connected cells have matching voltages. When arrays are uniformly illuminated and have cells of nearly identical performance, these matching conditions result in maximized output power. To mitigate problems arising from mismatches, bypass diodes can be used at the cell level to shunt current around underperforming cells that would otherwise currently limit a string. Blocking diodes can be used at the string level to prevent current flow into strings that are underperforming. These time-honored and widely used array techniques are further discussed in [41, p. 5.1–1]. For power beaming, where the illumination conditions on the receiver array might change rapidly over relatively short time scales due to scintillation and other atmospheric effects, there may be value in adding capacitance at the cell or string level to stabilize the array's output and minimize mismatch losses. Active electronic control of received power at the cell and string level can also be employed to compensate for changing illumination conditions, but this can add cost and complexity.

6.4.3.3   *Ohmic Losses*

PV cells and arrays use grid lines, bus bars, and other interconnects to gather and consolidate the generated electrical current. As all these features have finite conductivity, ohmic losses are associated with each of them. These losses can be mitigated through judicious design and sizing, but typically at the cost of adding more mass in the form of heavier gauge conducting elements.

6.4.3.4   *Conversion and Regulation*

The receiver's output voltage and current should ideally be matched to the application, but often the application or applications might vary. Voltage conversion and regulation are likely necessary to satisfy the user's load requirements, and these will incur losses arising from their inefficiencies.

6.4.3.5   *Additional Losses*

If the receiver employs processing, data acquisition, communication, power management electronics, or other subsystems, these will draw power from the receiver, further decreasing the amount of power that can be sent to the receiver's output. Though this functionality cannot be omitted, power beaming system designers should be cognizant of their power consumption. In addition, round-trip energy losses for local energy storage (charging and discharging of batteries or capacitors) can also impact net efficiency.

6.4.4   *Receiver Optics*

Optics can be used to tailor the beam and minimize wasted light, thereby reducing the number of PV cells required. Because the light only needs to be approximately uniformly spread across each PV cell, non-imaging optical designs with less stringent wavefront error specifications are adequate. Receiver optics generally act as concentrators, and so concentrator designs can be leveraged from concentrated solar PV systems and LED illuminators. A full treatment of non-imaging optics can be found in [42].

Conservation of étendue demands a trade-off between the acceptance angle range $\pm\theta$ of incoming collimated light and the concentration ratio C. When air or vacuum is the only transport medium, and we limit the maximum angle for light to strike the PV cell (because large angles are more likely to be reflected) as $\pm\alpha$, then this limit is

$$C \le \frac{\sin^2 \alpha}{\sin^2 \theta} \tag{6.2}$$

For example, if a PV cell can reasonably accept light from $\pm 45°$ and the design requires a concentration ratio of $25\times$, that implies the acceptance angle is $\pm 8.1°$.

6.4.5 *Thermal Management*

Heat sinks or fans powered by the receiver are frequently used to manage the temperature of the PV cells and other components on power beaming receiver arrays. As any additional energy required for cooling is generally subtracted from the power available at the output of the receiver, lower power, and lower-complexity thermal control methods are typically favored.

6.4.5.1 *Efficiency Effect of Temperature*

For simplicity, assume that all the cells in an array are at the same temperature. The temperature rise depends on the thermal resistance between the PV cells and the cooling medium, typically air, and on the waste heat:

$$\Delta T = \Theta P_{\text{heat}} \tag{6.3}$$

where Θ is the thermal resistance (measured in $°C/W$) between the PV cell and the cooling medium. Continuing this simple model, the waste heat is simply due to the inefficiency of the PV cells:

$$P_{\text{heat}} = P_{\text{in}}(1 - \eta_{\text{PV}}) \tag{6.4}$$

where P_{in} is the optical input power, and η_{PV} is the PV efficiency. But as seen above, PV efficiency depends on temperature:

$$\eta_{\text{RX}} = \eta_{\text{PV}} - c_{\text{T}}\Delta T \tag{6.5}$$

where η_{PV} is the nominal PV efficiency (normally measured at 25°C, and it is assumed that ΔT is relative to that temperature), c_T is the PV cell temperature coefficient (normally around 0.18%/C), and we are otherwise assuming the receiver design efficiency $H = 1$. We can solve these equations to see that the temperature rise is

$$\Delta T = \frac{\Theta P_{in}(1 - \eta_{PV})}{1 - \Theta P_{in} c_T} \tag{6.6}$$

And therefore the actual efficiency is

$$\eta_{RX} = \eta_{PV} - c_T \frac{\Theta P_{in}(1 - \eta_{PV})}{1 - \Theta P_{in} c_T} = \frac{\eta_{PV} - \Theta P_{in} c_T}{1 - \Theta P_{in} c_T} \tag{6.7}$$

An example can help show the potential impact even for this simple model. Figure 6.9 highlights the importance of efficiently removing heat. It assumes $c_T = 0.0018$ and $P_{in} = 1,000\,\mathrm{W}$, and compares the overall efficiency vs. thermal resistance of the heatsink, for cases of PV efficiencies of 45% and 60%. The lower axis shows the temperature rise.

The real picture is more complicated due to factors such as varying temperatures across an array, heat from ohmic losses and electronics,

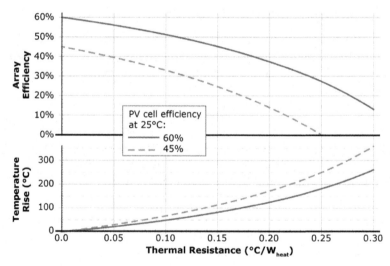

Figure 6.9.   Example impact of thermal resistance on PV temperature and array efficiency.

and power consumption of cooling fans and electronics as described previously.

If the output power is allowed to sag, then temperature increases only have the effect of reducing efficiency. But if a specific output power is required, then the operator could increase the input optical power to compensate for the thermal hit to efficiency. This eventually becomes a losing game, as more input power is added, which increases the heat, which decreases the efficiency, which requires more input power, until the cells get so hot that the target output power cannot be achieved no matter how much power is provided. Practical experience suggests that it's generally preferable to keep the PV cell temperature below about 90°C, with lower temperatures being favored. Figure 6.10 continues the example from above, showing the total output power (as well as temperature rise) vs. input power for the 45% and 60% efficient PV cells. Ohmic losses are not included here and would make the result even worse. Not only is there a point of diminishing returns of increasing optical power, but a negative effect, as modeled here.

6.4.5.2 *Types of Heat Sinks*

There are different regimes for thermal management. When the beam intensity is low enough, no heat sink is needed if surface convection

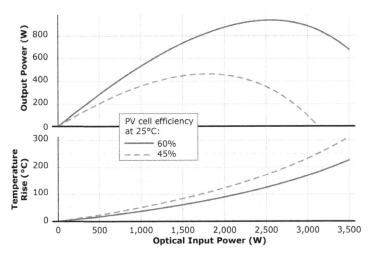

Figure 6.10. Impact of optical input power on output power and PV temperature.

and radiative heat rejection are adequate. The threshold for "low enough" depends on the PV efficiency and any ambient airflow. In some circumstances, such as on an aerial vehicle, the ambient airflow can be significant. As the beam intensity increases, cooling fins become necessary. Heat spreaders may also be needed if the cells are separated by a significant distance, or if a cooling area larger than the PV array area is required. Fans to force air through the heat sink fins become required if ambient airflow proves insufficient. In space, radiative heat transfer may be the only option.

Typical heat spreaders can be some combination of a metal plate, embedded graphite metals, heat pipes, and vapor chambers. Thermal conductivities for metals range between 90 and 400 W/m K. Heat pipes and vapor chambers, on the other hand, can have thermal conductivities up to 300,000 W/m K and beyond. Embedded-graphite metals are in the middle, around 900–1,200 W/m K.

Further exploration of thermal control methods for electronics can be found in [18,43].

6.4.6   *Power Management and Distribution*

Though the particulars of PMAD will be tailored to the power system's application, they may include DC/DC conversion, maximum power point tracking (MPPT), and energy storage. The implementation should be driven by the concept of operations, with attention paid to the possibility and likelihood of outages of the power beam, and the load demands of the application. In some cases, it may make the most sense to present the output of the power beaming receiver via a standard interface and offload aspects of power quality and energy storage to external elements, such as uninterruptable power systems.

6.4.7   *Receiver System Examples*

Once each of the considerations above has been balanced with the demands of the application and its constraints, the design can be undertaken. Shown in Figure 6.11 is an example of an integrated optical power beaming receiver.

This receiver was used in PowerLight's winning entry in the 2009 Space Elevator Power Beaming Challenge held by NASA [33]. It was

Figure 6.11.   Optical power beaming receiver system made by LaserMotive (Pow-erLight) for NASA's Space Elevator Power Beaming Challenge. Photo courtesy PowerLight.

part of a tether climbing robot that climbed a 1 km long cable attached to a helicopter. The power received was on the order of 1 kW when the climber was within a few hundred meters above the ground.

Figure 6.12 shows two receivers. The one on the left is from demonstrations conducted in 2019: one at the Port of Seattle, WA, and the other in Bethesda, MD. At each demonstration, more than 400 W of electrical power was delivered, and the system included a fully integrated safety interlock system [35].

The smaller receiver on the right side of Figure 6.12 was used in a demonstration in 2021 in Kent, WA that used the same transmitter shown in Figure 6.6. It also incorporated a full safety interlock system, and output more than 170 W electrical. A video showing the system in operation is available at [45].

Figure 6.12.   Laser power beaming receivers from demonstrations conducted in 2019 (left) and 2021 (right) [44]. *Photo courtesy PowerLight Technologies.*

6.5   Notional Link Examples

The notional examples below outline technical considerations and approaches for meeting requirements for several different instances where power beaming might be used. For the development of real-world links, additional consideration of factors such as regulations, mass producibility, marketability, and economics would likely be needed. System developers should consult systems engineering [46] and product design [47] resources as appropriate.

6.5.1   *Indoor Consumer Device*

In this example, the following top-level requirements are assumed:

- The system shall comply with all applicable safety and regulatory requirements.
- The transmitter input shall plug into a standard electrical outlet.
- The receiver output shall provide 5 W electrical output.
- The link shall operate at distances up to $10\,\mathrm{m}$.
- The receiver PV area shall not exceed $5\,\mathrm{cm}^2$.

If efficiency is the primary success metric, then 808 nm is probably the best choice. Allowing for a receiver design efficiency of 0.9 and a 59% PV efficiency at the operating temperature for a net 53% receiver efficiency, 5 W out implies 9.4 W of light into the receiver. Lasers come rated for specific, discrete power levels, so a 10 W laser is the likely choice. As a side benefit, under-driving lasers can extend their lifetime. End-to-end efficiency, including 2 W of other power consumed in the transmitter, can likely be just above 26%, requiring just under 20 W of input to the transmitter. Future improvements in lasers, PV cells, and controls are expected to improve efficiency to 40% or higher.

This low power example is one of the few cases when the lower efficiency of so-called "eye safer" wavelengths around 1,550 nm could be acceptable because the power consumption is so low that poor efficiency doesn't add a significant operating cost for the electricity. While PV efficiencies reported in the literature have exceeded 40% for 1,550 nm wavelength [48], commercially available cells are likely lower. A net receiver efficiency of 33% would require a 15 W laser, and allowing for some margin and laser lifetime extension would recommend the purchase of a laser rated for 18–20 W. 1,550 nm lasers have been demonstrated with 45% wall-plug efficiency [49], but ones with adequate power output available for sale appear to be less than 30% efficient. End-to-end efficiency is probably below 10%, requiring more than 50 W of system input power.

6.5.1.1 *Safety*

According to the ANSI Z136.1 standard for the safe use of lasers, the maximum permissible exposure (MPE) for 1,550 nm light for eyes and skin is 0.1 W/cm^2 (1,000 W/m^2) for long-duration exposure, defined as up to 30,000 s (more than 8 h) [50, pp. 80, 88]. The MPE for 808 nm light is about 60 times lower, driven by a more restrictive exposure level for eyes [50, pp. 79, 87].

Even a 100% efficient 5 cm^2 receiver would need to be illuminated with 1.0 W/cm^2, exceeding those safety thresholds by one or more orders of magnitude. An actual receiver will require a higher optical power density to offset the receiver losses described above (3.7 W/cm^2 for 808 nm, and more than 7 W/cm^2 for 1,550 nm). This demonstrates clearly that there will need to be robust methods for detecting objects approaching the beam to prevent exceeding the exposure limit. Given that these safety systems will need to be implemented regardless, there does not appear to be any significant benefit to choosing the eye-safer 1,550 nm wavelength. As the laser safety MPE decreases, more attention must be paid to the possibility of hazards from reflections or other indirect exposures.

In the US, the Food and Drug Administration (FDA)'s Center for Device & Radiological Health (CDRH) is the agency that approves laser products for consumer use [51], as further elaborated in Chapter 4. The system will need to pass muster with these or similar regulators in other countries. Safety approval considerations are critically important drivers in the design process.

The remainder of this link example will assume an 808 nm-based system.

6.5.1.2 *Optics*

Using 808 nm and assuming constraints on product cost, it might only be possible to get about half the maximum available radiance at this wavelength, which would be 2×10^{10} W $\cdot$ m^{-2} $\cdot$ sr^{-1} (0.02 TW $\cdot$ m^{-2} $\cdot$ sr^{-1}) and implies the aperture product $d_T d_R$ is then $\sim 2.9 \times 10^{-4}$ m^2. If the transmit and receive apertures are the same size, they could each be as small as 1.7 cm in diameter to reach 10 m if the beam shaping, tracking, and other factors are close to optimal.

6.5.1.3 *Tracking and Pointing*

Assuming a maximum allowed beam wander of 10% of the receiver beam aperture diameter implies that the beam should not wander more than about 2 mm from being centered, which implies a tracking accuracy of 2 mm/10 m = 0.2 mrad (200 μrad). That might be challenging in a consumer device due to the cost pressure on components, although the short distance and stable transmitter mounting could make it economical at scale. The 5 cm^2 receiver area limit implies a 2.5 cm diameter, so a 1.7 cm beam could have up to ± 4 mm of allowable wander. Doubling that accuracy then doubles the required tracking accuracy to 0.4 mrad.

The slew rate to track a nearby moving object, assuming around 20 m/s at a distance of 2 m, would require the transmitter to reach 10 rad/s = 570 deg/s = 95 rpm. No requirement was stated to allow for a moving receiver, so perhaps it should be unaddressed in the interest of simplicity. Tracking is much simpler for a stationary device, such as a wireless speaker or a smartphone sitting on a table.

6.5.1.4 *Thermal*

A 55% efficient transmitter will likely need about 20 W of input power to generate up to 10 W optical output, implying about 10 W of waste heat at the transmitter end. This amount of heat should be relatively easy to dissipate with conventional methods, such as fans or passive heat sinks. If spread over a 10 cm × 10 cm heat sink, the average heat flux is only 0.1 W/cm^2, which is close to the heat from the sun near noon on a summer day.

Because the receiver is roughly $5\,cm^2$ and 53% efficient, the waste heat power density will be roughly 0.88 W/cm^2, or lower if the heat sink is allowed to encompass a larger area than the light collection aperture). An appropriate pin-fin heat sink can reject this heat without the need for a fan. For any consumer device, care must be taken to ensure that users can't come into direct contact with a dangerously hot surface, but the accessible part of heatsink fins will be within normal temperature limits.

6.5.1.5 *Summary*

The indoor consumer device case is one that has been envisioned and discussed by many and determined to have promise [52,53]. In 2023, perhaps the closest company to having a product poised for widespread adoption is Wi-Charge [54], which has worked to develop and refine its hardware for many years, and which has also secured needed safety certification [55] although its choice of wavelength has not been publicly disclosed. Researchers have also demonstrated similar systems [56], but the formidable safety approval process has limited entrants in this area. For this reason, there has also been an investigation into the use of LEDs rather than lasers for this type of application [8].

6.5.2   *Outdoor Horizontal Point-to-Point*

In this example, the following top-level requirements are assumed:

• The system shall comply with all applicable safety and regulatory requirements.

- The transmitter input shall plug into a standard high-power electrical outlet.
- The receiver output shall provide 500 W electrical output.
- The link shall operate at distances up to 400 m.
- The receiver PV area shall not exceed $0.25\,\mathrm{m}^2$ ($2{,}500\,\mathrm{cm}^2$).

Potential use cases for a system meeting these parameters include sensors, telecom nodes, electric drones, and temporary power links.

Longer distances can push the design toward wavelengths with higher radiance laser sources, even though the component efficiencies tend to decrease for those wavelengths. Laser PV cell efficiencies will be assumed to be in the 45% to 65% range.

6.5.2.1 *Safety*

As with the indoor consumer device case, it is almost certain that continuous exposure limits according to ANSI Z136.1 for candidate wavelengths will be exceeded. This again will necessitate active and robust safety systems. A further consideration is that this link could potentially expose aircraft or satellites to the beam, depending on the circumstances of the receiver and what is behind it. Attention needs to be paid to the potential for light spilling past the receiver and how this could affect safety and beam exposure. In the US, the Federal Aviation Administration oversees laser operations that could affect aviation [57]. Though only US Department of Defense affiliated efforts are required to coordinate with the Joint Space Operations Center's Laser Clearinghouse (JSpOC LCH) [58], it is strongly recommended for all operators [59].

6.5.2.2 *Optics*

Allowing for atmospheric losses and likely receiver efficiencies, about 1.3–1.7 kW of light from the transmitter will be needed, depending on wavelength, which implies the following aperture sizes (see Equation (3.20)) for various wavelengths, assuming transmit and receive apertures are the same size:

- 808 nm ($0.04\ \mathrm{TW}\cdot\mathrm{m}^{-2}\cdot\mathrm{sr}^{-1}$): aperture product is $0.075\ \mathrm{m}^2$, implying $>31\,\mathrm{cm}$ for each aperture.

- 976 nm (0.16 TW $\cdot$ m^{-2} $\cdot$ sr^{-1}): aperture product is 0.037 m^2, implying >21.8 cm for each aperture.
- 1,075 nm (at least 15 TW $\cdot$ m^{-2} $\cdot$ sr^{-1}): is 0.004 m^2, implying >7.4 cm for each aperture.

If the transmit aperture is limited to no larger than 20 cm for reasons of cost, then the available radiance for each wavelength means the receiver diameters would be at least 48 cm at 808 nm, 24 cm at 976 nm, or 2.6 cm at 1,075 nm. Setting a maximum average incident light intensity at the receiver[4] of less than 100 suns (10 W/cm^2) will push the diameter for 1,075 nm up to at least 14.3 cm.

A reasonable trade-off between aperture sizes and efficiency suggests a choice of 976 nm. At this wavelength, PV efficiencies are up to 48% in 2023 and poised for improvement. A net receiver efficiency of 45% implies a need for 1,120 W of light. If we limit the average heat intensity at the receiver to under 1 W/cm^2, which can be managed with relatively low-power fans, then the light-collecting aperture could be 28 cm in diameter or larger.

6.5.2.3 *Tracking and Pointing*

If we assume we want the beam to wander no more than ± 2 cm from its nominal center, the pointing system needs to have an accuracy of ± 50 μrad. Turbulence-induced scintillation can add up to 30 μrad or more of beam wander, driving a need for active fast beam steering with a bandwidth on the order of 100 Hz.

6.5.2.4 *Thermal*

With a 976 nm laser having 50% efficiency and atmospheric loss of 10%/km, the transmitter needs to output nearly 1.2 kW of light, and so the cooling system will need to dissipate at least 1,200 W of waste heat. Since this may be generated within a small volume, a chiller or

---

[4]This choice is driven primarily by average heat flux rejection limits but is somewhat arbitrary. It could be increased as receiver efficiencies increase or when the thermal management system reduces thermal resistance, or both. At some point, beam intensity is limited because of the possibility of thermal blooming or lensing effects.

similar thermal management system will be needed. This subsystem will consume its own power, degrading the overall system efficiency. A chiller with a CoP of 3.5, for example, will consume more than 340 W.

An average heat intensity of around 1 W/cm^2 can be managed to keep the thermal resistance from PV to ambient air below 45 C with relatively low airspeed (about 2 m/s), which can be generated using low-power fans consuming approximately 15 W.

6.5.2.5 *Summary*

The link described here is like that demonstrated in 2019 by PowerLight Technologies in Seattle, WA and Bethesda, MD. This link was a functional demonstration and was not optimized for efficiency. It also used an earlier generation of 976 nm lasers with a lower radiance than the example above. Future links with similar parameters could be accomplished with greater efficiency and more compact transmitter and receiver systems. More information on the 2019 demonstration is in Chapter 2.

6.5.3 *Medium-Sized Uncrewed Air Vehicle (UAV)*

In this example, the following top-level requirements are assumed:

- The system shall comply with all applicable safety and regulatory requirements.
- The receiver output shall provide 3 kW electrical output.
- The transmitter input shall be able to run off a 20 kW portable generator or comparable power source.
- The receiver shall be able to integrate into a fixed-wing uncrewed aerial vehicle.
- The link shall operate at distances up to 3 km.

The goal of the system is to deliver power at a distance to an uncrewed aerial vehicle for the propulsion system and onboard electrical systems. These may include motors, batteries, imaging systems, communications nodes, and other systems. It is desirable that the aircraft would have enough energy storage so that it could fly for some period of time if the beam is interrupted, or to enable it to fly to areas beyond the transmitter's view before returning for inflight recharging.

6.5.3.1 *Safety*

Unlike the previous examples of indoor and horizontal point-to-point links, the potential effects of the system on other aircraft will need careful consideration due to the need to point the beam above the horizon. Spacecraft may be affected as well. A thorough investigation of applicable safety and regulatory requirements should be undertaken. In the United States, this will include coordination with the FAA and the Laser Clearinghouse, as described in the previous example.

Potential effects on the UAV itself and its payload(s) should also be considered. Receiver waste heat could affect hydrocarbon-based materials, such as carbon-fiber structural elements, paint, airplane skin material, and other materials integrated with the UAV body. A beam wandering off the receiver collection area could directly heat parts of the UAV, and high-intensity light from the beam could potentially damage any unprotected cameras.

6.5.3.2 *Optics*

Multi-mode fiber lasers are currently available with radiance up to $170 \text{ TW} \cdot \text{m}^{-2} \cdot \text{sr}^{-1}$, although more commonly they are below $50 \text{ TW} \cdot \text{m}^{-2} \cdot \text{sr}^{-1}$, primarily for use in industrial materials processing. For this case, slightly under 11 kW of laser output is needed, and reasonable aperture sizes can be accommodated with $35 \text{ TW} \cdot \text{m}^{-2} \cdot \text{sr}^{-1}$.

A beam size at the receiver of 50 cm, resulting in average waste heat intensity around $3 \pm 0.5 \text{ W/cm}^2$, would therefore require a transmit aperture of at least 15 cm. At that size, however, the average beam intensity at the exit window would be more than 62 W/cm^2, which might cause problems from any contamination on the exit window. If the exit optical intensity is limited to 40 W/cm^2, then the transmit aperture size would grow to almost 19 cm.

6.5.3.3 *Tracking and Pointing*

If the beam is permitted to wander ± 3 cm on the receiver, then the beam tracking and pointing system must provide a pointing accuracy of about 10 μrad. That level of accuracy can be challenging without major efforts in sensing, controls, fast steering hardware, and possibly transmitter vibration reduction.

6.5.3.4 *Thermal*

Assuming a receiver design efficiency of 80%, including temperature effects on PV, and a baseline PV efficiency of 50% (which is higher than currently commercially available, but reasonable to expect within a few years), the average waste heat on the receiver would be about 2.6 W/cm^2, which would require continuous airflow to aid with dissipation. Fortunately, the receiver is likely to be mounted on the underside of a wing or fuselage, and airflow is readily available as part of normal flight operations. The receiver design team should negotiate with the UAV team about minimizing the impact on UAV aerodynamics.

6.5.3.5 *Summary*

Given the many existing and emerging applications of drones, this example is of prime interest. An initial demonstration of laser power beaming to an uncrewed fixed-wing aircraft was executed by Lockheed Martin and PowerLight Technologies (when it was known as LaserMotive) in 2012 [60], as shown in Figure 2.28(b). Being able to power this type of aircraft indefinitely removes or relaxes limitations of flight time and payload mass, enabling new capabilities.

6.5.4 *Space to Earth*

In this example, the following top-level requirements are assumed:

- The system shall comply with all applicable safety and regulatory requirements.
- The transmitter input shall come from sunlight collected outside Earth's atmosphere.
- The receiver on the Earth shall provide at least 20 kW of electrical output.
- The receiver area shall not exceed 30 m in diameter.

The goal of this system is to deliver as much output power from the ground receiver as possible while staying within the continual exposure limits. To help derive the amount of energy that needs to be collected in orbit, it will be assumed that, at the time of system implementation, PV laser receivers are near 50% net efficiency, and that atmospheric losses are 30%. Obviously, atmospheric losses will vary with weather conditions and geographic location. Generating 20 kW of electricity from the receiver output would require more than 40 kW of light at the receiver and nearly 60 kW of light generated in orbit. With today's technology, near-single-mode fiber lasers in the 1,070 nm region are the only practical choice for reasonable aperture sizes. Assuming the laser transmitters have a net efficiency of 45%, the input power required is 127 kW, which we'll round up to 130 kW for some margin.

The solar array on orbit required to generate the electric power to deliver 20 kW output on Earth, assuming space solar PV array efficiency of 25%,[5] would need to capture 660 kW. That output requires less than 400 m^2 which for a single square array is less than 20 m on a side. The eight solar arrays on the International Space Station, for comparison, are more than 6 times as large, at 2,500 m^2 [62]. For reasons detailed below, an orbital altitude of 10,000 km will be selected.

6.5.4.1 *Safety*

Safety is considered the prime driving constraint. As with other links and as described in Section 4.4, it is possible to use a safety light curtain, but in this case, there is a significant difference. The speed of light delay in detecting approaching objects will be measured in the relatively long tens of milliseconds, as it takes about 67 ms for round-trip travel across 10,000 km. Other methods of sensing those objects should be considered as well.

The average beam intensity at the receiver is 5.7 mW/cm^2, but for a Gaussian profile, it will be roughly double that in the middle and much lower ($<$1 mW/cm^2) near the edges. "Continual" (classified as up to 3×10^4 s, or 8 h + 20 min) maximum permissible exposure limit for eye safety around 1,070 nm is 5.3 mW/cm^2, but beam jitter and scintillation will both move the beam around and change the intensity

---

[5]While some space PV cells exceed 30% [61, p. 27], there will be some losses associated with the array, as described in Section 6.4.1.

profile rapidly. Shorter duration exposures have higher MPEs (at this wavelength, the 0.25 s exposure limit is 14 mW/cm^2, and the 1 ms exposure limit is 55 mW/cm^2), but these are not the driving safety requirements.

With active safety sensors detecting people, birds, aircraft, and other objects, the safety system would need to provide separation between the beam and the first detection of an approaching object. It is relatively easy to detect passenger aircraft from long distances. Even if only detected when 1 km away, a plane traveling just below Mach 1 would still offer almost 3 s to detect and respond. Birds and people move much slower, so a 5-m border would allow 100 ms to detect and respond. Except for small birds at high altitudes, birds and people could also be readily detected within such a short stand-off distance. Spacecraft present a special case, as their orbital velocities relative to the power beam will generally exceed several kilometers per second, depending on the specific orbit's parameters. This great speed means that they will only be in the beam for a very short period.

Those considerations suggest that the allowed beam intensity at the receiver might be increased. However, consider the case of an event in orbit that could rapidly perturb the transmitter without dispersing the beam. This might arise from a debris collision or similar occurrence. Additional safety measures, such as onboard accelerometers, would need to be included. These would not be subject to the light travel time constraint, and could respond much more quickly [63].

6.5.4.2 *Optics*

For a medium Earth orbit (MEO) at 10,000 km altitude, the orbital period is about 5.8 h. A constellation of satellites would be needed for continuous receiver coverage for the ground receiver, ignoring weather outages. A high-altitude receiver would avoid weather outages but would require an additional link to send power to the ground. Limiting the beam size at the $1/e^4$ point at the receiver to 30 m and assuming the laser can have an M^2 of about 1.3 implies that the transmit aperture diameter needs to be at least 1.47 m. The primary mirror in the Hubble telescope, for comparison, is 2.4 m in diameter.

6.5.4.3   *Tracking and Pointing*

Constraining the beam to wander no more than 3 m requires a pointing accuracy of 0.3 μrad. If the 1.5 m mirror is an f/5, its focal length is then 7.5 meters. Achieving 0.3 μrad tracking requires positioning the equivalent point source within 2.3 microns. This is achievable with modern microelectromechanical systems (MEMSs).

On the ground, reflective or diffractive concentrators could focus the laser light from its approximately incident 0.015 W/cm^2 by a factor of 10$\times$ (resulting in 0.15 W/cm^2) up to 300$\times$ (4.5 W/cm^2). This level of concentration would put the intensity in the range that laser PV cells are often designed to (1–50 W/cm^2) while still having an acceptance angle of $\pm 3.3°$ up to $\pm 18.4°$, which would not unduly tax pointing of receiver optics.

6.5.4.4   *Thermal*

A 50% efficient receiver implies that the average waste heat would be 7.5 mW/cm^2, which is less than 10% of the intensity of sunlight. While the more concentrated heat from PV cells would need to be spread out, rejecting the average heat could be accomplished with passive cooling.

6.5.4.5   *Summary*

Solar power satellites have been examined for decades, and laser-based space solar has received a fair amount of attention [64,65]. Because of the scale of such a system, many unknowns remain to be resolved. These are discussed further in Section 8.4.3.

6.6   Conclusion

The selection of components and subsystems in the design and implementation of power beaming systems will be driven by the modality utilized. This chapter reviewed options and considerations for systems operating at optical wavelengths and explored notional example applications.

6.7   Further Reading

Many resources exist for more information on lasers, PVs, thermal management, beam control, and other topics introduced in this chapter. In addition to the specific chapter references, readers may wish to consult the following for further guidance. Additional sources can be found in this book's back matter.

B. E. A. Saleh and M. C. Teich, *Fundamentals of Photonics*, Third edition, 2 vols. Hoboken, New Jersey: Wiley, 2019.

J. L. Miller and E. Friedman, *Photonics Rules of Thumb: Optics, Electro-Optics, Fiber Optics, and Lasers*, Third edition. Bellingham, WA: SPIE Press, 2020.

Larry C. Andrews and Ronald L. Phillips, Laser Beam Propagation Through Random Media, 2nd ed. Bellingham, Washington, USA: International Society for Optics and Photonics (SPIE), 2005.

A. Luque López and S. Hegedus, Eds., *Handbook of Photovoltaic Science and Engineering*, Repr. Chichester, England: Wiley, 2009.

Hagop Injeyan and Gregory D. Goodno, Eds., *High-Power Laser Handbook*. New York: McGraw-Hill, 2011.

References

[1] R. McKie, "My Bright Idea: Robert Winston," *The Guardian*, February 27, 2010. [Online]. Available at: https://www.theguardian.co m/technology/2010/feb/27/robert-winston-my-bright-idea (accessed September 12, 2022).

[2] D. R. Paschotta, "Wall-Plug Efficiency." [Online]. Available at: https://www.rp-photonics.com/wall_plug_efficiency.html (accessed May 16, 2020).

[3] "nLIGHT Demonstrates 73% Wall-Plug Efficiency," nLIGHT. [Online]. Available at: https://www.nlight.net/press-releases-content/ nlight-demonstrates-73-wall-plug-efficiency (accessed September 19, 2022).

[4] P. Crump *et al.*, "SHEDs Funding Enables Power Conversion Efficiency up to 85% at High Powers from 975-nm Broad Area Diode

Lasers," Accessed: Oct. 21, 2023. [Online]. Available: https://www.academia.edu/11448147/SHEDs_Funding_Enables_Power_Conversion_Efficiency_up_to_85_at_High_Powers_from_975_nm_Broad_Area_Diode_Lasers.

[5] "High Power CW Fiber Lasers, 1 − 100+ kW | IPG Photonics," https://www.ipgphotonics.com/. [Online]. Available at: https://www.ipgphotonics.com/en/products/lasers/high-power-cw-fiber-lasers (accessed May 5, 2022).

[6] M. C. A. York and S. Fafard, "High Efficiency Phototransducers Based on A Novel Vertical Epitaxial Heterostructure Architecture (VEHSA) with Thin p/n Junctions," *J. Phys. Appl. Phys.*, vol. 50, no. 17, p. 173003, May 2017. doi: 10.1088/1361-6463/aa60a6.

[7] S. Jarvis, J. Mukherjee, M. Perren, and S. J. Sweeney, "On the Fundamental Efficiency Limits of Photovoltaic Converters For Optical Power Transfer Applications," in *2013 IEEE 39th Photovoltaic Specialists Conference (PVSC)*. Tampa, FL: IEEE, June 2013, pp. 1031–1035. doi: 10.1109/PVSC.2013.6744317.

[8] M. Zhao and T. Miyamoto, "1 W High Performance LED-Array Based Optical Wireless Power Transmission System for IoT Terminals," *Photonics*, vol. 9, no. 8, p. 576, August 2022. doi: 10.3390/photonics9080576.

[9] Anthony E. Siegman, *Lasers*. Mill Valley, CA: University Science Books, 1986.

[10] C. C. Davis, *Lasers and Electro-Optics: Fundamentals and Engineering*. Cambridge [England]; New York, NY: Cambridge University Press, 1996.

[11] B. E. A. Saleh and M. C. Teich, *Fundamentals of Photonics*, Third edition, 2 vols, in Wiley Series in Pure and Applied Optics. Hoboken, New Jersey: Wiley, 2019.

[12] H. Zhou, H. Stange, and M. Kneier, "How to Choose a Laser: How to Choose a Laser Diode for Your Application," Laser Focus World. [Online]. Available at: https://www.laserfocusworld.com/lasers-sources/article/16555361/how-to-choose-a-laser-how-to-choose-a-laser-diode-for-your-application (accessed September 25, 2022).

[13] L. Roberts, R. Ward, C. Smith, and D. Shaddock, "Coherent Beam Combining Using an Internally Sensed Optical Phased Array of Frequency-Offset Phase Locked Lasers," *Photonics*, vol. 7, no. 4, p. 118, November 2020. doi: 10.3390/photonics7040118.

[14] P. Sprangle, B. Hafizi, A. Ting, and R. Fischer, "High-Power Lasers for Directed-Energy Applications," *Appl. Opt.*, vol. 54, no. 31, p. F201, November 2015. doi: 10.1364/AO.54.00F201.

[15] F. Chen *et al.*, "10 kW-Level Spectral Beam Combination of Two High Power Broad-Linewidth Fiber Lasers by Means of Edge Filters," *Opt. Express*, vol. 25, no. 26, p. 32783, December 2017. doi: 10.1364/OE.25.032783.

[16] M. Vorontsov *et al.*, "Comparative Efficiency Analysis of Fiber-Array and Conventional Beam Director Systems in Volume Turbulence," *Appl. Opt.*, vol. 55, no. 15, p. 4170, May 2016. doi: 10.1364/AO.55.004170.

[17] "Temperature Control and Mount Selection," Arroyo Instruments. [Online]. Available at: https://www.arroyoinstruments.com/wp-content/uploads/2021/04/Arroyo-TemperatureControl.pdf (acce ssed September 29, 2022).

[18] A. D. Kraus and A. Bar-Cohen, *Thermal Analysis and Control of Electronic Equipment*. Washington: New York: Hemisphere Pub. Corp.; McGraw-Hill, 1983.

[19] R. J. Martin, *Thermal Systems Design: Fundamentals and Projects*, Second edition. Hoboken, NJ: Wiley, 2022.

[20] "Beam Quality and Strehl Ratio | Edmund Optics." [Online]. Available at: https://www.edmundoptics.com/knowledge-center/application-notes/lasers/beam-quality-and-strehl-ratio/ (accessed December 7, 2022).

[21] C. Palmer and E. Loewen, *Diffraction Grating Handbook*, Seventh edition. Rochester, New York: Newport Corporation, 2014.

[22] R. K. Tyson and B. W. Frazier, *Principles of Adaptive Optics*, Fifth edition. Boca Raton, London, New York: CRC Press, Taylor & Francis Group, 2022.

[23] "Adaptive Optics Technologies," Dynamic Optics. [Online]. Available at: https://www.dynamic-optics.it/ (accessed October 21, 2023).

[24] F. M. Dickey, Ed., *Laser Beam Shaping: Theory and Techniques*, Second edition. Boca Raton, FL: CRC Press, Taylor & Francis Group, 2014.

[25] F. M. Dickey and T. E. Lizotte, Eds., *Laser Beam Shaping Applications*, Second edition. Boca Raton, FL: CRC Press, Taylor & Francis Group, 2017.

[26] "Laser Beam Shaping Overview |Edmund Optics." [Online]. Available at: https://www.edmundoptics.com/resource-page/application-notes/optics/laser-beam-shaping-overview/ (accessed October 8, 2022).

[27] "Beam Shaping," Holo Or. [Online]. Available at: https://www.holoor.co.il/beam-shaping/ (accessed October 15, 2022).

[28] D. R. Paschotta, "Beam Shapers." [Online]. Available at: https://www.rp-photonics.com/beam_shapers.html (accessed October 15, 2022).

[29] T. Wescott, "Sampling: What Nyquist Didn't Say, and What to Do About It," Wescott Design Services, Aug. 2018. Accessed: Dec. 08, 2022. [Online]. Available: https://www.wescottdesign.com/articles/Sampling/sampling.pdf.

[30] W. Kester, "What the Nyquist Criterion Means to Your Sampled Data System Design," Analog Devices, Tutorial MT-002, October 2008. [Online]. Available at: https://www.analog.com/media/en/training-seminars/tutorials/MT-002.pdf (accessed December 8, 2022).

[31] K. J. Åström and R. M. Murray, *Feedback Systems: An Introduction for Scientists and Engineers*. Princeton, New Jersey: Princeton University Press, 2008.

[32] B. Friedland, *Control System Design: An Introduction to State-Space Methods*, Dover edition. Mineola, NY: Dover Publications, 2005.

[33] T. Talbert, "NASA — LaserMotive LLC Wins Prize in Power Beaming Challenge." [Online]. Available at: https://www.nasa.gov/directorates/spacetech/centennial_challenges/cc_pb_feature_11_10_09.html (accessed December 10, 2022).

[34] B. Boen, "After the Challenge: LaserMotive," NASA. [Online]. Available at: http://www.nasa.gov/directorates/spacetech/centennial_challenges/after_challenge/lasermotive.html (accessed October 27, 2022).

[35] *Energy Transmitted by Laser In 'Historic' Power Beaming Demonstration*, October 22, 2019. [Online Video]. Available at: https://www.youtube.com/watch?v=Xb9THqrXd4I (accessed July 21, 2023).

[36] M. A. Green, *Solar Cells: Operating Principles, Technology and System Applications*, Repr. [der Ausg.] Englewood Cliffs, New Jersey, 1982. Kensington, NSW: Univ. of New South Wales, 1998.

[37] D. L. Chubb, *Fundamentals of Thermophotovoltaic Energy Conversion*. Amsterdam, Netherlands; Boston; Oxford, UK: Elsevier, 2007.

[38] T. Nugent, "Improving Performance Metrics for Power Beaming," presented at the 3rd Optical Wireless Power Transfer Conference, Yokohama, Japan, Apr. 2021.

[39] J. R. Wertz and W. J. Larson, Eds., *Space Mission Analysis and Design*, Third edition, in Space technology library. El Segundo, CA: Dordrecht; Boston: Microcosm; Kluwer, 1999.

[40] W. Storr, "Kirchhoff's Current Law, (KCL) and Junction Rule," Basic Electronics Tutorials. Available at: https://www.electronics-tutorials.ws/dccircuits/kirchhoffs-current-law.html (accessed December 9, 2022).

[41] *Solar Cell Array Design Handbook*, Vol I. Jet Propulsion Laboratory, California Institute of Technology, Pasadena, CA, United States, NASA-CR-149364, October 1976. [Online]. Available at: https://ntrs.nasa.gov/api/citations/19770007250/downloads/19770007250.pdf (accessed October 15, 2022).

[42] J. Chaves, *Introduction to Nonimaging Optics*, Second edition. CRC Press, Taylor & Francis Group, 2016.

[43] Y. Shabany, *Heat Transfer: Thermal Management of Electronics*. Boca Raton, FL: CRC Press, 2010.

[44] P. Jaffe, "Power Beaming & Space Solar," presented at the *SERDP & ESTCP and OE-Innovation Symposium*, Arlington, VA, USA, December 1, 2022. [Online]. Available at: https://www.symposium.ser dp-estcp.org/event/a06cd5ff-a780-4fff-8cc7-75bfc70c32a6/summary. (accessed December 10, 2022)

[45] *Power Beaming-Phase 3*, April 19, 2021. [Online Video]. Available at: https://www.youtube.com/watch?v$=$9MI2ph9jptM (accessed December 10, 2022).

[46] *NASA Systems Engineering Handbook*. NASA Center for AeroSpace Information, 2007. [Online]. Available at: https://www.nasa.gov/ sites/default/files/atoms/files/nasa_systems_engineering_handbook. pdf (accessed October 25, 2022).

[47] K. T. Ulrich and S. D. Eppinger, *Product Design and Development*, Fifth edition. New York: McGraw-Hill/Irwin, 2012.

[48] S. D. Jarvis, J. Mukherjee, M. Perren, and S. J. Sweeney, "Development and Characterisation of Laser Power Converters for Optical Power Transfer Applications," *IET Optoelectron.*, vol. 8, no. 2, pp. 64–70, April 2014. doi: 10.1049/iet-opt.2013.0066.

[49] P. O. Leisher *et al.*, "Watt-Class 1550 nm Tapered Lasers with 45% Wallplug Efficiency For Free-Space Optical Communication," in *2016 International Semiconductor Laser Conference (ISLC)*, September 2016, pp. 1–2.

[50] Laser Institute of America, "ANSI Z136.1-2014, American National Standard For Safe Use of Lasers." 2014.

[51] "Laser Products and Instruments," FDA Center for Devices and Radiological Health. [Online]. Available at: https://www.fda.gov/ radiation-emitting-products/home-business-and-entertainment-produ cts/laser-products-and-instruments (accessed September 15, 2022).

[52] L. Olvitz, D. Vinko, and T. Švedek, "Wireless Power Transfer For Mobile Phone Charging Device," in *2012 Proceedings of the 35th International Convention MIPRO*, 2012, pp. 141–145.

[53] Q. Liu *et al.*, "Charging Unplugged: Will Distributed Laser Charging for Mobile Wireless Power Transfer Work?," *IEEE Veh. Technol. Mag.*, vol. 11, no. 4, pp. 36–45, 2016. doi: 10.1109/MVT.2016.2594944.

[54] "WI-CHARGE LTD.," Wi-Charge. [Online]. Available at: https:// wi-harge.com/ (accessed July 1, 2020).

[55] "Wi-Charge In-Room Wireless Charging Approved for Use in US," Digital Trends. [Online]. Available at: https://www.digitaltrends. com/mobile/wi-charge-in-room-wireless-charging-approved-fda/ (acce ssed October 23, 2022).

[56] V. Iyer, E. Bayati, R. Nandakumar, A. Majumdar, and S. Gollakota, "Charging a Smartphone Across a Room Using Lasers," *Proc. ACM Interact. Mob. Wearable Ubiquitous Technol.*, vol. 1, no. 4, pp. 1–21, January 2018. doi: 10.1145/3161163.

[57] "Laser Safety | Federal Aviation Administration." [Online]. Available at: https://www.faa.gov/about/initiatives/lasers (accessed October 27, 2022).

[58] "Laser Tagged — How the JSpOC Manages Laser Deconfliction." [Online]. Available at: https://www.vandenberg.spaceforce.mil/News/ Features/Display/Article/920559/laser-tagged-how-the-jspoc-manages- laser-deconfliction/https%3A%2F%2Fwww.vandenberg.spaceforce. mil%2FNews%2FFeatures%2FDisplay%2FArticle%2F920559%2Flaser- tagged-how-the-jspoc-manages-laser-deconfliction%2F (accessed October 27, 2022).

[59] R. Lafon, J. Wu, and B. Edwards, "Regulatory Considerations: Laser Safety and the Emerging Technology of Laser Communication," presented at the Commercial Laser Communications Interoperability and Regulatory Workshop, Washington, DC, Jun. 12, 2017.

[60] "Laser Powers Lockheed Martin's Stalker UAS For 48 Hours," Media — Lockheed Martin. [Online]. Available at: https://news.lock heedmartin.com/2012-07-11-Laser-Powers-Lockheed-Martins-Stalker- UAS-For-48-Hours (accessed October 27, 2022).

[61] "State-of-the-Art Small Spacecraft Technology," NASA Ames Research Center, Moffett Field, CA, NASA/TP—20210021263, October 2021. [Online]. Available at: https://www.nasa.gov/sites/de fault/files/atoms/files/soa_2021_1.pdf (accessed December 11, 2022).

[62] M. Garcia, "About the Space Station Solar Arrays," NASA. [Online]. Available at: http://www.nasa.gov/mission_pages/station/structure/ elements/solar_arrays-about.html (accessed December 11, 2022).

[63] J. Zhao, J. Jia, H. Wang, and W. Li, "A Novel Threshold Accelerometer with Postbuckling Structures for Airbag Restraint Systems," *IEEE Sens. J.*, vol. 7, no. 8, pp. 1102–1109, August 2007. doi: 10.1109/ JSEN.2007.897936.

[64] A. Rubenchik, J. Parker, R. Beach, and R. Yamamoto, "Solar Power Beaming: From Space to Earth," LLNL-TR-412782, 952766, April 2009. doi: 10.2172/952766.

[65] S. J. Mobilia, "Detailed Design of a Space Based Solar Power System," Master of Science, San Jose State University, San Jose, CA, 2009. doi: 10.31979/setd.eytn-2wht.

Chapter 7

Link Characterization

Quis, quid, ubi, quibus auxiliis, cur, quomodo, quando
Who, what, where, by what aids, why, how, and when

— Cicero via Aquinas [1, p. 109]

7.1 Introduction

As interest grows in considering power beaming as a means of moving energy, it will be increasingly important to accurately characterize the performance and particulars of power beaming links. This chapter shows one paradigm that might be utilized to measure such links and references a historical example to illustrate elements of the characterization approach. Appendix A contains a hands-on example suitable for reproduction by students and others wishing to develop first-hand experience in using the approach outlined. The general methodology is intended to be modality agnostic and is applicable to straightforward point-to-point power beaming links. It is also extensible to power beaming systems of other varieties and greater complexity.

A rudimentary power beaming link's functional blocks and selected measurable parameters are depicted in Figure 7.1.

The 15 measurable parameters called out in Figure 7.1 are defined and grouped in Table 7.1 by measurement type: length, power, power density, mass, volume, and time. Within each major grouping, measurable parameters proceed from the input side to the output.

251

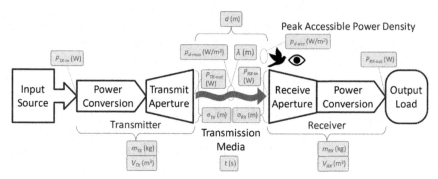

Figure 7.1.   Power beaming link functional blocks with measurable parameters of interest.

These parameters are by no means exhaustive and are oriented with a single, simple, and effectively stationary link in mind. Scenarios with multiple links, multiple transmitters, multiple receivers, or cases where link elements are in motion may include other parameters of interest. Sensitivity to aperture or polarization misalignment and component cost contributions could also be added.

The grouping of the parameters could be ordered to align with the interests of the demonstration performer to better reflect the goals of the system. They could proceed in order of interest priority along the energy path from the input to the transmitter to the output of the receiver (e.g., power delivered then link distance, or vice versa), or by measurement difficulty. The scheme selected here is convenient for later association with calculated parameters.

Reporting of any demonstrations or links should include basic identification information such as the responsible parties, title, location, date, contact information, and reference callouts. *At a bare minimum, reports of power beaming link demonstration should include the input power, link distance, and output power.*

7.2   Defining the Power Beaming Link System

For meaningful reporting, it is critical to be as clear as possible in defining what elements comprise the link and their boundaries. This can be accomplished by mapping the functional blocks in Figure 7.1 to the hardware employed for the link. Clear demarcation of the system's start and end points is paramount. It may also

Table 7.1. Parameter names, symbols, and descriptions for power beaming link measurements.

Parameter	Symbol	Description
Largest transmit aperture dimension	$\varnothing_{\text{TX}}$ (m)	The largest dimension of the transmitter aperture, often the diameter
Wavelength	λ (m)	The wavelength corresponding to the frequency of operation
Link distance	d (m)	The distance between the transmit and receive apertures
Largest receive aperture dimension	$\varnothing_{\text{RX}}$ (m)	The largest dimension of the receiver aperture, often the diameter
Transmitter power input	$P_{\text{TX-in}}$ (W)	The power from all input sources to the transmitter
Transmitter power output	$P_{\text{TX-out}}$ (W)	The power output of the transmitter at the wavelength of operation
Receiver power input	$P_{\text{RX-in}}$ (W)	The power incident on the receive aperture from the transmitter
Receiver power output	$P_{\text{RX-out}}$ (W)	The power from the receiver to the output load
Maximum power density	$p_{d\text{-max}}$ (W/m^2)	The maximum power density along the beam's path
Accessible power density	$p_{d\text{-acc}}$ (W/m^2)	The maximum power density accessible to people, animals, aircraft, etc.
Transmitter mass	m_{TX} (kg)	The mass of the transmitter, including the transmit aperture
Receiver mass	m_{RX} (kg)	The mass of the receiver, including the receive aperture
Transmitter volume	V_{TX} (m^3)	The volume of the transmitter, including the transmit aperture
Receiver volume	V_{RX} (m^3)	The volume of the receiver, including the receiver aperture
Link duration	t (s)	The duration over which the link was operated

be helpful to acknowledge potentially legitimate alternative interpretations. For instance, should the boundary between the input source and the transmitter be placed where a power supply is plugged into an outlet or be defined at the output of the power supply? Should equipment that performs cooling functions be considered as part of the transmitter or receiver? Different rationales could justify accounting in one way or another. The key is to achieve clarity in what has been assumed. This can be accomplished through the employment of diagrams, photographs, and narratives.

7.3 Measurable Parameters

Metrology has a robust history, and its measurement practices are described in various sources. Depending on the purpose of the link and the level of precision in reporting, link demonstrators may opt for different approaches. Many best practices are found in publications affiliated with quality or standards organizations, such as those derived from the International Organization for Standardization (ISO) and the like [2,3]. One exemplar for measuring and documenting power beaming links was written by Dickinson and Brown [4]. This source uses independent quality assurance staff for verification of measurements. A more recent publication that captures the majority of these parameters clearly is by He *et al.* [5].

7.3.1 *Reporting Uncertainty*

Demonstrations that seek to establish new records for one or more parameters should adhere to accepted measurement principles like those outlined by Taylor [6] and report uncertainty clearly. Measurement devices should have unexpired calibration certificates and measurement processes should follow guidelines outlined by a reputable national or international metrology laboratory, such as the National Institute of Standards and Technology (NIST) in the United States. Being able to independently verify a measurement with multiple different calibrated devices is recommended. For informal power beaming link demonstrations, less rigorous approaches may be sufficient. In general, reporting absolute uncertainty with significant figures indicated is preferable to reporting that uses fractional uncertainty. However, the circumstances of the measurement may drive the decision in choosing between absolute or fractional uncertainty reporting.

7.3.2 *Link Operating Conditions*

For the duration of the link's being active, environmental conditions that could affect the operation or measurements should be logged.

Some of these conditions could include ambient temperature, humidity, wind speed, pressure, sound level, lighting, and electromagnetic energy. A link demonstrated near external sources of energy to which the receiver is sensitive could skew measurements and results. Two examples of this are a rectenna receiver located near a mobile phone tower operating at the same frequency, or a photovoltaic receiver in sunlight collecting ambiently harvested energy in addition to beamed power received from the transmitter. While this may be a desirable feature of a given system, sources of the received energy should be clearly distinguished. In some cases, the temperatures of the transmitter, receiver, their components, or the transmission media may have a dramatic effect on link performance [7]. For such instances, denotation of these effects in the measurement should be reported appropriately.

7.3.3 *Length Measurements*

7.3.3.1 *Largest Transmit Aperture Dimension, $\emptyset_{TX}$ (m), and*
Largest Receive Aperture Dimension, $\emptyset_{RX}$ (m)

Structures that radiate electromagnetic waves or other forms of energy come in a variety of shapes and sizes. Here, the term "aperture" is used even though the radiating or transducing structure may not resemble a conventional aperture. To be clear, "aperture" refers to that part of the unit (transmitter or receiver) that is conveying the electromagnetic waves; housing and support structures will likely make the overall unit dimensions larger. The largest aperture dimension is used in lieu of aperture diameter, width, or area to accommodate the range of possible volumetric implementations. Though it may obscure aperture details, the largest dimension provides a starting point in establishing the link's operating envelope, particularly for the case of uniformly illuminated circular or square apertures [8]. When other dimensions are available, they should also be reported, especially if they help to define the physical or effective area of the aperture. This is particularly needed if the effective area differs significantly from estimates computed using the largest aperture dimension alone.

The rationale behind reporting the largest receive and transmit aperture dimensions is to provide a linear reference measurement that is often easy to measure and which can facilitate first-order calculations that can help to identify the link's geometric limitations.

Reporting the largest aperture dimensions should not discourage performers from identifying other dimensions and measurable quantities that assist in characterizing the transmitter or receiver, some of which may be referenced to the operating wavelength. These additional quantities could include the physical area, effective aperture, directivity, gain, number of array elements, aperture power distribution, aperture phase distribution, focal length, beam quality, beam parameter product, and others.

7.3.3.2 *Wavelength,* λ (m)

The wavelength or frequency of operation along with its associated center and range may be measured using a spectrum analyzer suitable for the modality employed. Typically, power beaming calls for the transmission of energy within a very narrow bandwidth or bandwidths, which can be challenging. The motivation for implementing narrow transmission bandwidths may stem from a desire to maximize the energy to the receiver's region of greatest sensitivity, minimize interference or incursion into neighboring spectral regions, optimize overall system efficiency, or a combination of these. Combining multiple narrow bandwidths might be employed to avoid deleterious destructive interference at the receiver, or perhaps as a route to maximize delivered power. Since the bandwidth of the transmitting source may not match the radiating aperture's bandwidth, wavelength measurements should be collected from the output of the transmitting aperture if possible.

The width, center frequency, full-width half-maximum value, coherence length, and spectral flux energy distributions for each significant contributing component of the transmitted energy may vary over time. Some might vary due to the media between the transmitter and receiver or show effects that are a function of distance from the transmitter. The degree to which each of these is measured and characterized statistically may vary depending on how much they influence the energy delivered by the link.

7.3.3.3 *Link Distance,* d (m)

The distance between the transmit and receive apertures defines the link distance, although in some cases, system implementations may call for other geometric interpretations of link distance. For instance, if a transmit or receive aperture is deeply recessed within a structure, or hardware is employed at any point between the transmitter and receiver for means of safety, re-direction of the beam, or for other purposes, then the impact of these should be clearly conveyed. If a given link employs active or passive elements to enhance transmission between the transmitter and receiver, such as in the case of a beam waveguide, the elements should be described appropriately.

7.3.4 *Power Measurements*

Power measurements, especially those focused on achieving record power transfer, should give attention to four considerations which deal with variations and account for sources and sinks of energy. These considerations are as follows:

- **Stored energy:** Elements within the link may exist in a state without any stored energy or be started and ended with no net change in stored energy. Stored energy in the system creates the possibility that there could be confusion about the performance of the link. This confusion arises from the uncertainty of whether the delivered energy received at the output traversed the entirety of the link, or whether some of it came from a storage element located somewhere in the system. Likewise, ambiguity could also depend on how energy is buffered within the system. Reporting data for links that have been in operation for a significant time period may aid in reducing or averaging the storage element effects observed.
- **Link modes:** Any different system modes that might substantively affect power measurements should be reported. Some links may simply be on or off with no distinctions for other possible power consumption modes. More sophisticated links and transmitters could quantify different power consumption modes such as stand-by mode, internal transmitter energy storage charging mode, receiver acquisition mode, safety interlock triggered mode, and a host of others. All of these modes could affect the amount of energy delivered and the practical efficiency observed over the link's operation, so their contributions should be accounted for.
- **Steady-state condition:** Hardware for power beaming links may take time to reach a steady-state condition. Power values and losses throughout the system may vary until thermal and other

steady-state response equilibrium conditions are reached. Indications should be provided on whether system tests were conducted under such steady-state conditions, and if so, what the wait duration was.

- **External power sources:** At any point in the link, there exists the possibility for the introduction of power from external sources. Power might be inadvertently provided by external temperature control systems or by other sources. Link implementers should look closely for external power sources that might distort collected and reported values.

These are the power measurements of principal interest for a simple point-to-point power beaming link.

7.3.4.1 *Transmitter Power Input, $P_{TX-\text{in}}$ (W)*

The total input power to the link is represented by $P_{\text{TX-in}}$. Techniques described by Malarić [9] and Witte [10] can be used to properly define this input power. If an electrical power source provides alternating current (AC) rather than direct current (DC), it will be necessary to measure the current and the voltage along with their corresponding phase difference in order to ascertain the contributions of the real and reactive components of the apparent power.

7.3.4.2 *Transmitter Power Output, $P_{TX-\text{out}}$ (W)*

The total transmitter output power, $P_{\text{TX-out}}$, represents all of the power that exits the transmit aperture at the associated operating wavelength λ and its statistically defined range. How measurements of the transmitter's output power are taken will depend on the modality employed, possibly requiring RF power meters, laser power meters, or other devices. It may not be practical to measure the total transmitter output power in cases where the transmit aperture is large or when the transmit power is high. In such cases, the output power distribution might be sampled at one or more points to calculate or extrapolate the total transmitter output power. Methodologies used should be articulated. The spectral power distribution might vary as well so power measurements may also need to be associated with the statistics collected for the wavelength measurement.

7.3.4.3 *Receiver Power Input, $P_{RX-\text{in}}$ (W)*

The total amount of transmit power that is incident upon the receive aperture defines $P_{RX-\text{in}}$. Much like the transmitter output measurement, the receiver power input measurement approach will depend on the modality employed. This measurement might need to account for transmission media effects that could introduce time-varying incident power fluctuations. Some of the transmitted power may never reach the receiver, and some of the power that is incident will be reflected by the receiver aperture rather than absorbed. Power that is reflected from the receiver aperture should be included as part of $P_{RX-\text{in}}$ since improved future iterations could capture this lost energy. Minimizing the reflected power could be achieved through the employment of matching networks, antireflective coatings, or other means, depending on the system's modality. The amount of the receiver's incident reflected power should be reported as a distinct subset of $P_{RX-\text{in}}$, as in [4]. This could be represented by $P_{RX-\text{refl}}$. Another potential impediment to fully utilizing the incident power is polarization mismatch if the receiver's conversion capability is sensitive to polarity. Loss contributions from polarity mismatch should be included in $P_{RX-\text{in}}$ and could be represented by $P_{RX-\text{pol}\neq}$. Improvements to this polarization mismatch could improve the receiver's performance. Losses from reflection and polarization mismatch might be interrelated, and care should be taken to avoid duplication in reporting.

7.3.4.4 *Receiver Power Output, $P_{RX-\text{out}}$ (W)*

The receiver power output, $P_{RX-\text{out}}$, is the usable power available to the output load. It may be measured in the same fashion as the transmitter power input. An implicit assumption in the reporting of the receiver output power is that it comes entirely from the energy that has been beamed to the receiver and that any energy storage elements within the receiver system (batteries, capacitors, etc.) have all been properly accounted for. The existence or usage of any energy storage elements in the system, particularly the receiver, should be clearly disclosed in the link description. Selecting or adjusting the load to maximize output power is desirable, and any approaches or efforts to accomplish this are worth mentioning.

An interesting possibility arises in the prospect of using multiple forms of output power from the receiver. For instance, if a given scenario requires both electricity and thermal energy to be delivered to the output load or loads, power that might otherwise be categorized as contributing to waste heat or inefficiency might instead be utilized. Such scenarios could be envisioned for space or lunar applications, where systems must contend with environments featuring very low temperatures. In such cases, all usable forms of energy provided by the receiver to the output loads should be accounted for, and the forms of power delivered (electrical, thermal, etc.) identified.

7.3.5   *Power Density Measurements*

7.3.5.1   *Maximum Power Density, $p_{d-\text{max}}$* (W/m^2)

The maximum power density is the point or points of highest power density between the transmit and receive apertures that result from the energy radiated by the transmitter. This notionally sets a prospective upper limit for power density exposure risk, except in instances where power might be concentrated by external factors, such as scintillation. Ideally, the power density should be directly measured at numerous positions located at, between, and near the transmitter and receiver apertures. These should include locations on and off the beam path's longitudinal axis connecting the centers of the apertures in order to form a complete volumetric picture of the power density profile. This aids in verifying actual radiation conditions against theory and any simulations that have been developed. Such measurements will help to positively identify the location and character of the beam waist if one exists. Measuring the maximum and minimum power densities at each aperture will envelop their corresponding achieved amplitude tapers. These could be reported as $p_{d-\text{TXmax}}$, $p_{d-\text{TXmin}}$, $p_{d-\text{RXmax}}$, and $p_{d-\text{RXmin}}$. More power density measurements will afford more data points to correspond the link's performance to modeling, and in the verification of the locations of the reactive near field, radiating near field, and far field (or Fresnel and Fraunhofer regions) as applicable. If a range of demonstration or link operating conditions are anticipated, power density measurements should be collected both for the nominal and for the extreme ranges of these conditions. Depending on the link's modality and power density, other units such as W/cm^2, mW/cm^2, or equivalent suns of concentration may be preferred to W/m^2.

7.3.5.2 *Accessible Power Density, p_{d-acc} (W/m^2)*

The accessible power density is somewhat subjective. It may depend on usage scenario assumptions or on the reliability of access restriction measures. The intent is to determine what hazards the power beam may pose to people, animals, devices, or other entities. Hazard classes could include interference, disruption, or damage to systems outside the link. Active or passive means may be employed to limit access to one or more parts of the power link to mitigate or eliminate hazards. Considerations might include analysis of usage scenarios and probabilistic assessments. Intrinsic qualities of the power beaming system should be considered in determining the accessible power density. For instance, a system operating in the visible range with a laser transmitter may elicit a natural blink reflex in people who might be exposed to the beam, whereas one operating in the infrared would not. Implementers should convey an awareness of and design within safety standards, limits, and practices relevant to the modality employed. As with p_{d-max}, units other than W/m^2 may be preferred for p_{d-acc}. Perhaps more than any other parameter, accessible power density should be measured and assessed by a reputable independent outside party in addition to any measurements made by the link implementer.

7.3.6 *Mass Measurements*

7.3.6.1 *Transmitter Mass, m_{TX} (kg), and Receiver Mass, m_{RX} (kg)*

The mass of the transmitter and receiver can be determined by weighing them as integrated wholes or in separate parts on scales of appropriate capacity and precision. This measurement requires that the transmitter and receiver be clearly defined. Elements that contribute to alignment, stability, thermal control, power conversion, structural reinforcement, and the like should be included or a rationale for their exclusion be presented. Discussion of the usage of alternate means that could result in lower masses is encouraged, as are caveats that may indicate that the mass of the apparatus used is not indicative of the likely mass of a subsequent system. In any case, reporting and making transparent the difference between the means that *could* have been used and those *actually* used is key.

7.3.7 *Volume Measurements*

7.3.7.1 *Transmitter Volume, V_{TX} (m^3), and Receiver Volume, V_{RX} (m^3)*

Volume determination is subject to challenges, especially for items with complex geometries. Volume determination may be of limited utility since it is perhaps most useful as a proxy for integrability or transportability. It may be clear that an item exceeding a certain volume cannot fit within a particular platform's payload area or inside a given shipping container, but a smaller item still may not fit if it has one or more dimensions that are greater than the intended accommodations. This means that even relatively intensive methods to measure volume like submerging an irregular shape in a fluid to find its displacement may not be very instructive. It may make the most sense to find and report the approximate volume of a rectangular box or the smallest standardized container in which the transmitter would fit. If parts of the system are collapsible or have several different physical configurations for operation or transport, it should be clearly noted.

7.3.8 *Time Measurements*

7.3.8.1 *Link Duration, t (s)*

The link duration is the difference between the link's start and stop times. It defines the period for which a run has been logged to bind the validity of other reported measurements. As needed, discussion should be devoted to considerations for duty cycles, operating modes, and limitations of the link that might not be immediately evident from the link duration alone. If either the transmitter or receiver has not reached thermal equilibrium or will require a cooling, recharging, or other recuperation period prior to re-energizing the link, this should be made clear. Since one of the goals of reporting t is to provide a sense of the energy delivered over the course of the operating period, factors that might complicate or confound this should be disclosed.

7.4 Calculated and Estimated Parameters

The foregoing section concerned quantities that could be subject to direct measurement or found by straightforward calculations using these measurements. In practice, equipment or other limitations might prevent the employment of direct measurements. By contrast, the quantities in this section are calculated from the measurements in the previous sections or through estimation.

Aside from the amount of power transmitted, there are three general types of factors that affect a power beaming link's performance and characteristics, after [11, p. 15]:

(1) **Geometric factors:** Those resulting from physical geometry, which are: the separation between the transmitter and receiver, the size and shape of the transmit and receive apertures, their orientation and alignment, and the operating wavelength.
(2) **Implementation factors:** Those resulting from the physical implementation, apart from those having to do with physical geometry, such as the transmitter aperture's amplitude and phase distributions and the device efficiencies of the components within the transmitter and receiver.
(3) **Propagation factors:** Those arising from media or objects between the transmit and receive apertures, such as the effects of the atmospheric gases, particles, spatiotemporal temperature variations, and weather; each of which can cause different combinations of reflection, refraction, diffraction, absorption, or scattering; and aperture-medium coupling loss. In some instances, the channel between the transmit and receive apertures might act as a gain medium.

Depending on the circumstances, developers of power beaming links may have some control or influence over each of the above factors.

The three principal calculated or estimated parameters that are generally not considered as figures of merit are summarized in Table 7.2 and explained further below.

Table 7.2. Calculated or estimated parameters that are generally not considered figures of merit.

Parameter	Symbol	Description
Beam collection efficiency	η_{BC}	Approximate link efficiency as bounded by geometry
Path loss	L_P	Losses arising between the transmit and receive apertures
Energy delivered	$E_{\text{RX–out}}$ (J)	The energy delivered via the output of the receiver

7.4.1 *Beam Collection Efficiency,* η_{BC}

A simplified expression approximating the beam collection efficiency (BCE) performance of a power beaming link, which neglects implementation and propagation factors, can be established by using the geometric factors considered in the Goubau relationship. BCE is referred to as beam efficiency (BE) by Shinohara and others [12, p. 23]. The Goubau relationship was introduced in Chapter 3 and is reiterated here for convenience. For optimally illuminated and aligned ideal apertures, the parameter τ from [14] is expressed as

$$\tau = \frac{\sqrt{A_{\text{TX}} A_{\text{RX}}}}{\lambda d} \qquad (7.1)$$

where A_{TX}, A_{RX}, λ, and d are the transmit aperture area, receive aperture area, the wavelength of operation, and the distance between the apertures, respectively.

Then τ can be used in the following expression to find the BCE, as in [12, p. 24]

$$\eta_{\text{BC}} = 1 - e^{-\tau^2} \qquad (7.2)$$

Additional context and caveats can be found in Chapter 3.

7.4.2 *Path Loss,* L_P

The losses in power that occur between the transmitter and receiver can be caused by a multitude of contributions. Some of these contributions have been previously enumerated as propagation factors. With a sense of the loss contributions due to the BCE, further losses can be attributed to either implementation or propagation factors. Selected implementation factors will be addressed in the next section which focuses on figures of merit.

Attenuation contributions from air molecules, aerosol absorption, and scattering can be expressed as shown by [15, p. 12]. These loss contributions might vary significantly with changing atmospheric conditions [16]. Certain operating wavelengths and link circumstances may cause refraction, diffraction, or ducting.

Some link configurations might place discrete objects directly in the beam path. This might block, refract, or diffract the beam. Multipath enhancement may increase the amount of energy falling on the receiver due to reflection or scattering effects coming from off-axis objects located near the longitudinal transmission path.

Comparing the difference between the transmitter power output, $P_{\text{TX}-\text{out}}$, and the receiver power input, $P_{\text{RX}-\text{in}}$, yields an expression for the approximate path loss, L_P:

$$L_P = 1 - \frac{P_{\text{RX}-\text{in}}}{P_{\text{TX}-\text{out}}} \tag{7.3}$$

The combination of Lp and BCE makes it possible to prospectively corroborate estimates of other loss sources.

7.4.3   *Energy Delivered, $E_{\text{RX}-\text{out}}$* (J)

The energy delivered by the link is merely the receiver power output over the duration of the link's operation. A simple expression for the energy delivered, assuming time-averaged receiver output power, is

$$E_{\text{RX}-\text{out}} = P_{\text{RX}-\text{out}} \times t \tag{7.4}$$

While total energy delivered is of interest, in some cases it might not be considered a figure of merit because of the possibility of operating the link over a longer period of time to deliver more energy.

7.5   Figures of Merit

In addition to the all-important measurable parameters such as transmitter power input $P_{\text{TX}-\text{in}}$, link distance d, received power output $P_{\text{RX}-\text{out}}$, and accessible power density, $p_{d-\text{acc}}$, there are several

Table 7.3.	*Power beaming link figures of merit.*

Parameter	Symbol	Description
Transmitter-specific power	SP_{TX} (W/kg)	The mass-specific power of the transmitter
Transmitter volume-specific power	VSP_{TX} (W/m^3)	The volume-specific power of the transmitter
Receiver-specific power	SP_{RX} (W/kg)	The mass-specific power of the receiver
Receiver volume-specific power	VSP_{RX} (W/m^3)	The volume-specific power of the receiver
Transmitter efficiency	η_{TX}	The combined efficiency of the transmitter's power conversion and aperture
Receiver efficiency	η_{RX}	The combined efficiency of the receiver's power conversion and aperture
End-to-end link efficiency	η_{Link}	Efficiency from the input to the transmitter to the output of the receiver

figures of merit that are beneficial in characterizing practical power beaming links. These are summarized in Table 7.3 and defined further in the later sections.

7.5.1 *Transmitter Specific Powers, SP_{TX} (W/kg) and VSP_{TX} (W/m^3)*

If the transmitter is to be on an airborne, spaceborne, or mobile platform of any kind, its mass and volume are often key considerations. Power-to-weight ratio, mass-specific power, or simply "specific power" are roughly equivalent ways of expressing the transmitter's mass.

The value for the transmitter's specific power SP_{TX} is easily found by dividing the transmitter's power output, P_{TX-out}, by the transmitter's mass, m_{TX}, giving

$$SP_{TX} = \frac{P_{TX-out}}{m_{TX}} \tag{7.5}$$

In some circumstances, volume might be an equal or greater consideration than mass. In such instances, a figure for volume-specific power could be employed. This figure denoted as VSP_{TX} is calculated using V_{TX} in the denominator to give

$$VSP_{TX} = \frac{P_{TX-out}}{V_{TX}} \qquad (7.6)$$

7.5.2 *Receiver Specific Powers, SP_{RX} (W/kg) and VSP_{RX} (W/m^3)*

Like the transmitter, the receiver might be integrated with a platform that has mobility or transportability expectations or constraints. This makes mass-specific and volume-specific powers prime figures of merit. The receiver's mass-specific power is found in a similar fashion to that of the transmitter:

$$SP_{RX} = \frac{P_{RX-out}}{m_{RX}} \qquad (7.7)$$

It is keenly important to recognize that the value of SP_{RX} might be correlated only to a particular link configuration. It should not be thought of as an intrinsic quality of the receiver in the same way that SP_{TX} is considered an intrinsic quality of the transmitter. The observed SP_{RX} might be dramatically increased or decreased with small link configuration changes. These include reorienting the receiver, changing the distance over which the link operates, changing the load presented to the receiver, or altering any of many other variables. Given this, maintaining the association of SP_{RX} with a particular link instance is important. Under favorable conditions, it can show what is achievable. It may make sense to find the SP_{RX} of a given receiver using idealized conditions, perhaps reported as SP_{RX-max}. These idealized conditions set an upper bound for performance expectations.

Similar to the transmitter, volume might be an equal or greater consideration than mass. Again, a figure for volume-specific power for VSP_{RX} can be expressed using V_{RX} in the denominator:

$$VSP_{RX} = \frac{P_{RX-out}}{V_{RX}} \qquad (7.8)$$

The same caveat above for the validity of VSP_{RX} applies regarding operating under different conditions. The maximum realistically achievable VSP_{RX} can be reported as VSP_{RX-max}.

7.5.3   *Transmitter Efficiency, η_{TX}*

The transmitter-side "Power Conversion" and "Transmit Aperture" functional blocks from the link depiction in Figure 7.1 can be used to represent the whole of the transmitter. The power output of the "Transmit Aperture" can be divided by the power input to the transmitter-side "Power Conversion" block to find transmitter efficiency as

$$\eta_{TX} = \frac{P_{TX-out}}{P_{TX-in}} \qquad (7.9)$$

Different transmitter operating conditions may be explored to find the peak achievable P_{TX-out} and η_{TX}. It is realistic to expect that these two quantities are not simultaneously maximized under the same operating conditions. Depending on the needs of a given application, more weight might be given to optimizing SP_{TX} over η_{TX}, or vice versa.

Manufacturers that supply transmitters as complete units may provide specifications or test data showing transmitter output and efficiency performance as a function of varying voltage or current input. Link implementers should make such information available in their reporting.

7.5.4   *Receiver Efficiency, η_{RX}*

Like the approach describing the transmitter, the "Receive Aperture" and receiver-side "Power Conversion" functional blocks from the link depiction in Figure 7.1 can be used to represent the whole of the receiver. The power input to the "Receive Aperture" can be divided by the power output of the receiver-side "Power Conversion" block to find receiver efficiency:

$$\eta_{RX} = \frac{P_{RX-out}}{P_{RX-in}} \qquad (7.10)$$

As with SP_{RX} and VSP_{RX}, η_{RX} should be viewed as specific to the particular link implementation. Different incident power densities, receiver temperatures, or other factors could dramatically affect the performance of the receiver and the resulting value for η_{RX}.

7.5.5 *End-to-End Link Efficiency,* η_{Link}

Historically, end-to-end efficiency is the primary figure of merit in determining which power beaming technologies are adopted. Depending on the circumstances, this may not actually be truly the most important figure of merit. For instance, for solar power satellite systems, two of the most frequently asked questions are "How efficient is it?" and "Will this fry birds?" The efficiency question is an inquiry into cost while the frying birds question speaks to system safety. As another example, the specific power of a receiver might be the most important parameter for powering aerial vehicles. Power beaming link implementers are often reluctant to advertise the end-to-end link efficiency because values achieved to date are generally low, with a few notable exceptions, such as in [4]. Some have reported in manners that, while perhaps not intentionally misleading, have nonetheless sown confusion as to the actual performance achieved. To forestall this lack of clarity, end-to-end efficiency, η_{Link}, should be reported in terms of the ratio of both the input to the transmitter-side "Power Conversion" and the output of the receiver-side "Power Conversion" blocks from Figure 7.1 or

$$\eta_{Link} = \frac{P_{RX-out}}{P_{TX-in}} \qquad (7.11)$$

It is critical to report clearly how the quantities for P_{TX-in} and P_{RX-out} were determined, and what assumptions were made in establishing the boundaries for the power input and the power output of the link. Dickinson and Brown are exemplars of this in their reporting of their landmark high-efficiency demonstration in 1975 [4]. For their DC-RF source, they clarify that the power for the cathode heater and the blower power for air cooling is not accounted for, stating that these quantities "by an alternate more efficient design they could be minimized or eliminated" [4, p. 2].

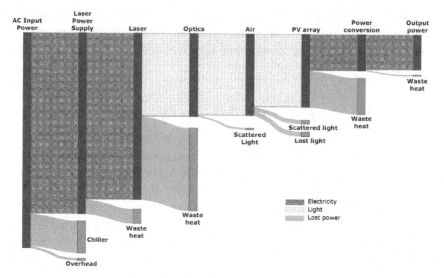

Figure 7.2.   A Sankey diagram for a notional laser power beaming link. Image courtesy PowerLight Technologies.

7.5.6   *Depiction of Sources of Losses*

A Sankey diagram [17] can be used to depict sources of loss and their relative magnitude. These diagrams provide confidence that reporters know that the link is subject to the law of conservation of energy and that all significant sources or sinks of energy are accounted for. An example Sankey diagram for a notional laser power beaming link is shown in Figure 7.2.

In Figure 7.3, an example Sankey diagram for Dickinson and Brown's 1975 record-efficiency microwave power beaming demonstration in Waltham, MA [4] is shown.

The categories of losses could be further subdivided depending on the implementation of the link and the ability to quantify loss sources. For instance, losses within the transmitter arising from thermal management systems, electrical power conversion, and optical or microwave component inefficiencies could be enumerated and grouped separately.

Depicting each loss source in the context of either a generic functional block diagram or one specific to the implementation will further clarify where and how the greatest contributions originate. In principle, any point-to-point power beaming link's loss sources should

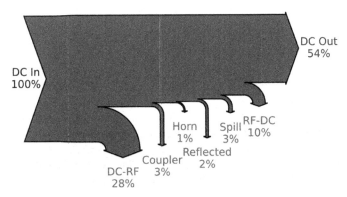

Figure 7.3. Sources of loss in Dickinson and Brown's 1975 high-efficiency microwave power beaming demonstration conducted in Waltham, MA. Outflows sum to 101% due to rounding.

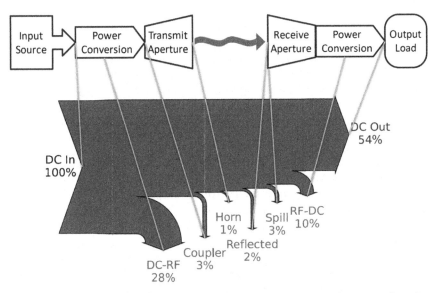

Figure 7.4. Mapping of loss sources to the generic point-to-point power beaming functional block diagram for Dickinson and Brown's 1975 Waltham demonstration. Outflows sum to 101% due to rounding.

be mappable to the generic functional block diagram introduced previously. A mapping of Dickinson and Brown's 1975 Waltham demonstration is shown in Figure 7.4.

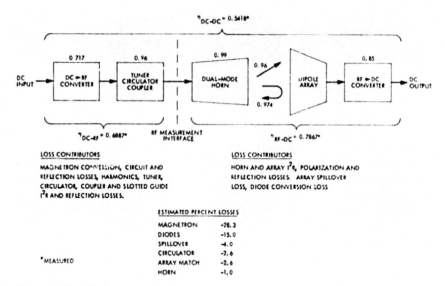

Figure 7.5. Dickinson and Brown's "Distribution of system efficiencies (measured and estimated)" depiction for their 1975 Waltham demonstration, from [4, p. 38].

Dickinson and Brown realized the value of doing something like this over 45 years ago. Losses associated with their demonstration are seen cascaded in stages in Figure 7.5.

Showing the relative power levels at each stage of a power beaming link is instructive. Table 7.4 captures the Dickinson and Brown Waltham demonstration from this perspective by reporting the power level present at each stage, and by showing the component efficiencies as well as their contribution to the total system efficiency.

The second column from the right of Table 7.4 lists the loss percentages represented in the Sankey diagram of Figure 7.3. The figure shows the percentages rounded to the nearest whole digit. This is also true for the figure's end-to-end link efficiency.

Each power measurement has an uncertainty associated with it. Dickinson and Brown did an admirable job capturing and reporting these uncertainties, though most uncertainties were reported fractionally. These are not shown in Table 7.4 but can be readily found in [4]. For reference, they reported a $P_{\text{TX}-\text{in}}$ of 914.76 W $\pm$ 0.61%,

Table 7.4. Power dissipation and loss contributions for Dickinson and Brown's 1975 Waltham demonstration, source data derived from [4].

	Power through link (W)	Power dissipated by subsystem (W)	Subsystem efficiency (%)	Subsystem loss (%)	Subsystem's contribution to total link loss (%)	Cumulative link efficiency to subsystem output (%)
DC input	914.76	—	—	—	—	—
DC→RF converter	655.88	258.88	71.7	28.3	28.3	71.7
Tuner circulator coupler	629.65	26.24	96.0	4.0	2.9	68.8
Dual-mode horn	623.35	6.30	99.0	1.0	0.7	68.1
Spillover	598.42	24.93	96.0	4.0	2.7	65.4
Array match	582.86	15.56	97.4	2.6	1.7	63.7
RF→DC converter	495.43	87.43	85.0	15.0	9.6	54.2

a $P_{\text{RX-out}}$ of 495.62 W $\pm$ 0.72%, and a corresponding η_{Link} of 54.18% $\pm$ 0.94%.[1]

7.6 Other System Characteristics of Interest

Many other system or component characteristics and figures of merit are possible and may gain prominence for particular applications. Those might include transmitter or receiver area-specific mass (ASm_{TX} or ASm_{RX} in kg/m^2) [18, p. 1434], transmitter or receiver average area-specific power (ASP_{TX} or ASP_{RX} in W/m^2), transmit efficiency versus steering angle ($\eta_{\text{TX}}(\theta)$ in %), or a host of others.

Certain power beaming systems may feature qualities that could have compelling utility for potential users. Some of these qualities could include the ability to address multiple receivers simultaneously

---

[1]There exists a very slight discrepancy between their reported $P_{\text{RX-out}}$ and that listed in Table 7.4. This difference arises from combining their reported fractional uncertainties pertaining to other parts of the link to find the losses in wattage for each subsystem. The same slight discrepancy, which is much less than 1 W, occurs for other values in the "Power through link" column of Table 7.4.

or in a time-shared manner, the ability to propagate in the absence
of a clear line of sight, the ability to employ energy harvesting when
a transmitter is unavailable, the ability to operate in a beam waveguide mode with phase reconstruction nodes [19], transmit efficiency
versus steering angle for receivers that move, or any of a number
of other novel features. These aspects should be described at least
qualitatively, and quantitatively if possible.

Power density can vary in many different ways in different parts of
the link. Numerous characteristics related to power density might be
of interest such as those that describe the power densities on different
parts of the apertures as well as the link's required "keep out" areas or
buffer regions to assure conformance with RF, optical, or other power
density safety guidelines under nominal and anomalous operating
conditions.

For many applications, cost concerns may be paramount or
present driving significance. This has been explored by Dickinson
[20], with the quantity $/MW-km introduced as a means of com-
parison. Depending on the circumstances, $/W, $/kWh, $/kg, $/m^2,
$/m^3 or other cost factors or combinations of cost factors may be
of principal concern. These could apply on the transmitter side,
receiver side, or both. As with almost any engineered system, the end
result will likely need to balance the competing considerations of size,
weight, power, and cost. Even more broadly, system implementations
will depend on feasibility evaluations stemming from technical, economic, legal, operational, and schedule factors [21, p. 176]. These are
further addressed in Chapter 8.

7.7   Comparison with Other Links

There are a range of means for comparison with previously demonstrated and proposed links. Links can be compared using measured,
calculated, or estimated parameters that are based on either a variety
of figures of merit or on a collection of any of these or values derived
from them. In many cases, extraordinarily complex systems can be
abstracted and compared based on the performance of one narrowly
defined metric. This approach is not without pitfalls and is not appropriate for all power beaming demonstrations, given the diversity of
their aims and goals. Nevertheless, uniform bases of comparison can
provide an impression of what has been achieved, where progress can

be made, and where research, development, and engineering efforts should be focused.

Plots employed to compare previous links include power versus distance as seen in Figure 1.5 and normalized efficiency from the transmitter to the receiver as seen in Figure 1.6. In some instances, log scales will provide the greatest utility for comparison. Insights gleaned from demonstrated and proposed links can guide technology goals and assessment of utility for given applications.

7.8 Link Reporting

How to report power beaming performance should be considered prior to ensuring that measurable parameters have been collected and calculated, estimated parameters have been determined, figures of merit have been found, and other characteristics of interest have been identified. The audience for the reporting should be assessed, as has been previously explored by Nugent [22]. Possible audiences include the following:

- The academic, research, and science/engineering communities.
- Corporate or organizational management.
- Government or other sponsors.
- Industrial partners or potential partners.
- Prospective end-users.
- A general audience.

The options for the form of the reporting itself are similarly broad and may be driven by the target audience selected. Some include the following:

- Journal or conference paper.
- Technical report.
- Chart presentation.
- Marketing literature.
- Video.

Finally, the critical question of the goal of reporting should be considered. Possibilities include the following:

- Disseminating a notable performance result.
- Establishing the achievability of a given aspect of practicality.

Table 7.5. Rudimentary demonstration identification information.

Item	Description
Date (YYYY-MM-DD)	The date the demonstration occurred, or the first day of operation.
Location (city/country)	The location where the demonstration occurred.
Title	A short, descriptive title to distinguish the demonstration from others.
Responsible party	The organization or person responsible for the demonstration.
Contact information	Email, phone, address, or other means of contact for additional information.
Reference(s)	Citable pointer to additional documentation.

- Garnering support for future technological investment.
- Satisfying a contractual requirement.
- Marketing a product or capability.
- Raising general awareness about a technological capability.

The goals of the reporting, the reporting forms selected, and the audiences to be reached will frame the approach. In each case, information that is to be captured describing the link will form the foundation for the documentation. In addition to the technical results, rudimentary identification information should be reported to substantiate the demonstration and provide a conduit for inquiries. An example of elements for basic identification information is seen in Table 7.5.

Some parties that implement power beaming links will have legitimate motivations to limit the disclosure of certain details. Such motivations might include the protection of company proprietary information or national security considerations. The measurements and practices suggested herein endeavor to respect those concerns, while simultaneously permitting both independent verification and clarity on the current state-of-the-art to prospective power beaming users. Ultimately, it is at the discretion of the link implementers as to what they are willing to disclose.

Table 7.6 shows a power beaming link summary (PBLS, pronounced "pebbles") table for reporting basic identifying information, link parameters, and figures of merit.

Table 7.6. Power beaming link summary (PBLS).

Power Beaming Link Summary	Item	Value
	Date (YYYY-MM-DD)	
	Location (city/country)	
	Title	
	Responsible party	
	Contact information	
	Reference(s)	

Parameter	Symbol	Value
Largest transmit aperture dimension	$\varnothing_{TX}$ (m)	
Wavelength	λ (m)	
Link distance	d (m)	
Largest receive aperture dimension	$\varnothing_{RX}$ (m)	
Transmitter power input	$P_{TX-\text{in}}$ (W)	
Transmitter power output	$P_{TX-\text{out}}$ (W)	
Receiver power input	$P_{RX-\text{in}}$ (W)	
Receiver power output	$P_{RX-\text{out}}$ (W)	
Maximum power density	$p_{d\text{-max}}$ (W/m^2)	
Accessible power density	$p_{d\text{-acc}}$ (W/m^2)	
Transmitter mass	m_{TX} (kg)	
Receiver mass	m_{RX} (kg)	
Transmitter volume	V_{TX} (m^3)	
Receiver volume	V_{RX} (m^3)	
Link duration	t (s)	
Beam collection efficiency	η_{BC}	
Path loss	L_P	
Energy delivered	$E_{RX-\text{out}}$ (J)	
Transmitter specific power	SP_{TX} (W/kg)	
Transmitter volume-specific power	VSP_{TX} (W/m^3)	
Receiver specific power	SP_{RX} (W/kg)	
Receiver volume-specific power	VSP_{RX} (W/m^3)	
Transmitter efficiency	η_{TX}	
Receiver efficiency	η_{RX}	
End-to-end link efficiency	η_{Link}	

7.9   Summary of Best Practices for Power Beaming Demonstrations and Link Reporting

1. **Safety first:** Ensure that anyone who may be exposed to the link is cognizant of potential hazards and appropriately trained to understand and mitigate safety risks.
2. **Specify equipment:** Include references to manufacturer documentation and calibration certifications for the link hardware and for the equipment used to characterize it.
3. **Document carefully:** Employ recordkeeping and documentation methods such as configuration management of test procedures and taking notes and photographs.
4. **Claim clearly:** Explicitly prepare for and report the "run for the record" demonstration distinctly from any others, stating all assumptions, demarcations, and definitions.
5. **Verify independently:** Enlist the participation of a dispassionate outside party to independently conduct or verify critical measurements.

7.10   Resources

A simple link example suitable for implementation by students to practice and understand power beaming link characterization methods described in this chapter is included in Appendix A. The measurement reporting table and template for demonstration documentation are freely available at the following links:

Power Beaming Link Summary (PBLS) v1d.pdf. https://drive. google.com/file/d/18OOsBwSkxU3APlsXenY09sqAKOFcLWER/ view?usp=sharing.

Power Beaming Link Summary (PBLS) TEMPLATE v1d.xlsx. https://docs.google.com/spreadsheets/d/19JRu9hwtzxIHzAESo 54VOHldygCfo-4L/edit?usp=sharing&ouid=10020991152972107 7529&rtpof=true&sd=true.

For characterizing microwave links specifically, Wang *et al.* [23] and Shiv [24] have outlined methodologies.

7.11 Conclusion

This chapter has sought to explore the particulars of measuring and characterizing power beaming links in a way that can serve both as an exemplar for seasoned implementers and as an accessible introduction for those new to power beaming. The methods outlined are intended to be extensible to point-to-point power beaming links of virtually any modality or application. Even though the link implementation hardware and the measurement equipment will vary, the underlying principles should still apply.

References

[1] S. Thomas (Aquinas), *The "Summa Theologica."* Burns Oates & Washbourne, 1914.

[2] J. L. Bucher, *The Metrology Handbook,* 2nd edition. ASQ Quality Press, 2012.

[3] Barry N. Taylor and Chris E. Kuyatt, "Guidelines for Evaluating and Expressing the Uncertainty of NIST Measurement Results," National Institute of Standards and Technology, Gaithersburg, MD, NIST Technical Note 1297, September 1994. [Online]. Available at: https://nvlpubs.nist.gov/nistpubs/Legacy/TN/nbstechnicalnote1 297.pdf (accessed July 27, 2020).

[4] R. M. Dickinson and W. C. Brown, "Radiated Microwave Power Transmission System Efficiency Measurements," Jet Propulsion Lab., California Inst. of Tech., Pasadena, CA, Technical Report NASA-CR-142986, JPL-TM-33-727, May 1975. [Online]. Available at: https://ntrs.nasa.gov/search.jsp?print=yes&R=19750018422 (accessed April 25, 2020).

[5] T. He *et al.*, "High-Power High-Efficiency Laser Power Transmission at 100 m Using Optimized Multi-Cell GaAs Converter," *Chin. Phys. Lett.*, vol. 31, no. 10, p. 104203, October 2014. doi: 10.1088/0256-307X/31/10/104203.

[6] J. R. Taylor, "An Introduction to Error Analysis: The Study of Uncertainties in Physical Measurements," in *ASMSU/Spartans.4.Spartans Textbook.* University Science Books, 1997. [Online]. Available at: https://books.google.com/books?id=ypNnQgAACAAJ.

[7] S. Jarvis, J. Mukherjee, M. Perren, and S. J. Sweeney, "On the Fundamental Efficiency Limits of Photovoltaic Converters for Optical Power Transfer Applications," in *2013 IEEE 39th Photovoltaic Specialists Conference (PVSC)*, Tampa, FL: IEEE, June 2013, pp. 1031–1035. doi: 10.1109/PVSC.2013.6744317.

[8] G. Goubau and F. Schwering, "Free Space Beam Transmission," in *Microwave Power Engineering: Generation, Transmission, Rectification*, E. C. Okress, Ed., in Electrical Science Series, Vol. 1. Academic Press, 1968. [Online]. Available at: https://books.google.com/books?id=Obw3BQAAQBAJ.

[9] R. Malarić, *Instrumentation and Measurement in Electrical Engineering*. Boca Raton, FL: BrownWalker Press, 2011.

[10] R. A. Witte, *Electronic Test Instruments*, 2nd edition. Upper Saddle River, NJ: Prentice Hall, 2002.

[11] P. Jaffe *et al.*, "Opportunities and Challenges for Space Solar for Remote Installations," U.S. Naval Research Laboratory, Washington, DC, Memo Report NRL/MR/8243–19-9813, October 2019. [Online]. Available at: https://apps.dtic.mil/sti/pdfs/AD1082903.pdf (accessed April 21, 2020).

[12] N. Shinohara, *Wireless Power Transfer via Radiowaves*, in Iste Series. Wiley, 2014. [Online]. Available at: https://books.google.com/books?id=pJqOAgAAQBAJ.

[13] G. Borgiotti, "Maximum Power Transfer Between Two Planar Apertures in the Fresnel Zone," *IEEE Trans. Antennas Propag.*, vol. 14, no. 2, pp. 158–163, March 1966. doi: 10.1109/TAP.1966.1138660.

[14] G. Goubau, "Microwave Power Transmission from an Orbiting Solar Power Station," *J. Microw. Power*, vol. 5, no. 4, pp. 223–231, 1970.

[15] H. Weichel, *Laser Beam Propagation in the Atmosphere*. Bellingham, WA: The International Society for Optical Engineering, 1990.

[16] K. Ardon-Dryer, Y.-W. Huang, and D. J. Cziczo, "Laboratory Studies of Collection Efficiency of Sub-Micrometer Aerosol Particles by Cloud Droplets on a Single-Droplet Basis," *Atmospheric Chem. Phys.*, vol. 15, no. 16, pp. 9159–9171, August 2015. doi: 10.5194/acp-15-9159-2015.

[17] A. B. W. Kennedy and H. R. Sankey, "The Thermal Efficiency of Steam Engines. Report of the Committee Appointed to the Council upon the Subject of the Definition of a Standard or Standards of Thermal Efficiency for Steam Engines: With an Introductory Note (Including Appendixes and Plate at Back of Volume)," January 1898. doi: 10.1680/imotp.1898.19100.

[18] P. Jaffe and J. McSpadden, "Energy Conversion and Transmission Modules for Space Solar Power," *Proc. IEEE*, vol. 101, no. 6, pp. 1424–1437, June 2013. doi: 10.1109/JPROC.2013.2252591.

[19] G. Goubau and P. D. Coleman, "Beam Waveguides," in *Microwave Power Engineering: Generation, Transmission, Rectification*, E. C. Okress, Ed., in Electrical Science Series, vol. 1. Academic Press, 1968. [Online]. Available at: https://books.google.com/books?id=Obw3BQAAQBAJ.

[20] R. M. Dickinson and O. Maynard, "Ground Based Wireless and Wired Power Transmission Cost Comparison," presented at the *International Energy Conversion Engineering Conference (IECEC)*, Vancouver, British Columbia, Canada, August 1999. [Online]. Available at: http://hdl.handle.net/2014/17841 (accessed April 25, 2020).

[21] P. M. Heathcote, *"A" Level Computing*. Ipswich: Payne-Gallway Publishers, 2001.

[22] T. Nugent, "Uniform Comparisons of Power Beaming Efficiency," presented at the *20th Annual Directed Energy Science and Technology Symposium*, Oxnard, CA, February 2018.

[23] C. Wang, W. Xu, C. Zhang, M. Wang, and X. Wang, "Microwave Wireless Power Transmission Technology Index System and Test Evaluation Methods," *EURASIP J. Adv. Signal Process.*, vol. 2022, no. 1, p. 16, December 2022. doi: 10.1186/s13634-022-00846-7.

[24] Shiv Prasad Tripathy, *RF & μWave Measurements: For Design, Verification and Quality Control*. Independently published, 2019.

Chapter 8

Power Beaming Applications

Celebrating innovation for its own sake is in bad taste. For technology truly to augment reality, its designers and engineers should get a better idea of the complex practices that our reality is composed of.

— Evgeny Morozov [1, p. 13]

8.1 Introduction

Often when a new phenomenon or technology emerges, it is not clear how it might be used or whether it will have any utility whatsoever. Even the confirmation of Maxwell's theory in 1888 by Heinrich Hertz reportedly aroused little excitement initially:

> *Hertz's students were impressed and wondered what use might be made of this marvelous phenomenon. But Hertz thought his discoveries were no more practical than Maxwell's.*
>
> *"It's of no use whatsoever," he replied. "This is just an experiment that proves Maestro Maxwell was right — we just have these mysterious electromagnetic waves that we cannot see with the naked eye. But they are there."*
>
> *"So, what next?" asked one of his students. Hertz shrugged. He was a modest man, of no pretensions and, apparently, little ambition.*
>
> *"Nothing, I guess."* [2, p. 5]

What would Hertz make of our world today, where the electromagnetic spectrum is employed for a panoply of diverse purposes, ranging from communications and navigation to imaging and sensing?

In 1891, Oliver Heaviside remarked somewhat more presciently *Three years ago, electromagnetic waves were nowhere. Shortly afterward, they were everywhere* [2, p. 5].

Still, it would take many years for applications to resolve themselves and enter the realm of practicality. In the case of power beaming, the fundamentals have long been established, but the technology is only beginning to enter mainstream, practical development.

Though much of the historical focus for power beaming development has been on space solar (a concept alternately known as solar power satellites, space solar power, space-based solar power, and many other names), there exist an increasing number of applications that are already being realized. This chapter explores how limitations arising from constraints on current means of power provision and storage could be addressed by power beaming, and what classes of applications are likely to be targeted. Applications are broken into two groups: those for terrestrial and space uses. Within each group, applications are arranged in approximate order of increasing power demand, though such demand could vary widely depending on the context specifics. First, a method for establishing a basis of power beaming's suitability for a given case is explored.

8.2   Suitability Considerations

Power beaming is obviously not appropriate for every application or circumstance. One first-order process for analyzing the characteristics and suitability of an application's energy requirements to be satisfied via wireless energy delivery for a given moment and location is shown in Figure 8.1.

The process shown in Figure 8.1 is far from comprehensive. Though it addresses functional alternatives to power beaming, many interrelated technical, economic, and operating constraints will heavily influence whether power beaming makes sense in a given context. Some of these factors were explored and quantified in Chapters 4 and 7. Determining the thresholds of suitability for the most relevant factors could be decisive in whether power beaming is attractive for

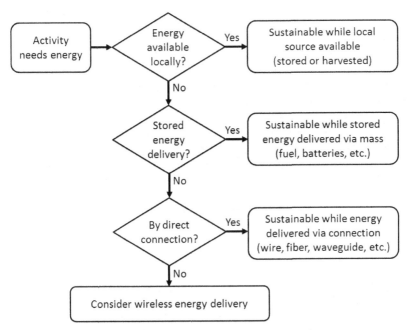

Figure 8.1. Analysis of steps for activities requiring energy at a given time and location.

a particular application. In each case, it will be helpful to estimate the size, weight, power, and likely cost of the system by properly combining the mass-specific, volume-specific, and efficiency metrics for the proposed scenario. Peak and accessible power densities will be important drivers as well. Finally, many of the terms in Figure 8.1 are subjective or situationally dependent (such as "local", "sustainable", and "direct"), leading to some ambiguity when assessing a particular real-world circumstance.

Economic consideration for a given application will include weighing the costs and benefits of satisfying the technical requirements using power beaming and determining market suitability. Key questions include: Is the technology proven? Is it affordable? Is it suitable for production at an appropriate scale? Are there externalities that work for or against it, such as regulatory, political, or public perception challenges? Many of these same questions are similar to those evaluated when considering ways to mitigate or offset climate change, as addressed by Smil [3]. The user's needs and sensitivities could vary widely, whether the application is consumer, commercial, civil, military, scientific, or otherwise. Each may have application-specific metrics or key performance parameters with different weights.

Energy storage deserves special attention in the design and implementation of power beaming systems and in application suitability assessment. Since one of the principal envisioned benefits of using power beaming is often the reduced need for storage, the question of how great that reduction might be is of prime importance. In nearly every scenario, the sizing of storage will be driven by the application itself and the concept of operations. The duration that a system element can be without externally provided power is of key concern, such as in the case when a power beaming link is interrupted.

Factors to consider in storage sizing include the duration of anticipated outages or other unavailability of energy resupply via power beam, limitations on the recharge rate of the storage system, and peak power demands of the application. All of these might affect not just the energy storage elements themselves, but also the entire system's power management and distribution architecture.

8.3 Terrestrial Applications

8.3.1 *Consumer Electronics*

The proliferation of portable consumer electronics has created possibilities for recharging or powering via power beaming. Smartphones, watches, smart speakers, virtual reality headsets, earphones, tablet computers, notebook computers, and other devices are most convenient when they don't need a wired power connection. Most of these have internal energy storage in the form of a battery which needs periodic recharging. Ensuring devices are charged when they need to impose a mental load on those who use or maintain them. Being freed of this imperative will likely increase quality of life and enhance the users' experience. In addition to portable consumer electronics, devices in places that would otherwise require routing of electrical wiring to supply power have been powered via power beaming. Door locks, security cameras, lighting, and motion-activated faucets are a few examples of these devices. Opportunities for deploying power beaming for consumer electronics recharging present themselves in residential, workplace, retail, conference, medical, government, transportation, and hospitality settings.

8.3.2 *Industrial and Commercial Equipment*

In industrial and commercial contexts, there are often collections of distributed devices that require both data connectivity and power. These might be electronic shelf labels in a grocery store, environmental sensors or robots in a warehouse, strain measurement sensors on a bridge, lighting units in a factory, information kiosks in a terminal, telecommunications radios such as for 5G almost anywhere, or a myriad of other devices in similar situations. Examples of Powercast's wirelessly powered, batteryless shelf labels are shown in Figure 8.2.

Historically, many of these devices have had both wired data and power connections, but with the advent of Wi-Fi and other wireless communication standards, the data needs have in many cases been met wirelessly for years. For power, a connection to facility power or a replaceable or rechargeable battery might have been used. Power beaming permits energy to be provided to these devices wirelessly, avoiding the need to run cabling or periodically change numerous batteries, some of which may be difficult to access. This also can

Figure 8.2. Wirelessly charged batteryless electronic shelf labels. Image courtesy and © Powercast. Used with permission.

allow rapid reconfigurability or redeployment of assets within an area served by beamed power without the constraints imposed by wired power distribution, such as outlet availability and proximity. For mobile warehouse robots, power beaming also may obviate the need for charging stations and the corresponding downtime needed for recharging, since robots could be charged via a power beam while in motion. Some companies have elected to focus on these industrial and commercial use cases rather than those for consumer electronics.

8.3.3 *Sensors*

For scientific and military purposes, it is often necessary to deploy sensors to record or relay information about an environment. This application is similar to the industrial and commercial equipment application described above but distinguished by the likelihood that the area to be monitored will be larger than a single building or complex. Sensors may need to operate in vast and harsh outdoor environments in which it is impossible or impractical to employ energy harvesting means because of mission constraints such as concealability or higher power demand that can be effectively satisfied by energy harvesting. Depending on the specifics of the situation, it may be possible to "bathe" an area with a low level of electromagnetic energy that powers many sensors continually over time, in the manner of prototypical energy harvesting scenarios, or to employ a "lawn sprinkler" approach in which sensors are addressed sequentially in a time-shared manner from a single or small number of power transmitters. Limitations stemming from the wavelength used, power density requirements and constraints, and line of sight needs could drive the approach taken. The anticipated required operating time and whether the sensor needs to be retrieved at a future point will factor into the tradeoff study between using power beaming, embedded long-term energy storage, or other means for energy provision.

8.3.4 *Ground and Marine Vehicles*

By virtue of their intrinsic mobility, outdoor vehicles pose a particular challenge for power provision. In most cases, they can't be easily tethered or otherwise provided with energy through direct physical contact. These limitations can be overcome for vehicles operating within confined areas, or instances where rails or catenaries can be used to supply energy. But most of the time, vehicles need to carry their own stored energy source and employ an energy resupply scheme that depends on some form of refueling or recharging by means that requires at least periodic proximity to the resupply source. Power beaming potentially increases this distance dramatically.

Vehicles come in a range of sizes and may carry people or not. Power beaming might apply to situations as varied as delivery robots, passenger vehicles, or trucking. One vision from a US patent for power beaming for passenger vehicle recharging is shown in Figure 8.3.

In addition to the benefits from the scenarios described in previous sections, power beaming could eliminate the need for an outdoor vehicle to return to a specific refueling or recharging station. This could permit it to be resupplied with energy wherever it is, either in situ or on the move. In certain situations, this could have tremendous value. If a robot is contaminated during its work in bomb disposal or reactor cleanup, it might be unsafe, undesirable, or impossible to have it return to an energy resupply point, despite the need to keep it operational. Power beaming could extend indefinitely its ability to operate in a harmful environment.

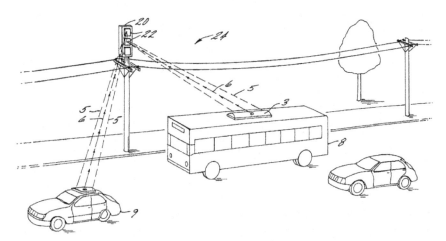

Figure 8.3. A depiction of vehicle recharge via power beaming [4]. *Uncopyrighted image from US patent.*

Autonomous and remotely operated vehicles, especially those that are uncrewed, stand to gain from the possibility of energy replenishment at a distance. Without human operators or passengers, limitations driven by people's need to eat, sleep, and relieve themselves are lessened or eliminated. Vehicles wouldn't need to stop or have onboard facilities to provide for these human needs. With these reduced needs, vehicles could be smaller, and might more easily be employed in groups or swarms.

Application categories for ground and marine vehicles that might benefit from power beaming include agriculture, communications, security, surveillance (for wildlife, environmental, infrastructure, military, law enforcement, mapping, prospecting, disaster response, search and rescue, photography, or other monitoring purposes), transportation, logistics, marketing, and entertainment.

8.3.5 *Aircraft*

Flying vehicles present a profound opportunity for power beaming, since in most cases they require significant and continuous energy to stay aloft. Because they are often accessible via direct lines of sight, power beaming offers a potentially compelling means of energy resupply. One vision of how this could be realized is shown in Figure 8.4.

Some hardware demonstrations of power beaming to aircraft can be found in [5–7].

The attractiveness of power beaming for flying vehicles is likely to be driven by the application constraints and limits of the technology. For stationary and quasi-stationary air platforms, the demands on a ground transmitter for tracking will be reduced, although the receiver-specific power requirements increase due to the inefficiency of hovering. If the application requires all-weather operation, longer electromagnetic wavelengths that penetrate rain could be employed. However, these would either need larger transmit and receive apertures or would have limited range versus shorter distances. For fast-moving and dynamic platforms, transmitter tracking of the receiver on the platform would be an important consideration. While the tracking challenge has been solved in directed energy contexts, it requires expensive and precise hardware. Tracking for power beaming is a more tractable problem in some ways since the receiver can potentially provide feedback and host tracking aids.

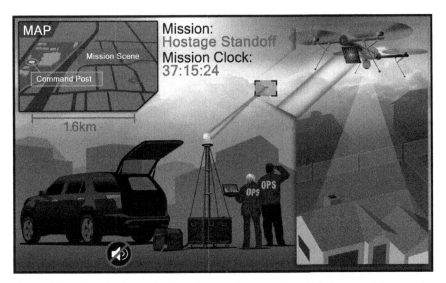

Figure 8.4. Power beaming for a law enforcement application requiring persistent drone coverage. Image courtesy and © PowerLight Technologies. Used with permission.

Air vehicles share many of the advantages, challenges, and opportunities as ground and marine vehicles as explored in the previous section. The need to remain aloft has implications for each of these. Air vehicles can be less concerned about terrain, but must either have buoyancy, suitable aerodynamics, or expend considerable energy to remain in the air. These can entail demands on stored energy sources, like batteries and fuel supplies. In turn, this can cascade into constraints on payload capacity, flight performance, and flight duration. In circumstances where power beaming can relieve these constraints, applications become unbound in new and profound ways.

For instance, imagine a delivery system in which drones never need to land and can form a sort of continuous "air conveyor" or dynamic web for shipping logistics. Material, supplies, cargo, or shipments could be launched by a catapult or similar device to a flying drone or drones and could then bring it to the desired delivery destination. A line or swarm of such drones could rapidly move material from place to place or disperse it as needed. Depending on the number of power beaming stations needed to supply such a network, they might be positioned in conjunction with existing infrastructure that already has established access to electricity, such as power lines or cell towers. In-flight recharging might be done piecemeal along the delivery route, or in demarcated zones.

The applications for air vehicles, especially for those that are autonomous and uncrewed, include those described in previous sections but expand to address them in ways that simply aren't easily achievable without the benefit of air mobility. The clear byways of the sky could simplify navigation and transcend impediments arising from the terrain. However, the challenge of air traffic control may be significant, and regulatory hurdles will likely need to be overcome.

8.3.6 *Power Distribution*

Terrestrial power distribution employing power beaming has been proposed for both the infrastructure-disadvantaged developing world and for developed countries with circumstances adverse to conventional power distribution. Power beaming links for electrical grid distribution have been proposed for locations in Africa [8], Canada [9], and New Zealand [10]. These types of links could offer the advantages of rapid reconfigurability and resilience to environmental and geographical challenges that make using power lines unattractive. In most non-emergency circumstances, the suitability of power beaming links as a viable alternative to power lines will likely hinge on cost factors. New Zealand's Emrod has explored a number of utility power use cases, including for enhancing grid resilience, distribution to and within islands, "last mile" distribution, outage response, cellular base station power, and mining and remote facility power [11].

To date, most of the proposed power beaming links for carrying grid-scale amounts of power employ microwaves. Rain and fog concerns associated with optical power beaming dampen enthusiasm for using grid-scale links, though its performance under typical conditions for many locations can be excellent. In either case, close attention needs to be given to safety because of the large amounts of power and likely high-power densities involved.

Conventional transmission and distribution lines address safety in large part by elevating wires on poles or burying them underground. Elevation of a power beaming link might be one element of a safety system but would almost certainly need to be accompanied by other means. A sensing system to monitor the volume over which the transmission occurs and the surrounding area would likely be needed since the power levels for this application are almost certain to exceed exposure limits. Such safety systems have already been implemented [12] and have been discussed in more detail in Chapter 4. A key difference between power line safety scenarios and those for power beaming is that there would not be the intrinsic visual cue of a physical power line to alert those nearby to the potential hazard. Indeed, power line safety standards can be based on the distance from the power line [13], whereas in a power beaming transmission circumstance, the mere presence of the beam might be difficult to detect, much less the distance from it.

8.4   Space Applications

On Earth, the presence of the atmosphere and its characteristic features, like weather, must almost always be considered in the design, deployment, and use of a power beaming link. This typically makes transmission more challenging, especially for optical power beaming. In space, the lack of atmosphere can radically reshape or eliminate considerations for power beaming, as the terrestrial challenges of scattering and losses from the weather are absent. In this section, applications are arranged notionally in approximate order of increasing likely link distance, though particular application cases may vary in their characteristics. The proposed nomenclature for space power beaming link distances is shown in Table 8.1.

8.4.1   *Lunar and Planetary Body Applications*

The Moon, planets, and similar bodies often present distinct surfaces on which operations can be conducted. On the Earth's surface,

Table 8.1.   *Nomenclature for space power beaming link distances* [14].

Link distance	Link distance range	Example application
"Short"	$d \leq 0.01\,\text{km}$	Inter- or intra-satellite power links
"Medium"	$d \leq 100\,\text{km}$	Lunar power beaming networks
"Long"	$d \leq 100{,}000\,\text{km}$	Solar power satellites
"Very long"	$d > 100{,}000\,\text{km}$	Beamed energy propulsion

power distribution is typically accomplished via wires and grid infrastructure. The characteristics of the demands of power distribution on those planetary-like surfaces beyond the Earth may be different enough that power beaming would offer compelling advantages over traditional approaches in many circumstances.

It is anticipated that intensive operations in craters of the Moon, especially those in permanent shadow, will require large amounts of power. This power might be consumed by mobile platforms, like rovers, that are prospecting for or mining water ice, minerals, or other commodities [15]. Options for powering these platforms include using stored energy in batteries or fuel cells, nuclear sources, reflected solar power, power provided via tether, and power beaming. Some of these options have been explored by Landis for over 30 years [16–18]. Each has advantages and disadvantages. Challenges are compounded by the fact that the Moon's day and night periods each last approximately 2 weeks. Thermal considerations are of special concern since the measured temperature on the Moon ranges from 120° centigrade at the lunar equator during the day to −250° centigrade on crater floors at night [19]. Maintaining suitable operational temperatures for electronics and equipment is likely to require considerable expenditure of energy. To reduce the mass and volume of a given platform, it may be desirable to deliver energy to it instead of incorporating the source in it directly. While tethers are one option for doing this, they are likely to have a loss and require management of cabling to avoid snags and breaks resulting from lunar surface features. In circumstances where a clear line-of-sight along the link path exists, power beaming may be the most attractive option, as it would likely offer flexibility in the provided power density that reflected sunlight could not. However, important questions about challenges posed by operations in the lunar environment, including dust buildup and regolith effects, need to be investigated and resolved. NASA has funded efforts that have begun exploring some of these through the Watts on the Moon and BIG Idea Challenges [20].

Having the transmitter and receiver of a power link both attached to the same lunar or planetary surface might simplify pointing and closed-loop link control, though seismic activity or changes in the surface due to thermal variation or settlement activities could introduce challenges. Depending on the circumstances, it may be practical or necessary to decouple one or both of the transmitter and receiver from the surface. Other scenarios include having either the transmitter or receiver in orbit or otherwise in motion around the body, with the other end of the link on the surface. This would levy increased pointing complexity, likely mostly on the transmitter side, to effectively maintain the power beaming link. Putting the transmitter in orbit could allow it greater access to sunlight versus the surface.

An interesting application for orbit-to-surface power beaming was explored by Costanzo *et al.* [21]. The authors examined how the European Space Agency's Rosetta spacecraft might have sent power from orbit to the Philae probe, which had landed on Comet 67P/Churyumov–Gerasimenko, but in a shaded region in which it could not collect sunlight to recharge. Had the orbiting spacecraft been able to transmit power to the probe on the surface of the comet, its mission might have been extended substantially.

The Moon, comets, and asteroids are not the only places where power beaming might play a role in space exploration. Power satellite systems similar to those long proposed for Earth have also been proposed for Mars [22], and novel exploration architectures employing power beaming have been proposed for Venus [23]. Kare envisioned an extensive optical power beaming infrastructure that could serve the bulk of cislunar space [24].

8.4.2 *Between Spacecraft*

Without an atmosphere or terrain features to interfere with line of sight, the appeal of beaming power between spacecraft in space has enticed technologists for many years. In a 2006 paper that explored the possible value of fractionated space architectures, Brown and Eremenko discussed both microwave and laser power beaming as potential means for sharing energy between spacecraft [25]. Prior to that, Bekey devoted an entire chapter in his landmark and visionary 2003 book *Advanced Space System Concepts and Technologies* [26] to *Power and Energy Beaming Concepts*. Even earlier, in 1993, the ISY-METS experiment demonstrated microwave power beaming in suborbital space. Given the years that have passed since the recognition of these ideas, one might reasonably ask why there hasn't been greater employment of power beaming between spacecraft.

Perhaps the fundamental obstacle to power beaming between spacecraft is that in nearly every scenario of potential interest, the sun presents a more attractive alternative as an energy source than a power beam. This is for several reasons:

(1) Sunlight is provided without cost and complexity, unlike a power beaming transmitter.
(2) For missions that operate at a distance closer to the sun than the orbit of Mars, the power density of sunlight is likely sufficient to meet most objectives.

For a spacecraft that relies solely on receiving beamed power versus using sunlight, there is a reduction in robustness and fault tolerance. If something goes wrong with a conventional spacecraft that causes a loss of orientation control, its solar array or arrays might be positioned so as to keep at least a minimum of energy flowing into the power system. For a spacecraft that relies on beamed power, the options may be more limited. If something has gone wrong with the transmitter, there's not much the receiver can do. Even if the transmitter is operating normally, the receiver will still need to be able to orient itself properly within the power beam to keep energy flowing into the power system.

Certainly, there do exist scenarios in which power beaming between spacecraft might be a more attractive option. For instance:

(1) An instance where a spacecraft needs power when it is in shadow.
(2) Circumstances where sunlight doesn't provide a high enough power density, or where having solar arrays of the required size would impede mission success.

Because power beaming could permit much greater control over the available power density, it could be attractive for powering very small spacecraft.

8.4.3 *Space Solar*

Though included in the space applications section of this chapter, space solar as usually conceived spans space and terrestrial domains. It is classified here as a space application because a significant amount of the infrastructure for implementation would be in space.

Space solar is inextricably tied to power beaming, both conceptually and historically. As noted in the introduction to this chapter, it has gone by many names, but in nearly every case it refers to the collection of sunlight in space for utilization on Earth. In almost every instance, power beaming is how the energy is delivered from space to the Earth. An overview of the history of space solar's development is largely covered in Chapter 2 as part of the history of power beaming technology. Many studies and other major works exploring space solar's history, concepts, and feasibility have been published over the years. A partial listing appears in Table 8.2. Additional publications can be found in the National Space Society's Space Solar Power Library [27].

Examinations of space solar have addressed issues related to technology, economics, societal impacts, environmental effects, legal implications, and other topics. The considerations for each of these areas change depending on the particulars of the technology employed and the application context. For instance, supplying energy via space solar with optical power beaming to a remote place devoid of infrastructure is likely to be very different than supplying energy to a national power grid using microwave power beaming.

The aspects of the power beam to be used will have a huge effect on the issues listed above. The power density at the receiver site is of great interest, as it will have many implications for safety and economic feasibility. Section 5.5.3 includes several proposed power beaming links for space solar and their associated parameters.

The space segments used to create the power beams vary widely, and dozens of different architectures for space solar have been proposed. The Solar Power Satellite by means of Arbitrarily Large PHased Array (SPS-ALPHA) concept has been developed by John Mankins, and is detailed in his book *The Case for Space Solar Power* [40]. A key feature of the SPS-ALPHA design and other relatively recent approaches are their extensive use of modular elements, which could be manufactured with mass-production methods, reducing the system's cost. One version of the SPS-ALPHA space segment is shown in Figure 8.5.

Another proposed architecture of note is Ian Cash's Constant Aperture, Solid-State, Integrated, Orbital Phased Array (CAS-SIOPeiA) concept, detailed in several papers [50,51]. It addresses the need to collect sunlight from one direction while emanating a power beam from a different direction with novel phased array elements and is pictured in Figure 8.6.

Table 8.2. Selected major and summary works concerning space solar.

Year	Work	Type	Cite
1980	The Final Proceedings of the Solar Power Satellite Program Review	Report	[28]
1981	Electric Power From Orbit: A Critique of a Satellite Power System	Book	[29]
1981	Space Power	Book	[30]
1981	Solar Power Satellites	Book	[31]
1995	Sun Power: The Global Solution for the Coming Energy Crisis	Book	[32]
1998	Solar Power Satellites: A Space Energy System for Earth	Book	[33]
2001	Laying the Foundation for Space Solar Power	Book	[34]
2007	Report of the URSI Inter-Commission Working Group on SPS	Report	[35]
2007	Space-Based Solar Power As an Opportunity for Strategic Security	Report	[36]
2009	Space-based Solar Power: Possible Defense Applications and Opportunities	Report	[37]
2011	Space Solar Power: The First International Assessment of Space Solar Power	Book	[38]
2012	Solar Power Satellites	Book	[39]
2014	The Case for Space Solar Power	Book	[40]
2015	Space-Based Solar Power	Book	[41]
2015	Space-Based Solar Power: A Technical, Economic, and Operational Assessment	Report	[42]
2019	Opportunities and Challenges for Space Solar for Remote Installations	Report	[43]
2020	Introduction to Space Solar Power Plant (in Chinese)	Book	[44]
2020	Space-Based Solar Power: A Near-Term Investment Decision	Report	[45]
2021	Catching the Sun: A National Strategy for Space Solar Power	Report	[46]
2021	Space Based Solar Power: De-Risking the Pathway to Net Zero	Report	[47]
2022	Space-Based Solar Power: A Future Source of Energy For Europe?	Report	[48]
2022	Space-Based Solar Power: Can It Help to Decarbonize Europe and Make It More Energy Resilient?	Report	[49]

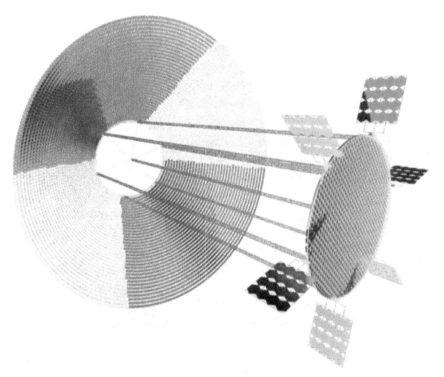

Figure 8.5. Spacecraft from an SPS-ALPHA solar power satellite system. Sunlight reflectors are shown on the structure on the left, and solar cells and microwave conversion occur in the disc-shaped structure on the right. Image courtesy and © John C. Mankins. Used with permission.

A selection of additional architectures also appears in Appendix D of [43], and new concepts emerge with some regularity. Some concepts employ microwave power beaming, some employ optical power beaming, and some use simple reflection of sunlight. Sunlight reflection has been investigated by Ehricke [52], Fraas and O'Neill [53], and Çelik [54] and is currently being pursued by Nowack's Reflect Orbital [55]. The original space reflector concept appears in Oberth's 1929 classic *Wege zur Raumschiffahrt* [56, p. 481], and is insightfully and humorously explored by Houghton [57]. Laser space solar approaches are summarized by Summerer and Purcell [58], including those employing solar-pumped lasers. Hybrid optical-microwave approaches have also been suggested [59,60, p. 46].

Figure 8.6.   A CASSIOPeiA solar power satellite system. Upper and lower reflectors redirect sunlight to the central conversion structure, which can steer the power beam to a ground receiver as the spacecraft orbits. Image courtesy and © Ian Cash. Used with permission.

Notable advantages of energy from space solar are that it could be clean, constant, unlimited, and globally transmissible. The degree to which these advantages were realized would depend in part on the implementation method and its effectiveness. It is the attraction of these features that has captured the attention and imagination of researchers, enthusiasts, and leaders for decades. However, the challenges of economic feasibility, regulatory and public perception readiness, along with technology immaturity have tempered progress and support for the concept in many circles. Expositions of both the opportunities and challenges are summarized in [43, Sec. 7.1] and [43, Sec. 7.2], respectively.

Perhaps the most decisive parameter for informing economic feasibility is specific power, expressed as watts transmitted from the space segment per kilogram of space segment. If a great amount of power can be transmitted by employing a small amount of mass, less mass will need to be put in space, and the cost of launch is reduced. Similarly, if the cost of the actual hardware to be put in space can be reduced via mass production, the overall cost of the system is likely to be lower as well. Thresholds and trends for these parameters and others are developed in appendices C and J of [43], and a generalized cost model is proposed in appendix M. Revolutionary changes in access to space, like the successful development and deployment of a functional and practical space elevator, could dramatically alter the space segment emplacement cost for space solar.

From a regulatory, safety, and public perception standpoint, discussions have focused on human exposure to energy in the power beam, especially for those who work at or near the receiver. While receivers could be designed to accept power densities within human exposure limits, this would necessitate large collection areas for large amounts of power, which might not be attractive in every scenario. Active safety systems using interlocks like those described in Chapter 4 have been successfully demonstrated at smaller scales and have yet to be operated on a space to ground power beaming link, but they may offer a path toward increasing the power density at the receiver. Effects on wildlife and aircraft will vary with the power beam implementation and would need to be addressed. The impact of reradiated harmonic energy arising from the rectenna arrays or reflected optical energy would need consideration, as would the prospect of interference with electronics and imaging devices. These and other factors would feed into the site selection criteria for the receiver.

A concern held by some astronomers is that solar power satellites would be of sufficient size to confound some astronomical observations, including those in the visible, infrared, and microwave portions of the spectrum. Another possible concern for astronomers is that a large object that reflects sunlight and has a significant apparent magnitude will appear in the night sky. This is likely a realistic concern, especially since a full-scale system could call for a constellation of very large satellites, each with a potentially significant apparent magnitude.

Many of the works from Table 8.2 include roadmaps outlining approaches to the numerous technology areas, political and regulatory hurdles, and economic challenges that need to be addressed for space solar to become a reality. As power beaming technology matures, it is likely that roadmaps will need to be evolved and iterated accordingly.

The underlying risk in ignoring space solar is that if a nation or corporate entity masters the technology in a cost-effective way, that country could be positioned to become an electricity provider for the world. As global per capita demand for energy continues to grow, this would be a geopolitically attractive position to occupy.

For further expositions on this expansive topic, the reader is urged to consult the resources in Table 8.2 or Chapter 24 of *Future Energy* [61].

8.4.4 *Beamed Energy Propulsion*

The challenge inherent in the "tyranny of the rocket equation" [62] in which increasing the amount of payload to be sent into space also dramatically increases the amount of propellant needed has motivated the investigation of other means of propulsion. One prospective method is to use a power beam, either to directly push on a spacecraft via radiation pressure or to provide energy to excite a propellant or otherwise power a propulsion mechanism. This section explores some scenarios where this method might be used.

8.4.4.1 *Earth to Orbit*

One of the greatest challenges in spaceflight is successfully and economically getting from the surface of the Earth to Low-Earth Orbit (LEO). This step of getting to LEO has been characterized as getting "halfway to anywhere" by the famed science fiction author Robert Heinlein [63]. To date, humans have only achieved establishing LEO by using rockets that employ chemical propulsion, and only with NASA's space shuttle in 1981 [64] were major portions of an orbital launch system reused. It took over three and a half more decades before meaningful progress was made in developing orbital launch systems with significant reusability when SpaceX successfully reused and recovered for a second time a Falcon 9 rocket in 2017 [65]. By July 2022, SpaceX had reused boosters on more than 100 occasions and had several boosters that had flown a dozen or more times [66].

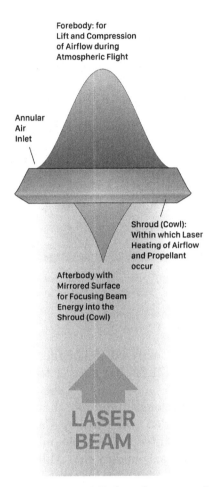

Figure 8.7. Myrabo's laser powered Lightcraft concept. Image by Tokamac — Own work, CC BY-SA 4.0, https://commons.wikimedia.org/w/index.php? curid=68054922.

Both optical wavelengths and millimeter waves have been investigated for Earth-to-Orbit propulsion [67,68]. Work by Myrabo *et al.* [69] culminated in October 2000 with an outdoor demonstration at White Sands Missile Range in New Mexico that employed a 10-kW pulsed carbon dioxide laser to propel 12.2-cm diameter craft over 70 m upwards into the air. The concept of operation of Myrabo's Lightcraft concept is shown in Figure 8.7.

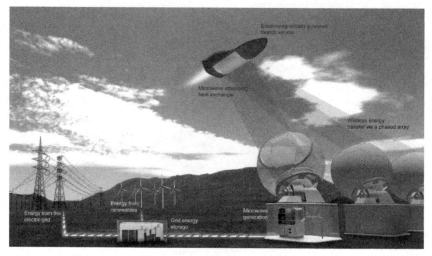

Figure 8.8. Overview of Escape Dynamics' planned microwave launch concept. Image courtesy and © Escape Dynamics. Used with permission.

Laser launch has been analyzed and further explored subsequently by Kare [70] and others [71,72] but has yet to be demonstrated at a larger scale than Myrabo.

On the longer wavelength end of the spectrum, an extensive DARPA/NASA-funded effort led by Carnegie Mellon University's Kevin Parkin resulted in notable millimeter-wave beamed propulsion demonstrations under the Millimeter-wave Thermal Launch System (MTLS) program, described in detail in [73]. The company Escape Dynamics also sought to use millimeter waves beamed from the ground to heat hydrogen in a vessel aboard a launcher to achieve orbit. An overview of the concept is shown in Figure 8.8.

Prior to suspending operations in 2015, Escape Dynamics reported performing testing that showed specific impulse results exceeding those possible with chemical rockets [68].

For further information on beamed energy propulsion, readers are urged to consult the forthcoming book by Leik Myrabo to be published by Elsevier, and Philip Lubin's *The Path to Transformational Space Exploration* [74].

Figure 8.9. Skylon launcher with enhanced propulsion from space-based lasers as conceived by Keith Henson. Base artwork by Reaction Engines, Ltd., modified by Barbara Graham, laser additions by Anna Nesterova. Used with permission.

8.4.4.2 Orbit Raising, Maneuverability, and In-space Propulsion

Many of the challenges of Earth-to-orbit launch with beamed power are absent or reduced when the spacecraft is already in space, such as the need to provide a very large amount of thrust for long enough to leave the Earth's atmosphere. The power transmitter might still be terrestrially based, or it could also be space-based. Henson has proposed power beaming from space to Reaction Engines' Skylon launcher while it is in orbit to help provide the propulsion needed to transport materials for building solar power satellites, as depicted in [75] (see Figure 8.9).

As the importance of cislunar space has gained recognition, the use of power beaming for energy distribution and propulsion enhancement has also garnered attention. The Cislunar Lighthouse Access and Moon Power (CLAMP) concept envisions creating resilient lunar transportation lines and addressing lunar outpost energy needs by

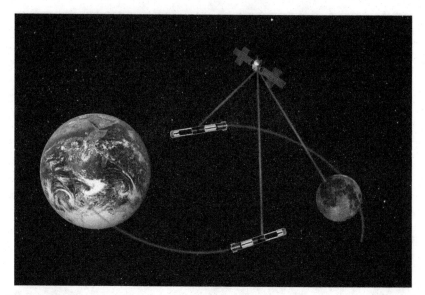

*Figure 8.10.   The Cislunar Lighthouse Access and Moon Power (CLAMP)
concept for providing beamed power for in-space propulsion and lunar usage.*

employing a power beaming architecture with a transmission station
at the Earth-Moon L4 or L5 stable equilibrium Lagrange point. An
overview of the concept, not to scale, is shown in Figure 8.10.

Cislunar space and within the solar system are only the beginning
of concepts for power beaming for energy distribution and propulsion.
The next section explores using this technology over much greater
distances.

8.4.4.3   *Interstellar Travel*

The enormous distance between the sun and nearby stars makes the
possibility of sending a probe to explore them extremely daunting.
Even a tiny probe would likely require immense amounts of energy
to accelerate to a speed that would permit completing the journey
within the span of a human lifetime. The prospect of carrying the
source of this energy on board is intimidating since any additional
mass would, in turn, require more mass to propel it, as described
at the beginning of this section. As a result, concepts for interstellar
travel have been proposed and developed to address these challenges.

As of 2022, few flight missions have truly started exploring this concept. Interest and enthusiasm were stimulated following JAXA's successful IKAROS mission in 2010, in which a solar sail was used for interplanetary travel to Venus [76], demonstrating the use of off-board energy for propulsion. In 2011, NASA and DARPA funded an extensive study on beamed energy propulsion that examined a range of potential applications, including those that advanced technology for interstellar travel [77]. Several long-standing concepts, such as SailBeam employing laser power beaming [78] and Starwisp using microwave power beaming continue to be of periodic interest [79]. More recent reviews by Bae [80,81], and Lubin [82] lay out potential paths forward. One current effort is the Breakthrough Starshot Initiative [83], which envisions "a ground-based light beamer pushing ultra-light nanocrafts" to "reach Alpha Centauri in just over 20 years from launch".

8.5 Companies

At the time of this writing, numerous start-up companies have been working for many years to bring power beaming to a range of applications. Some of these are listed in Table 8.3. This tabulation does not include governmental, academic, or non-startup corporate entities which are also doing work in this area, nor the numerous start-up companies that were engaged in work in different power beaming applications but that became defunct prior to October 2023.

A comprehensive listing of companies involved with power beaming for space applications can be found at https://www.factor iesinspace.com/space-utilities.

8.6 Standards

Much as short-distance wireless power has enjoyed widespread usage because of the adoption of standards, like Qi [84], similar campaigns have been undertaken for the standardization of power beaming. Among them is the AirFuel RF^{TM} [85] effort. Standards have historically played a pivotal role in the establishment of a technology's uptake. Examples like the Universal Serial Bus (USB),

Table 8.3.   *Power beaming application area start-up companies.*

Company name	Principal application area(s)	Modality	Website
Ossia	Consumer electronics	Microwave	www.ossia.com
Wi-Charge	Consumer electronics, commercial	Optical	www.wi-charge.com
Aquila	Infrastructure	Optical	www.aquila.earth
Energous	Consumer electronics, industrial	Microwave	energous.com
GuRu Wireless	Consumer electronics	Microwave	guru.inc
Reach	Industrial electronics	Microwave	reachpower.com
Aerocharge	Consumer electronics, industrial	Microwave	www.aerocharge.co
MH GoPower	Sensors, OEM	Optical	www.mhgopower.com
Powercast	Sensors, RFID, IoT	Microwave	www.powercastco.com
PowerLight Technologies	Telecom, defense	Optical	powerlighttech.com
VanWyn	Commercial, defense	Microwave	vanwyn.com
Emrod	Utility power distribution	Microwave	emrod.energy
Virtus Solis Technologies	Space solar	Microwave	www.virtussolis.space
Solar Space Technologies	Space solar	Microwave	www.solarspacetechnologies.com.au
Solaren	Space solar	Microwave	www.solarenspace.com
Overview Energy	Space solar	Undisclosed	overviewenergy.com
Reflect Orbital	Space solar	Reflection	www.reflectorbital.com

Bluetooth®, and Wi-Fi are but a few instances where standards for data connectivity were crucial in empowering stakeholders to specify, produce, and market technology. Though power standards have typically played a less visible role in recent decades, they too have often been instrumental in promulgating technology. The National

Electrical Manufacturers Association (NEMA) in North America and the Institute of Electrical and Electronics Engineers (IEEE) maintain collections of power standards in widespread use. It is likely that wide adoption of a power beaming standard will accelerate fielding and usage of the technology, especially if used in conjunction with other power standards. In instances where competing standards first resulted in a protracted battle for market dominance, such as with VHS and Betamax [86], or more recently with CHAdeMo versus CCS for electric vehicle fast charging [87], technology deployment was arguably delayed and consumers had greater hesitation and uncertainty in adoption. Coalescing around a standard or small group of standards that address different applications could be valuable in speeding power beaming technology acceptance. Such standards would likely be well-served by maximizing the use of existing safety and interface standards to the greatest extent practicable.

8.7 Conclusion

While power beaming is clearly not suited for every situation, there are numerous applications in which it offers advantages unattainable or difficult to achieve by other means. A survey of these has been reviewed in this chapter, ranging from those over short indoor distances at low power levels to interplanetary distances at huge power levels. In each case, the appropriateness of employing power beaming should be weighed against the available alternatives. It is likely that the number of new contexts in which power beaming can provide compelling value will increase as the technology further matures.

References

[1] E. Morozov, *To Save Everything, Click Here: The Folly of Technological Solutionism*, First edition. New York: PublicAffairs, 2013.

[2] A. Krawczyk and A. Byliniak, "Some Remarks on Life and Achievements of Heinrich Rudoplh Hertz on His 150th Birth Anniversary," in *Electromagnetic Field, Health and Environment Proceedings of EHE'07*. Amsterdam: IOS Press, 2008.

[3] V. Smil, "It'll Be Harder Than We Thought to Get The Carbon Out [Blueprints for a Miracle]," *IEEE Spectrum*, vol. 55, no. 6, pp. 72–75, June 2018. doi: 10.1109/MSPEC.2018.8362233.

[4] R. J. Parise, "Remote Charging System for a Vehicle," 5,982,139, November 9, 1999 [Online]. Available at: https://patentimages.storage. googleapis.com/70/a7/15/34fa03572fd52d/US5982139.pdf (accessed August 22, 2022).

[5] N. Kawashima and K. Takeda, "Laser Energy Transmission for a Wireless Energy Supply to Robots," in *Robotics and Automation in Construction*, 2008, p. 8. [Online]. Available at: https://www. intechopen.com/books/robotics_and_automation_in_construction/laser_ energy_transmission_for_a_wireless_energy_supply_to_robots (accessed July 02, 2020).

[6] T. Nugent, "Review of Laser Power Beaming Demonstrations by PowerLight Technologies (formerly LaserMotive)," presented at the *20th Annual Directed Energy Science and Technology Symposium*, Oxnard, CA, February 2018.

[7] Z. Tong, "Chinese Scientists Develop Laser-Powered Drone to Stay Aloft 'Forever,'" *South China Morning Post*. [Online]. Available at: https://www.scmp.com/news/china/science/article/3205885/chinese- scientists-develop-laser-powered-drone-stay-aloft-forever (accessed January 14, 2023).

[8] A. Douyère, G. Pignolet, E. Rochefeuille, F. Alicalapa, J. L. S. Luk, and J. Chabriat, "'Grand Bassin' Case Study: An Original Proof-of- Concept Prototype for Wireless Power Transportation," in *2018 IEEE Wireless Power Transfer Conference (WPTC)*, June 2018, pp. 1–4. doi:10.1109/WPT.2018.8639227.

[9] K. A. Carroll, "Feasibility of Supplying 1 GW of Power from Labrador to Newfoundland via Microwave Power Beaming Back- ground: Space Based Solar Power," Unpublished, 2009. [Online]. Available at: http://rgdoi.net/10.13140/2.1.1857.3445 (accessed November 24, 2020).

[10] M. D. P. Emilio, "EETimes — Emrod Enables Nikola Tesla's Dream," EETimes. [Online]. Available at: https://www.eetimes.com/emrod- enables-nikola-teslas-dream/ (accessed May 13, 2021).

[11] "Wireless Power Use Cases | Emrod," Emrod Energy. [Online]. Avail- able at: https://emrod.energy/use-cases/ (accessed January 13, 2023).

[12] T. J. Nugent, Jr., David Bashford, Thomas Bashford, Thomas J. Sayles, and Alex Hay, "Long-Range, Integrated, Safe Laser Power Beaming Demonstration," in *Technical Digest OWPT 2020*. Yokohama, Japan: Optical Wireless Power Transmission Committee, The Laser Society of Japan, April 2020, pp. 12–13.

[13] "1926.1408 — Power Line Safety (up to 350 kV) — Equipment Operations." *Occup. Safety Health Admin.* [Online]. Available at: https://www.osha.gov/laws-regs/regulations/standardnumber/1926/1926.1408 (accessed May 11, 2022).

[14] P. Jaffe, "Power Beaming and Space Applications," presented at the *3rd Optical Wireless Power Transmission (OWPT) Conference*, 2021.

[15] J. Grandidier *et al.*, "Laser Power Beaming for Lunar Night and Permanently Shadowed Regions," *Bulletin of the AAS*, vol. 53, no. 4, March 2021. doi: 10.3847/25c2cfeb.13f46900.

[16] G. A. Landis, "Solar Power for the Lunar Night," NASA-TM-102127, May 1989.

[17] G. A. Landis, "Applications for Space Power by Laser Transmission," presented at the *OE/LASE '94*, J. V. Walker and E. E. Montgomery IV, Eds., Los Angeles, CA, May 1994, pp. 252–255. doi: 10.1117/12.174188.

[18] G. A. Landis, "Laser Power Beaming for Lunar Polar Exploration," in *AIAA Propulsion and Energy 2020 Forum*, VIRTUAL EVENT: American Institute of Aeronautics and Astronautics, August 2020. doi: 10.2514/6.2020-3538.

[19] "Lunar Reconnaissance Orbiter: Temperature Variation on the Moon." NASA. [Online]. Available at: https://lunar.gsfc.nasa.gov/images/lithos/LROlitho7temperaturevariation27May2014.pdf (accessed May 17, 2022).

[20] J. St. Martin *et al.*, "Beaming of Energy via Laser for Lunar Exploration (BELLE)," University of Virginia, November 2020.

[21] A. Costanzo *et al.*, "Could the Space Probe Philae© Be Energized Remotely?," *Wirel Pow Transfer*, vol. 6, no. 2, pp. 154–160, September 2019. doi: 10.1017/wpt.2019.5.

[22] A. Baraskar, Y. Yoshimura, S. Nagasaki, and T. Hanada, "Space Solar Power Satellite for The Moon and Mars Mission," *J. Space Safety Eng.*, vol. 9, no. 1, pp. 96–105, March 2022. doi: 10.1016/j.jsse.2021.10.008.

[23] E. J. Brandon, R. Bugga, J. Grandidier, J. L. Hall, J. A. Schwartz, and S. Limaye, "Power Beaming for Long Life Venus Surface Missions," California Institute of Technology, Pasadena, CA, USA, NIAC Phase I Final Report, 2019. Accessed: May 19, 2022. [Online]. Available: https://www.nasa.gov/wp-content/uploads/2019/04/niac_2019_phi_brandon_powerbeaming_tagged.pdf?emrc=97da3d.

[24] J. T. Kare, "Laser Power Beaming Infrastructure for Space Power and Propulsion," presented at the *Defense and Security Symposium*, V. Korman, Ed., Orlando (Kissimmee), FL, May 2006, p. 62220W. doi: 10.1117/12.671943.

[25] O. Brown and P. Eremenko, "The Value Proposition for Fraction-ated Space Architectures," in *Space 2006*. San Jose, CA: American Institute of Aeronautics and Astronautics, September 2006. doi: 10.2514/6.2006-7506.

[26] I. Bekey, *Advanced Space System Concepts and Technologies, 2010–2030+*. El Segundo, CA: Reston, VA: Aerospace Press; American Institute of Aeronautics and Astronautics, 2003.

[27] "Space Solar Power — Library," National Space Society. [Online]. Available at: https://space.nss.org/space-solar-power-library/ (accessed April 11, 2020).

[28] "The Final Proceedings of the Solar Power Satellite Program Review," July 1980. [Online]. Available at: https://space.nss.org/wp-content/uploads/1981-DOE-SPS-Final-Proceedings-Of-The-Solar-Power-Satellite-Program-Review.pdf (accessed August 26, 2022).

[29] *Electric Power From Orbit: A Critique of a Satellite Power System*. Washington, DC: The National Academies Press, 1981. doi: 10.17226/19663.

[30] G. H. Stine, *Space Power*. New York: Ace Books, 1981.

[31] *Solar Power Satellites*. Washington, DC: U.S. Government Printing Office, 1981. [Online]. Available at: https://space.nss.org/wp-content/uploads/1981-OTA-Solar-Power-Satellites.pdf (accessed August 26, 2022).

[32] R. Nansen, *Sun Power: The Global Solution for the Coming Energy Crisis*. Ocean Shores, WA: Ocean Press, 1995.

[33] P. E. Glaser, F. P. Davidson, and K. I. Csigi, Eds., *Solar Power Satellites: A Space Energy System for Earth*, in Wiley-Praxis Series in Space Science and technology. Chichester, New York: Wiley published in association with Praxis Publishing, Chichester, 1998.

[34] *Laying the Foundation for Space Solar Power: An Assessment of NASA's Space Solar Power Investment Strategy*. Washington, DC: National Academies Press, 2001, p. 10202. doi: 10.17226/10202.

[35] "Report of the URSI Inter-Commission Working Group on SPS," 2007. [Online]. Available at: https://www.ursi.org/files/ICWGReport070 611.pdf. (accessed April 11, 2020).

[36] National Security Space Office, "Space-Based Solar Power As an Opportunity for Strategic Security – Phase 0 Architecture Feasibil-ity Study," October 2007. [Online]. Available at: https://space.nss.org/wp-content/uploads/Space-Based-Solar-Power-Opportunity-for-Strategic-Security-assessment.pdf (accessed April 11, 2020).

[37] N. W. Johnson *et al.*, "Space-based Solar Power: Possible Defense Applications and Opportunities for NRL Contributions," NRL/FR/7650–09-10,179, 2009.

[38] J. C. Mankins, Ed., *Space Solar Power: The First International Assessment of Space Solar Power: Opportunities, Issues and Potential Pathways Forward.* International Academy of Astronautics, 2011.

[39] D. M. Flournoy, *Solar Power Satellites*, in SpringerBriefs in space development. New York: Springer, 2012.

[40] J. Mankins, *The Case for Space Solar Power.* Houston, TX: Virginia Edition Publishing, 2014.

[41] Shri N. R. Sonkavday, *Space Based Solar Power.* Pune, India: Golden Page Publication, 2015.

[42] J. L. Caton, "Space-Based Solar Power: A Technical, Economic, and Operational Assessment," *Strategic Studies Institute*, 2015.

[43] P. Jaffe *et al.*, "Opportunities and Challenges for Space Solar for Remote Installations," U.S. Naval Research Laboratory, Washington, DC, U.S.A., Memo Report NRL/MR/8243–19-9813, October 2019. [Online]. Available at: https://apps.dtic.mil/sti/pdfs/AD1082903.pdf (accessed April 21, 2020).

[44] X. Hou, W. Li, and Z. Xinghua, *Introduction to Space Solar Power Plant.* Beijing, China: China Aerospace Press, 2020.

[45] J. A. Vedda and K. L. Jones, "Space-Based Solar Power: A Near-Term Investment Decision," 2020.

[46] "Catching the Sun: A National Strategy for Space Solar Power," Beyond Earth Institute, August 2021. [Online]. Available at: https://beyondearth.org/wp-content/uploads/2021/09/SSP-Report-Compressed.pdf (accessed December 10, 2021).

[47] "Space Based Solar Power — De-risking the Pathway to Net Zero," Frazer-Nash Consultancy, 2021.

[48] Frazer-Nash Consultancy, "Space-Based Solar Power: A Future Source of Energy for Europe?," European Space Agency, 2022.

[49] Roland Berger GMBH, "Space-based Solar Power: Can It Help to Decarbonize Europe and Make It More Energy Resilient?," European Space Agency, July 2022.

[50] I. Cash, "CASSIOPeiA — A New Paradigm for Space Solar Power," *Acta Astronautica*, vol. 159, pp. 170–178, June 2019. doi: 10.1016/j.actaastro.2019.03.063.

[51] I. Cash, "CASSIOPeiA Solar Power Satellite," in *2017 IEEE International Conference on Wireless for Space and Extreme Environments (WiSEE)*, 2017, pp. 144–149. doi: 10.1109/WiSEE.2017.8124908.

[52] K. A. Ehricke, "Space Light: Space Industrial Enhancement of the Solar Option," *Acta Astronautica*, vol. 6, no. 12, pp. 1515–1633, December 1979. doi: 10.1016/0094-5765(79)90003-1.

[53] L. M. Fraas and M. J. O'Neill, "Sunbeams from Space Mirrors for Terrestrial PV," in *Low-Cost Solar Electric Power*, Cham: Springer

International Publishing, 2023, pp. 163–176. doi: 10.1007/978-3-031-30812-3_12.

[54] O. Çelik, A. Viale, T. Oderinwale, L. Sulbhewar, and C. R. McInnes, "Enhancing Terrestrial Solar Power Using Orbiting Solar Reflectors," *Acta Astronautica*, vol. 195, pp. 276–286, June 2022. doi: 10.1016/j.actaastro.2022.03.015.

[55] "Reflect Orbital - Solar Energy, After Dark." [Online]. Available at: https://www.reflectorbital.com/ (accessed September 24, 2023).

[56] H. Oberth, *Ways to Spaceflight*. Munich-Berlin: R. Oldenbourg Verlag, 1929. [Online]. Available at: https://ia600304.us.archive.org/24/items/nasa_techdoc_19720008133/19720008133.pdf (accessed September 24, 2023).

[57] V. Houghton, *Nuking the Moon: And Other Intelligence Schemes and Military Plots Left on the Drawing Board*. New York: Penguin Books, 2019.

[58] L. Summerer and O. Purcell, "Concepts for Wireless Energy Transmission via Laser," presented at the *International Conference on Space Optical Systems and Applications (ICSOS)*, 2009, p. 10.

[59] C. A. Schafer and D. Gray, "Transmission Media Appropriate Laser-Microwave Solar Power Satellite System," *Acta Astronautica*, vol. 79, pp. 140–156, 2012. doi: http://dx.doi.org/10.1016/j.actaastro.2012.04.010.

[60] R. M. Dickinson, "Power in the Sky," *IEEE Microwave Magazine*, pp. 36–47, March 2013. doi: 10.1109/MMM.2012.2234632.

[61] T. M. Letcher, *Future Energy: Improved, Sustainable and Clean Options for Our Planet*. Amsterdam, Netherlands: Elsevier, 2020.

[62] C. Zur, "Escaping the Tyranny of the Rocket Equation," Scientific American Blog Network. [Online]. Available at: https://blogs.scientificamerican.com/observations/escaping-the-tyranny-of-the-rocket-equation/ (accessed July 4, 2022).

[63] J. P. Pournelle, "Halfway to Anywhere," *Galaxy*, vol. 34, no. 07, pp. 94–95, 1974.

[64] Y. Smith, "April 12, 1981: Launch of the First Shuttle Mission," NASA. [Online]. Available at: http://www.nasa.gov/image-feature/april-12-1981-launch-of-the-first-shuttle-mission (accessed July 5, 2022).

[65] "SpaceX Successfully Launches First Recycled Rocket Booster," *Reuters*, March 30, 2017. [Online]. Available at: https://www.reuters.com/article/us-space-spacex-launch-idUSKBN1711JY (accessed July 5, 2022).

[66] A. Trancart, "SpaceX Stats," SpaceX Stats. [Online]. Available at: https://www.spacexstats.xyz (accessed July 10, 2022).

[67] J. Kare, "Program and Applications for a Near-Term Laser Launch System," UCID-21718, 6907065, ON: DE90011760, June 1989. doi: 10.2172/6907065.

[68] "Advanced Space Propulsion Startup Shuts Down," *SpaceNews*. [Online]. Available at: https://spacenews.com/advanced-space-propulsion-startup-shuts-down/ (accessed July 10, 2022).

[69] L. Myrabo, "World Record Flights of Beam-Riding Rocket Lightcraft — Demonstration of 'Disruptive' Propulsion Technology," in *37th Joint Propulsion Conference and Exhibit*, Salt Lake City, UT: American Institute of Aeronautics and Astronautics, July 2001. doi: 10.2514/6.2001-3798.

[70] D. J. T. Kare, "Modular Laser Launch Architecture: Analysis and Beam Module Design," *Kare Technical Consulting*, Seattle, WA, Final Report, Apr. 2004. Accessed: Jul. 10, 2022. [Online]. Available: https://www.niac.usra.edu/files/studies/final_report/897Kare.pdf.

[71] M. Beam, "ToughSF: Laser Launch into Orbit," ToughSF. [Online]. Available at: http://toughsf.blogspot.com/2017/03/laser-launch-into-orbit.html (accessed July 10, 2022).

[72] K. Parkin, "Microwave and Laser Thermal Rockets - Parkin Research." [Online]. Available at: https://parkinresearch.com/microwave-thermal-rockets/ (accessed July 10, 2022).

[73] K. Parkin and T. Lambot, "Microwave Thermal Propulsion," NASA/TP—2017-219555, August 2017. [Online]. Available at: https://ntrs.nasa.gov/api/citations/20170009162/downloads/20170009162.pdf (accessed February 13, 2021).

[74] P. Lubin, *The Path to Transformational Space Exploration: Volume 2: Applications of Directed Energy*, vol. 2. Singapore: World Scientific, 2022. doi: 10.1142/11918-vol2.

[75] K. Henson, "Rays of Hope." [Online]. Available at: https://docs.google.com/file/d/1PHkFACumTHyfMPOfIDhAY46vPe_mt8zNmy3i2ZsOnHgqZqpGuMpSh3JaJsCO/edit?usp=sharing&usp=embed_facebook (accessed July 11, 2022).

[76] Y. Tsuda *et al.*, "Flight Status of IKAROS Deep Space Solar Sail Demonstrator," *Acta Astronautica*, vol. 69, no. 9–10, pp. 833–840, November 2011. doi: 10.1016/j.actaastro.2011.06.005.

[77] P. George and R. Beach, "Beamed-Energy Propulsion (BEP) Study," NASA Glenn Research Center, Cleveland, Ohio, NASA/TM—2012-217014, 2012. [Online]. Available at: https://ntrs.nasa.gov/citations/20120002761 (accessed July 18, 2020).

[78] J. T. Kare, "SailBeam: Space Propulsion by Macroscopic Sail-Type Projectiles," in *AIP Conference Proceedings*, Albuquerque, NM: AIP, 2001, pp. 402–406. doi: 10.1063/1.1357954.

[79] G. Landis, "Microwave Pushed Interstellar Sail – Starwisp Revisited," in *36th AIAA/ASME/SAE/ASEE Joint Propulsion Conference and Exhibit.* doi: 10.2514/6.2000-3337.

[80] Y. K. Bae, "Prospective of Photon Propulsion for Interstellar Flight," *Physics Procedia*, vol. 38, pp. 253–279, 2012. doi: 10.1016/j.phpro.2012.08.026.

[81] Y. K. Bae, "Photonic Laser Thruster: 100 Times Scaling-Up and Propulsion Demonstration," *J. Propul. Power*, pp. 1–8, December 2020. doi: 10.2514/1.B38144.

[82] P. Lubin, "A Roadmap to Interstellar Flight," *University of California Santa Barbara*, Santa Barbara, CA, Apr. 2016. [Online]. Available: https://www.nasa.gov/wp-content/uploads/2015/05/roadmap_to_interstellar_flight_tagged.pdf?emrc=7e698f (accessed April 05, 2021).

[83] "Breakthrough Initiatives." [Online]. Available at: https://breakthroughinitiatives.org/initiative/3 (accessed August 26, 2022).

[84] "Qi — Mobile Computing |Wireless Power Consortium." [Online]. Available at: https://www.wirelesspowerconsortium.com/qi/ (accessed April 27, 2022).

[85] "RF Wireless Power & Radio Frequency Charging," AirFuel Alliance. [Online]. Available at: https://airfuel.org/airfuel-rf/ (accessed April 27, 2022).

[86] "The Betamax vs VHS Format War." [Online]. Available at: https://www.mediacollege.com/video/format/compare/betamax-vhs.html (accessed April 28, 2022).

[87] "Is the 'War' of EV Fast Chargers Really Over? | Greenbiz." [Online]. Available at: https://www.greenbiz.com/article/war-ev-fast-chargers-really-over (accessed April 28, 2022).

Acknowledgments

Paul Jaffe's acknowledgments:

There were many events that shaped the path that led to this book. My interest in solar power satellites, one of the largest scale potential applications of power beaming, loomed large because of my long friendship with polymath and entrepreneur John Mankins. My focus on power beaming, which is a critical element for solar power satellites, resulted from enlightening communications I had with Steve Fetter in 2016. The opportunities to perform much of the work that has advanced power beaming technology in the United States since 2017 can be ultimately credited to the visionary leadership of RuthAnne Darling.

This book grew around the chapter on link characterization, which was written first. People who gave substantive feedback on that initial work and its precedents include Robert Winsor, James McSpadden, Hooman Kazemi, Bert Murray, Peter Schubert, Frank Little, Kevin Parkin, Keith Lofstrom, Yuval Boger, Gary Barnhard, Eric Conrad, Yei Wo, Mike Kelzenberg, Ali Hajimiri, Behrooz Abiri, Florian Bohn, and Avi Bar-Cohen.

My thinking on these topics has been influenced enormously over the years by my discussions with and the writings and technical advances of intelligent and passionate colleagues and innovators from around the world, too many to list here. I am awed by your determination — you are changing the world.

This book might have languished indefinitely had it not become a team effort. I am indebted to Tom Nugent, Berndie Strassner, and

Mitchel Szazynski for their willingness to join me in producing this volume. Throughout the process, the team at World Scientific has been patient and indispensable, especially Zvi Ruder, Steven Patt and Balasubramanian Shanmugam.

Few authors could labor effectively without the stalwart support of their families. This is certainly the case for me, and I am forever grateful to my family for permitting me the time required to complete this volume, especially to Jacinda, Jules, and Elliott.

Tom Nugent's acknowledgments:

So many things came together to lead to my contribution to this book. From an early interest in space development shaped by my friend and first mentor, the late Scott MacLaren, which led to my working in Japan for a year. While there, I attended the 1995 WPT conference in Kobe, where I met many researchers in the field. An interest in the space elevator concept led to a meeting in ~2006 with my friend, second mentor, and pioneer in laser power beaming, the late Jordin Kare. His vision and 30+ years of experience with lasers were critical to the founding of LaserMotive (now PowerLight Technologies) and our team's 2009 win of NASA's Centennial Challenge for Power Beaming.

A chance meeting with Richard Gustafson around 2011 eventually led to his joining PowerLight and taking on the burden of being CEO of our ambitious little start-up. And finally meeting Paul Jaffe in 2016 led us to the successful performance in OECIF's PTROL laser power beaming program. Our many discussions over the years helped inform the goal of this book to help bring together the microwave and laser power beaming communities by "translating" the different languages and perspectives to make it easier to discuss and compare different technologies. Thanks to Paul for inviting me to share those talks with a wider audience in this book.

Thank you to all of the private investors and employees of government agencies who have helped support us and others, and who pushed all of us to bring this transformative technology to reality.

In addition to the thanks to all of those people, I owe my wife Elizabeth the biggest debt of gratitude for believing in me, for her patience, as well as encouraging and supporting me in my non-traditional career path. I should also thank my funny, bright children for keeping me humble by pointing out my many mistakes.

Finally, I want to thank the entire PowerLight team especially those who provided feedback on my chapters in this book.

Berndie Strassner's acknowledgments:

I would like to first thank Paul Jaffe for inviting me to contribute to this book. I was pleased to be able to once again write on the historical accounts of power beaming. I owe thanks to numerous people who have helped guide me on my journey specifically in the space solar power arena. I would like to foremost thank Kai Chang who was my graduate advisor at Texas A&M University. He is the one who provided me the opportunity to research circular-polarized rectennas used for solar power collection here on Earth via microwaves. Others who were at Texas A&M who were involved in helping me with my rectenna work are Frank Little, Youngho Suh, and James McSpadden. I would also like to thank Dick Dickinson who influenced my research during his time at NASA's Jet Propulsion Laboratories. Many others have provided various forms of assistance for my work in space solar power, including my family.

Mitchel Szazynski's acknowledgments:

My thanks to Paul Jaffe for including me in this effort, and his willingness to provide advice and make connections over the years. I also want to thank John Bucknell and Ed Tate of Virtus Solis Technologies for their support, and the great deal I've learned from them. Finally, I want to thank Peter Schubert for inspiring me to pursue this field in the first place, and for his continued support thereafter, as my thesis advisor.

Appendix A

Simple Power Beaming Link Examples

For the things we have to learn before we can do them, we learn by doing them.

— Aristotle [1]

To illustrate the characterization methodology outlined in Chapter 7 in a straightforward and accessible way, a power beaming link entitled "Simple Flashlight" was constructed using relatively inexpensive and readily available materials for both the link itself and the measurement equipment. This permits the link to be replicated by students or anyone seeking to explore and understand the measurement and characterization considerations for power beaming. A short video showing the finished link in operation can be found at: https://youtu.be/OuOIRemLWU4.

The main elements employed were a flashlight and solar panel. These were selected in large part because the flashlight provided a square light field at its tightest focus setting, owing to its employment of a chip LED. This happened to closely match the shape and size of the chosen solar panel at a distance of about 2 m, minimizing the amount of light that missed the receiver. The matching of the light field and the solar panel is evident on the right side of Figure A.1, which shows the mapping of the functional blocks of the link to the way it was implemented. The link was demonstrated on April 18, 2020 in Springfield, Virginia in the United States during the initial stages of the global COVID-19 pandemic.

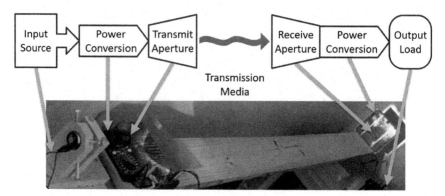

Figure A.1.   Mapping of the power beaming link functional block diagram to the configuration of the Simple Flashlight power beaming demonstration.

The input source was a universal AC adapter set to 4.5 V, to match the voltage the flashlight would have received if operated with the 3 AAA batteries specified. Note that the designation of this point in the larger power provision infrastructure is necessarily somewhat arbitrary. A case could be made that the AC wall outlet should instead be considered as the input source, rather than the output of the AC adapter since the AC adapter consists of conversion electronics, which have intrinsic inefficiency. Similarly, a counterargument can be made that the focus of power beaming link is more properly placed on the electrical-to-optical conversion element, in this case, the flashlight.

The flashlight employed as the transmitter was an LED flashlight, the "iKustar T6 LED Handheld Torch". Though this item appears to be currently unavailable from major retailers, it is essentially identical to a host of similar focusable "tactical" flashlights that employ the "CREE XML T6 LED chip". The flashlight used is shown in Figure A.2, held in the makeshift bracketry used for alignment.

The receiver consisted of a small 4.5 W solar panel, as shown in Figure A.3. The panel has multiple cells encapsulated in epoxy arranged in a series-parallel configuration to achieve the desired output voltage (6 V) and current (720 mA) under ideal solar illumination conditions.

The output of the receiver was fed into a resistor substitution box, also known as a "decade box". The Elenco RS-500 Resistance Substitution Box was used and can be seen in Figure A.4.

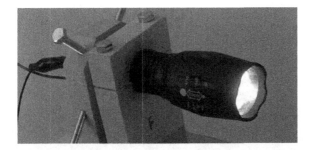

Figure A.2.   The LED flashlight used as the power transmitter.

Figure A.3.   Solar panel used as the receiver.

It is possible a potentiometer could have been used for the output load as well, to permit changing the resistance to trace the receiver's current–voltage $(I-V)$ curve. To maximize the link's efficiency, an effort was made to set the decade box to present the resistance that would result in operating at the peak power point.

Figure A.4. Resistance substitution box used as the output load.

A.1 Measurements

A.1.1 *Reporting Uncertainty*

In the case of "Simple Flashlight", all measurements were taken using commonly available, consumer-grade measurement devices, none of which had calibration certificates or the capability to perform NIST-traceable measurements. For many measurements, there was no effective means of performing an independent check with a different device intended to measure the same quantity. As such, the uncertainty of many of the measurements was difficult to quantify with confidence.

A.1.2 *Link Operating Conditions*

For the duration of the link's being active, the ambient temperature was approximately 17°C with an estimated uncertainty of ±3°C and the humidity was approximately 50% with an estimated uncertainty

Figure A.5. Thermometer/hygrometer used for measuring ambient conditions.

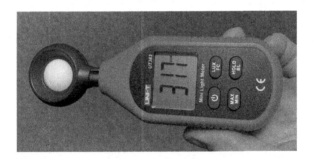

Figure A.6. Light meter used for roughly gauging illumination conditions.

of ±10%. Measurements were taken with the unbranded thermome-ter/ hygrometer seen in Figure A.5.

The ambient sound level was estimated at 40 dB(A) with an esti-mated uncertainty of ± 10 dB(A) but was not measured. It was antic-ipated that the sound level should not have any measurable effect on link measurements.

The ambient light level was measured both with a light meter and by measuring the output of the solar panel when the flashlight was off. The meter, shown in Figure A.6, read 0 lux under the conditions under which the link was operated while the link was inactive.

As lux are not directly convertible to watts per square meter, this and other lux measurements in this example should be treated with

caution, particularly since guidelines for such conversions typically
concern lux to solar irradiance flux [2], which has a different spectral
composition than the flashlight used. The output of the solar panel
in the test area without flashlight illumination with the decade box
presenting $1.60\,\mathrm{k\Omega} \pm 0.01\,\mathrm{k\Omega}$ was $9\,\mu\mathrm{A} \pm 1\,\mu\mathrm{A}$ and $10 \pm 1\,\mathrm{mV}$ or on
the order of $9\,\mathrm{nW}$. Uncertainties are estimated.

A.1.3 *Largest Transmit Aperture Dimension, $\varnothing_\mathrm{TX}$ (m)*

A digital caliper was used to measure the diameter of the optic of
the flashlight's output aperture, as shown in Figure A.7. The diam-
eter of the transmit aperture, $\varnothing_\mathrm{TX}$, is approximately $27\,\mathrm{mm}$, with
uncertainty estimated at $\pm\,0.5\,\mathrm{mm}$.

For many transmitters, a straightforward measurement of $\varnothing_\mathrm{TX}$
may be all that is needed to determine the transmit aperture area,
$\mathrm{A_{TX}}$, such as in the case of the circular transmit aperture for "Simple
Flashlight." Whenever possible though, $\mathrm{A_{TX}}$ should be reported as
well, particularly if it differs significantly from estimates that might
be found from $\varnothing_\mathrm{TX}$ alone.

A.1.4 *Wavelength, $\lambda(\mathrm{m})$*

As the power transmitter in this case was a flashlight, the transmission spectrum encompassed a wavelength range, rather than a

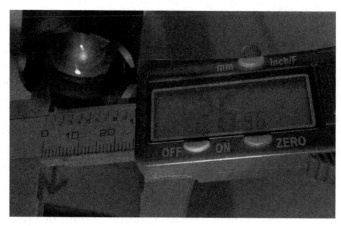

Figure A.7. Measuring the transmitter's aperture diameter with a digital caliper.

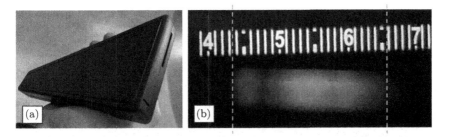

Figure A.8.   Spectroscope and measured flashlight spectrum.

single wavelength. While employing an optical spectrum analyzer would yield the most accurate characterization of the flashlight's output spectrum, it would be inconsistent with the demonstration link's philosophy of adhering to inexpensive and commonly available equipment. An alternative frequently used at the grade school level is the plastic spectroscope, a simple device that allows visualization of spectral content, and which is available for less than 10 USD. In this case, the "EISCO High Resolution Quantitative Spectroscope, 400–700 nm, 5 nm" was used. The device appears in Figure A.8(a), and the spectrum measured is shown in Figure A.8(b). The camera used to take the picture for both figures, and all pictures in this document, was an iPhone XR.

As seen in the figure, the spectral content of the flashlight output in the visual range appears to be predominantly between 430 and 660 nm, with the uncertainty of these boundaries estimated at ±10 nm. The energy distribution within the visible range is challenging to definitively determine with the simple instrument employed, though the dimmer region seen between about 470 and 490 nm is consistent with the relative spectral distribution plot from the datasheet for the Cree XLamp XM-L product family data sheet, shown in the overlay on the left side of Figure A.9.

The three separate data series shown on the overlay correspond to parts with different color temperature ranges within the XLamp family. As the flashlight manufacturer did not specify the particular part used, the spectral distribution from the Cree datasheet should not be interpreted as a definitive representation of what was present during the demonstration link's operation.

Though the spectroscope cannot effectively measure the ultraviolet or infrared regions bordering the visible spectrum, nor could

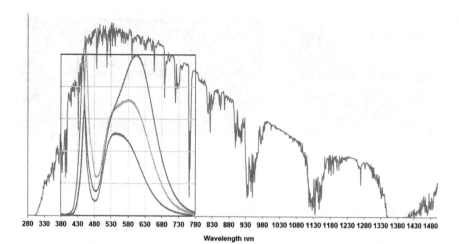

Figure A.9.　*Relative spectral power distribution of an LED family* [3] *either used or likely similar to the LED in the flashlight employed, overlaid on the ASTM G173-03 Reference Spectrum for Direct+circum solar irradiance* [4]. *Data for the LED family and solar irradiance do not share the same y-scale and are presented together to show relative spectral composition only. The three curves in the inset are for different LED components from the LED family.*

the author's camera capture the results, the overlay of the claimed LED optical output on the solar irradiance data suggests that energy present in these parts of the spectrum will be much lower from the LED than for sunlight, and almost certainly negligible.

A.1.5　*Link Distance, d* (m)

The distance between the closest parts of the transmitter and receiver was measured with a common tape measure, as seen in Figure A.10.

　　The link distance, d, was 2.048 m, with an estimated uncertainty of ± 0.002 m. While the actual transmit aperture was recessed about 7 mm from the front edge of the transmitter, and the receiver's photovoltaics were beneath a thin layer of epoxy encapsulation, the distance between the nearest parts transmitter and receiver was used. This is not only because it was more straightforward to measure, but because it would be this distance rather than another that would be of interest in ascertaining a system's suitability for a given application. Note this treatment deviates from the literal definition of d from Table 7.1, which specifies that d is the distance between the apertures themselves.

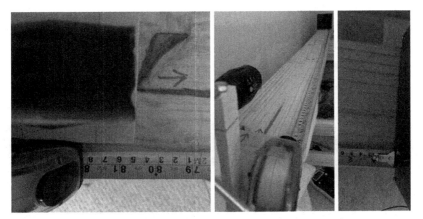

Figure A.10. Measurement of the distance between the transmitter and receiver.

A.1.6 Largest Receive Aperture Dimension, $\varnothing_{\mathrm{RX}}(\mathrm{m})$

The receiver for this link is unlike those of several previous power beaming links in that it is square, rather than circular. Thus, while the largest dimension of the receiver is the diagonal distance across the solar panel, rather than a diameter, the other dimensions are reported as well.

For the "Simple Flashlight" receive aperture, the diagonal, length, and width measurements are 213 mm, 156 mm, and 158 mm, respectively, each with an estimated measurement uncertainty of ±1 mm. Note that this is only the actual photovoltaic cell array dimensions and does not include the additional area that includes the border of the epoxy encapsulation. The aperture dimension measurements are shown in Figure A.11.

The area calculation performed to determine the receiver efficiency was $0.156\,\mathrm{m} \times 0.158\,\mathrm{m} = 0.0246\,\mathrm{m}^2$. The areas of the gaps between the cells and the cutoffs in the corners were not subtracted from this result in calculating the receiver efficiency. Regardless, the largest dimension of the receive aperture, $\varnothing_{\mathrm{RX}}$, is the diagonal, at $0.213 \pm 0.001\,\mathrm{m}$.

A.1.7 Transmitter Power Input, $P_{\mathrm{TX-in}}(\mathrm{W})$

For this and all electrical measurements, a collection of different multimeters was used. These meters were acquired at different times and

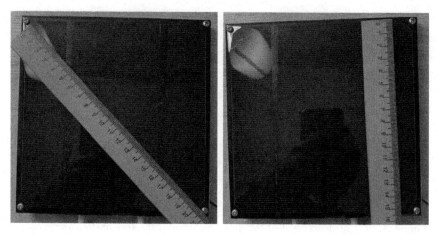

Figure A.11. Measuring the photovoltaic cell area of the panel.

from different places and thus may exhibit varying levels of precision. None had been formally calibrated, and none were capable at the time of the demonstration of making a NIST-traceable measurement. Estimated uncertainty was generated by comparing the readings of the 4 meters when subjected to similar test conditions, (measuring approximately 5.6 V, 10 mA, and 68 Ω) and they generally agreed within ±2% of each other.

The meter ensemble showing measurements recorded at the beginning and end of the link's period of being active is shown in Figure A.12.

The voltage and current were found to be essentially constant over the course of the approximately 5-minute link uptime, recorded as 3.62 V ± 0.02 V and 1.28 A ± 0.01 A, respectively, with estimated uncertainties. The product of these yields 4.63 W of input power, P_{TX-in}.

Because this was a very simple transmitter that is either on or off, there are no distinctions for other possible power consumption modes which might exist for more sophisticated transmitters and that could impact the practical efficiency observed over the operating duration.

A.1.8 *Transmitter Power Output, P_{TX-out} (W)*

The output power of the transmitter was difficult to measure to any degree of precision, given the lack of specialized equipment.

Figure A.12. Meter ensemble from left to right in each panel: input current, input voltage, output current, output voltage. The left panel is near the beginning of the link's active period and the right panel is near the end of the approximately 5-min link duration.

The range of the light meter used was billed as 0 to 199,999 lux, and positioning it as close as possible and aligned with the output of the flashlight gave a maximum of 123,800 lux. In the range $\geq$ 10,000 lux, the light meter documentation on the packaging claims an accuracy of $\pm$ 5%. Performing a linear extrapolation for lux to W/m^2 per [2] (which holds that "Solar Irradiance of 1 Sun (1,000 W/m^2) equals approximately 120,000 Lux") while recognizing the limitations of its applicability as previously discussed, this yields about 1,032 W/m^2, or nearly the equivalent of one sun under air mass 1.5 conditions. The area of the transmit aperture can be calculated thusly: $\varnothing_{TX} = 0.027\,m \rightarrow \pi r^2 = \pi(0.027/2)^2 = 5.7 \times 10^{-4}\,m^2$. Assuming a relatively uniform distribution across the transmit aperture, the resulting output power, P_{TX-out}, is $1,032\,W/m^2 * 5.7 \times 10^{-4}\,m^2 = 590\,mW$. Due to the many sources of potential error, the uncertainty is roughly estimated at $\pm20\%$.

A.1.9   *Receiver Power Input,* $P_{RX-in}(W)$

The challenge of quantifying the power that falls on the receiver aperture is similar in this case to measuring the transmitter's optical power output. In the absence of a beam profiler, the measurements taken were necessarily crude. An image of the light field immediately in front of the receiver was captured by putting a matte but

Figure A.13.   The light field as it appeared when a white piece of foam board was placed in front of the receiver.

non-Lambertian white piece of foam board in front of it, as seen in Figure A.13.

Most of the square field falls on the array. Inspection of the light field itself reveals the structure of the LED source and exhibits considerable variation. While image analysis software could be used to better characterize the relative optical intensity distribution, without an absolute known reference point it would be difficult to ascertain the total power in the field. To attempt to measure this, the light meter was slowly manually scanned across the front of the receiver in both minimum and maximum logging modes to capture the range of intensities falling on the receiver. The resulting minimum and maximum measurements were 1,647 lux and 3,276 lux respectively. Neglecting any weighting, the resulting arithmetic mean is 2,462 lux, which using the method previously outlined gives 20.52 W/m^2. The product of this quantity and the receiver aperture area of 0.0246 m^2 yields a receiver power input, P_{RX-in}, of 505 mW. Uncertainty is roughly estimated at ±20%.

A.1.10   *Receiver Power Output, P_{RX-out} (W)*

The receiver power output was measured in the same fashion as the transmitter power input, using two multimeters: one for measuring

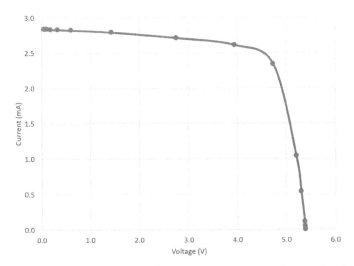

Figure A.14.   Current–voltage (I–V) plot collected prior to the run-for-the-record to locate the approximate position of the peak power point. The trend line is fitted to the data points.

voltage and the other for current. The multimeters used can be seen on the right side of each of the two panels in Figure A.12.

Prior to doing the "run-for-the-record" a current–voltage plot was generated by presenting a range of loads using the decade box to the receiver output to localize the peak power point. The resulting plot appears in Figure A.14.

A load of 1.60 kΩ with an estimated uncertainty of ±0.01 kΩ was determined to be near the peak power point under the operating conditions and was the load setting used for the run-for-the-record. The transmitter was realigned slightly to increase the optical flux falling on the receiver prior to the run-for-the-record, so the peak power point shown in Figure A.14 was improved upon. It is a near certainty that this change in conditions at least slightly altered the location of the peak power point, but it was deemed small enough so as not to necessitate collecting data to generate a second $I-V$ curve.

The voltage and current were found to decline slightly during the run-for-the-record link's being active, possibly due to slight heating of the flashlight or solar panel, but these temperatures were not specifically monitored. Within the first 30 seconds of link operation, the voltage and current were recorded as 4.96 V ± 0.02 V and 3.08 mA ± 0.01 mA, respectively. After approximately five minutes,

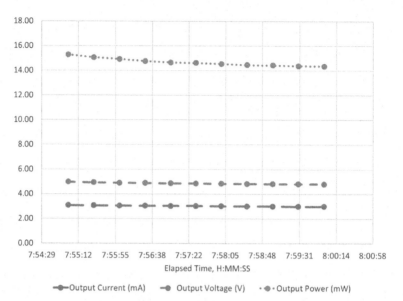

Figure A.15. Plot of current, voltage, and calculated output power over the course of the 5-min link's active period.

the readings had dropped to 4.80 V ± 0.02 V and 2.99 mA ± 0.01 mA. The meter readings can be seen in Figure A.12. These yield calculated power values of 15.3 mW at the start of the link operation and 14.4 mW at the end. The full set of 11 data collected for measured voltage, measured current, and calculated power is plotted in Figure A.15.

Finding the mean of the eleven measurements taken over the course of the link's active period yields 14.7 mW of output power, P_{RX-out}, with an uncertainty of ±0.2 mW.

This value would likely have declined slightly if the link had been permitted to run for a longer period, as it appears the solar panel had not quite yet reached thermal equilibrium.

An implicit assumption in the reporting of the receiver output power is that it comes in its entirety from energy that has been delivered by power beaming to the receiver and that any energy storage elements within the receiver system itself (batteries, capacitors, etc.) were empty or contained only negligible amounts of energy.

A.1.11 *Maximum Power Density, $p_{d-\text{max}}$ (W/m^2)*

The output from the transmitter appears to diverge at all points beyond the transmit aperture, indicating the beam waist is coincident with it. The largest lux measurements were recorded nearest the transmit aperture, supporting this observation. Thus, the maximum from the previous section is 1,032 W/m^2, with uncertainty roughly estimated at $\pm 20\%$.

A.1.12 *Accessible Power Density, $p_{d-\text{acc}}$ (W/m^2)*

In this scenario, no active or passive measures were taken to limit access to any part of the beam, other than physical access to the space where the demonstration was conducted. An imprudent experimenter or spectator would not be prevented from placing his or her eye right at the maximum output of the transmitter, other than by biological reflex. Because the output spectrum of the transmitter is in the visible range, the natural blink reflex would ostensibly offer some protection. Cree's application note on LED Eye Safety [5] suggests as much for the types of LEDs likely to have been used in the flashlight, which fall into the IEC 62471-2006 standard's risk group RG-2: "Moderate risk — Does not pose a hazard due to aversion response to bright light or thermal discomfort." The same might not be true for other demonstrations with sources outside the visible portion of the spectrum.

A.1.13 *Transmitter Mass, m_{TX} (kg)*

To determine the mass of the transmitter, it was weighed on an ordinary digital scale, as seen in Figure A.16.

In observing that both the flashlight itself and the wooden bracketry employed were considered as part of the transmitter mass, it might be asked whether it would be more appropriate to weigh the flashlight only. The flashlight itself, which was approximately 162 g when weighed separately, could not effectively be used to close the link because it does not have an effective aiming and

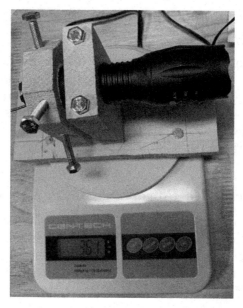

Figure A.16.   Weighing the transmitter with a digital scale.

stability mechanism. Though it would have been possible to simply lay it on another object or objects to achieve the desired pointing, the question might then become whether those objects should have been considered as part of the transmitter mass. It can also be postulated that alternate means of providing alignment and stability could have been employed (additively manufactured jigs, a tripod, or some combination of these or other means) that might have resulted in a reduction in total transmitter mass. While it is true that different means *could* have been used, the key is to report what was *actually* used. Thus, the mass of the transmitter, m_{Tx}, was considered to be 361 g, with uncertainty estimated at ±1 g.

A.1.14   *Receiver Mass,* m_{RX} (kg)

The receiver mass was measured in a similar manner to the transmitter, using a digital scale, as seen in Figure A.17.

The receiver mass, m_{RX}, was found to be 359 g with an estimated uncertainty of ±1 g.

Figure A.17. Weighing the receiver.

A.1.15 *Transmitter Volume,* $V_{TX}(m^3)$

For "Simple Flashlight" it was determined that it made the most sense to find the approximate volume of a rectangular box in which the transmitter would fit, neglecting minor protuberances such as the alignment bolts. The measured length, width, and height were 175 mm, 82 mm, and 120 mm, respectively. The measurement uncertainty for each is estimated at ± 1 mm. The transmitter volume, V_{TX}, was found as $0.175\,m \times 0.082\,m \times 0.120\,m = 1.7 \times 10^{-3}\,m^3$.

A.1.16 *Receiver Volume,* $V_{RX}(m^3)$

The receiver volume was calculated in a similar fashion to the transmitter volume: determining the volume of a box in which the bulk of the receiver would fit. Small protuberances, such as wiring, alignment pegs, staples, and screw heads were neglected. The length and width were each 165 mm with an estimated uncertainty of ± 1 mm. Thus, the receiver volume, V_{RX}, was: $0.165\,m \times 0.165\,m \times 0.014\,m = 3.8 \times 10^{-4}\,m^3$.

A.1.17 Link Duration, t(s)

The start and stop time of the links being active were measured approximately using an Android phone's stopwatch feature. The application of power to the transmitter, start of the stopwatch, stop of the stopwatch, and removal of power from the transmitter occurred manually in succession, so the link was actually active for a period marginally longer than indicated by the elapsed time. The duration the link was active, t, was at least 330 s, with an estimated uncertainty of up to +5 s. For "Simple Flashlight", the duty cycle was 100% for the duration of the link's operation.

A.2 Calculated and Estimated Parameters

A.2.1 Beam Collection Efficiency, η_{BC}

First finding the power beaming parameter τ, using the formulation from [6]:

$$\tau = \frac{\sqrt{A_{TX}A_{RX}}}{\lambda d} \tag{A.1}$$

with A_{TX}, A_{RX}, λ, and d as the transmit aperture area, receive aperture area, the wavelength of operation, and the distance between the apertures, respectively.

For this demonstration,

$$\tau = \frac{\sqrt{(5.7 \times 10^{-4} \text{ m}^2)(0.0246 \text{ m}^2)}}{(4.30 \times 10^{-7} \text{ m})(2.048 \text{ m})} = 4.3 \times 10^3 \tag{A.2}$$

The value for λ was selected at the shorter end of the wavelength range to be conservative, though the mean or another value could have been used instead. Using τ in turn with the following expression to find the BCE [6]

$$\eta_{BC} = 1 - e^{-\tau 2} \tag{A.3}$$

In this case, $\eta_{BC} \approx 1$, or 100%. For optical wavelengths over relatively short distances and apertures of reasonable size, this is typically the case. Thus, the beam collection efficiency is not a significant constraint on the link's theoretical maximum performance for "Simple Flashlight."

However, the energy emanating from the transmit aperture deviated from one with ideal directivity as evidenced by visual inspection. The receive aperture reflected incident energy by virtue of its not being perfectly absorptive. These are merely two of the sources of implementation loss that were beyond the scope of the equipment and techniques employed for the "Simple Flashlight" link to quantitatively characterize.

A.2.2 Path Loss, L_P

As it has been calculated that the BCE has a theoretical maximum of effectively 100%, losses can be attributed to either implementation factors or propagation factors.

For "Simple Flashlight," the propagation medium was ordinary air of approximately uniform temperature over a relatively short distance, and there was not a significant population of aerosols discernible by the naked eye, only occasional dust motes. Accordingly, no meaningful indications of atmospheric or aerosol absorption, scattering, refraction, diffraction, or ducting were observed.

In the link configuration, there were no solid objects directly in the beam path that might have blocked, refracted, or diffracted the beam. It is possible that there was some minuscule multipath enhancement to the amount of energy falling on the receiver from reflections off the board used to which the transmitter and receiver were affixed, or from the wall that ran parallel to the beam path, or other nearby objects.

Remaining are implementation factors originating from the non-idealized character of the transmit and receive apertures themselves, which can be superficially considered by inspecting the transmit aperture in Figure A.2 and the transmitted light's appearance at the distance of the receiver in Figure A.13, and the appearance of the receiver in Figure A.3. More in-depth probing and analysis of the non-idealized character of the apertures is beyond the scope of this investigation.

Simplistically, comparing the difference between the transmitter power output, P_{TX-out}, and receiver power input, P_{RX-in}, can be employed to approximate path loss, L_P:

$$L_P = 1 - \frac{P_{RX-in}}{P_{TX-out}} = 1 - \frac{0.505\,\text{W}}{0.590\,\text{W}} = 14.4\% \qquad (A.4)$$

This value fits with the intuitive, subjective result of inspecting the fraction of energy falling on the receiver.

A.2.3 Energy Delivered, E_{RX-out} (J)

The energy delivered by the link, the product of the receiver power output over the course of the link's operation:

$$E_{RX-out} = P_{RX-out} \times t \qquad (A.5)$$

For "Simple Flashlight" this was:

$$0.0147\,\text{W} \times 330\,\text{s} = 4.9\,\text{J} = 1.3 \times 10^{-6}\,\text{kWh} \qquad (A.6)$$

As a point of reference, a typical consumer AA cell contains about 7000 J at the beginning of its life [7].

A.3 Figures of Merit

A.3.1 Transmitter Specific Power, SP_{TX} (W/kg)

The value for SP_{TX} for Simple Flashlight:

$$SP_{TX} = \frac{P_{TX-out}}{m_{TX}} = \frac{0.590\,\text{W}}{0.361\,\text{kg}} = 1.63\,\text{W/kg} \qquad (A.7)$$

In this case, the transmitter was not operated over a range of different input voltages that might have been used to determine the maximum possible output power. Doing this may well have found an operating point that would have increased the SP_{TX}, though perhaps at the risk of damaging the transmitter or reducing its operating lifetime.

A.3.2 Receiver Specific Power, SP_{RX} (W/kg)

For this configuration of Simple Flashlight:

$$\mathrm{SP_{RX}} = \frac{P_{\mathrm{RX-out}}}{m_{\mathrm{RX}}} = \frac{0.0147\,\mathrm{W}}{0.359\,\mathrm{kg}} = 0.0409\,\mathrm{W/kg} \qquad (A.8)$$

A.3.3 Transmitter Efficiency, η_{TX}

For Simple Flashlight:

$$\eta_{\mathrm{TX}} = \frac{P_{\mathrm{TX-out}}}{P_{\mathrm{TX-in}}} = \frac{0.590\,\mathrm{W}}{4.63\,\mathrm{W}} = 12.7\% \qquad (A.9)$$

Compared to the assessed energy conversion efficiency of LED sources, which are reported as between 40% and 50% [8], this figure appears plausible, if on the low side. Principal losses in the transmitter likely a rise from diode and phosphor conversion inefficiency within the LED itself and the flashlight's transmission optics. Much as different transmitter input voltages were not explored to find the peak achievable $P_{\mathrm{TX-out}}$, neither was the voltage varied to find the peak η_{TX}. It is realistically expected that these two quantities would not be simultaneously maximized at the same input voltage.

A.3.4 Receiver Efficiency, η_{RX}

For Simple Flashlight:

$$\eta_{\mathrm{RX}} = \frac{P_{\mathrm{RX-out}}}{P_{\mathrm{RX-in}}} = \frac{0.0147\,\mathrm{W}}{0.505\,\mathrm{W}} = 2.91\% \qquad (A.10)$$

A.3.5 End-to-end Link Efficiency, η_{Link}

For Simple Flashlight:

$$\eta_{\mathrm{Link}} = \frac{P_{\mathrm{RX-out}}}{P_{\mathrm{TX-in}}} = \frac{0.0147\,\mathrm{W}}{4.63\,\mathrm{W}} = 0.317\% \qquad (A.11)$$

A.3.6 Depiction of Sources of Inefficiency

In Figure A.18, a Sankey diagram for Simple Flashlight with major categories of losses and power delivered is shown.

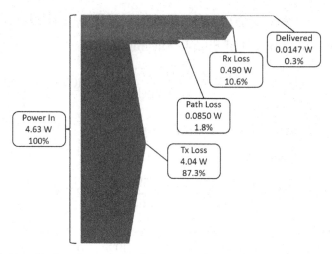

Figure A.18. Sankey diagram for Simple Flashlight with categories of losses and power delivered.

Though Simple Flashlight falls below the 1% end-to-end efficiency threshold put forth in this book as a minimum for defining a link as demonstrating power beaming, it nonetheless shows all of the elements of the link and can be used and characterized.

A.3.7 *Power Beaming Link Summary*

Concatenating all of the measurements and pertinent data from the demonstration in the Power Beaming Link Summary (PBLS, pronounced "pebbles") sheet gives the results shown in Table A.1.

A.4 Going Beyond Simple Flashlight

Paraphrasing thoughts and feedback on Simple Flashlight from Robert Winsor:

> On the receiver side, you covered the wavelength range, but I think it's also helpful to understand some physics of semiconductors and conversion efficiency. Most notable is that one photon of blue light at 440 nm will hit the photodiode and eject one single electron, just as one photon at 660 nm will also eject one single electron AND both photons create a photocurrent at the same voltage so they generate the same received power. Yet a photon at 440 nm requires 50% more power to emit than a

Table A.1.   *Power beaming link summary for a simple flashlight.*

Power Beaming Link Measurement Summary

Parameter		Description
Date (YYYY-MM-DD)	2020-04-18	The date the demonstration occured, For multi-day demonstrations, the first day of operation
Location [city, state, country]	Springfield VA, USA	The location the demonstration occurred
Title	Simple Flashlight	A short, descriptive title to distinguish the demonstration from others
λ(m)	430 nm–700 nm	The wavelength corresponding to the frequency of operation (or operating frequency in hertz)
$\varnothing_{TX}$(m)	0.027	The largest dimension of the transmitter aperture (typically the diameter)
m_{TX}(kg)	0.361	The mass of the transmitter, including power conversion elements and the transmit aperture
V_{TX}(m^3)	0.0017	The volume of the transmitter, including power conversion elements and the transmit aperture
$\varnothing_{RX}$(m)	0.205	The largest dimension of the receiver aperture, typically the diameter
m_{RX}(kg)	0.359	The mass of the receiver, including power conversion elements and the transmit aperture
V_{RX}(m^3)	0.00038	The volume of the receiver, including power conversion elements and the transmit aperture
d(m)	2.05	The distance between the transmit and receive apertures
$P_{TX-\text{in}}$(W)	4.63	The input source power to the transmitter
$P_{TX-\text{out}}$(W)	0.590	The power output of the transmitter at the frequency of operation

(*Continued*)

Table A.1.   (Continued)

Power Beaming Link Measurement Summary

Parameter		Description
$p_{d-\max}(\text{W/m}^2)$	1.03E+03	The maximum power density along the beam's path
$p_{d-\text{acc}}(\text{W/m}^2)$	1.03E+03	The maximum power density accessible to people, animals, aircraft, etc.
$P_{RX-\text{in}}(W)$	0.505	The power incident on the receive aperture
$P_{RX-\text{out}}(W)$	0.0147	The average power from the receiver to the output load during the demonstration
$t(\text{s})$	330	The duration over which the power link was active
$\text{SP}_{\text{TX}}(\text{W/kg})$	1.63	
$\text{SP}_{\text{TX}}(\text{W/kg})$	0.0409	
η_{TX}	12.7%	
L_{P}	14.4%	
η_{RX}	2.91%	
η_{Link}	0.317%	

photon at 660 nm (!!). So, let's say for example you repeated this exact demo but you used straight LED flashlights (not phosphor converted to generate white light like the one you used). You should find that a red LED flashlight power beam demo like this is more efficient. This is why the responsivity curve for silicon has a characteristic curve, and why you would normally want your transmitter wavelength to reside at the peak of the responsivity curve.

Another aspect of receiver efficiency is the degree of illumination and charge carrier density. A solar panel is generally intended for a much brighter source of light and may "leak" more electrons than you would like for your relatively lower intensity demo. Something you could try is a much smaller, more efficient photodiode but then use a large lens to concentrate the power into the much smaller photodiode. A Fresnel lens is a simple approach but has losses too. It may be cool to show how changing a thing or two here or there could change the link efficiency.

Given Robert Winsor's feedback, a second link was developed that addressed some of his points. This was "Simple Laser" and a short video summary of it can be found at this link: https://youtu.be/BtQcPHBMd8U

For Simple Laser, materials and equipment were again selected that were broadly available, but with a higher tolerance for cost than for Simple Flashlight. The core transmitter component was a 150 mW 650 nm Class 3B laser pointer, and the receiver was an array of monocrystalline silicon solar cells. The link distance was about 12 m, six times the length of Simple Flashlight, and the end-to-end link efficiency was improved by an order of magnitude to 3.0%. The link elements are shown in Figure A.19.

Because of the use of a Class 3B laser, laser safety eyewear and other safety practices were employed during the operation. Replicating this demonstration is strongly discouraged for those without laser safety training.

It is also possible to do similar simple demonstrations in the microwave regime. Though not quantified here, a range of links could be made with LEctennas or LEctenna-inspired derivatives and with Wi-Fi hardware. Videos with instructions for making LEctennas can be found at:

Easy Wireless LED–LEctenna
https://www.youtube.com/watch?v=V5SMF9p-4Q0.

and

How to Build a LEctenna
https://www.youtube.com/watch?v=3j7sAjWgySQ.

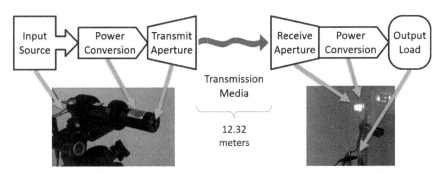

Figure A.19.   Mapping of the power beaming link functional block diagram to the configuration of the Simple Laser power beaming demonstration.

a demonstration of how a LEctenna can be illuminated from about 1 m away can be found at:

Light a LEctenna from a distance
https://www.youtube.com/watch?v=oTUCNQa8TFw.

A collection of resources for experimenting with LEctennas is freely available at:

LEctenna diodes, ideas, and more
https://docs.google.com/spreadsheets/d/1N-uxUcLifVlOrwhBJk59
Pop88IaBFhOHArOXSpjDmks/edit?usp=sharing.

LEctenna experimentation has been documented in peer-reviewed research, such as in Yu's "Quantifying the LEctenna: Measuring the Invisible Made Visible" [9].

A.5 Conclusion

Implementing a simple power beaming link is possible without great expense. Readily accessible items can be used to demonstrate the relevant principles and to characterize the link's performance. Hands-on experience in creating a simple link is helpful in understanding and approaching power beaming links of greater sophistication.

References

[1] Aristotle, "Nicomachean Ethics." [Online]. Available at: https://classics.mit.edu/Aristotle/nicomachaen.mb.txt (accessed October 14, 2023).

[2] P. Michael, "A Conversion Guide: Solar Irradiance and Lux Illuminance." IEEEDataport^{TM}, September 20, 2019. [Online]. Available at: https://ieee-dataport.org/open-access/conversion-guide-solar-irradiance-and-lux-illuminance (accessed April 18, 2020).

[3] "XLampXML.pdf." [Online]. Available at: https://www.cree.com/led-components/media/documents/XLampXML.pdf (accessed April 16, 2020).

[4] "Reference Air Mass 1.5 Spectra | Grid Modernization | NREL." [Online]. Available at: https://www.nrel.gov/grid/solar-resource/spectra-am1.5.html (accessed April 17, 2020).

[5] "Eye Safety with LED Components." [Online]. Available at: https://
www.cree.com/led-components/media/documents/XLamp_EyeSafety.
pdf (accessed April 18, 2020).

[6] N. Shinohara, *Wireless Power Transfer via Radiowaves*, in ISTE series.
Wiley, 2014. [Online]. Available at: https://books.google.com/books?
id=pJqOAgAAQBAJ (accessed April 21, 2020).

[7] "Product Datasheet Energizer E91." [Online]. Available at: https://
data.energizer.com/pdfs/e91-na.pdf (accessed April 23, 2020).

[8] D. GmbH, "Efficiency of LEDs: The Highest Luminous Efficacy of
A White LED," DIAL GmbH. [Online]. Available at: https://www.
dial.de/en/blog/article/efficiency-of-ledsthe-highest-luminous-efficacy-
of-a-white-led/ (accessed April 25, 2020).

[9] M. L. Yu, "Quantifying the LEctenna: Measuring the Invisible Made
Visible," *Int. J. High Sch. Res.*, vol. 3, no. 1, pp. 38–44, March 2021,
doi: 10.36838/v3i1.7.

Appendix B

Rectenna Varieties

Variety's the very spice of life, That gives it all its flavour.

—William Cowper[1]

In the many decades since the first rectennas were implemented, a multitude of types have arisen. These have been used not only for power beaming, but also for energy harvesting, radio frequency identification (RFID), and other applications. Both energy harvesting and RFID technologies have robust research and industrial communities. This section surveys some of the many types of rectennas used in all these areas.

B.1 Rectennas for Energy Harvesting

Over the last two decades, ambient power collection has been another popular application area for utilizing rectennas. In 2001, the University of Colorado, Boulder, CO introduced two rectenna arrays, seen in Figure B.1, for recycling ambient noise power over large bandwidths [2]. The grid rectenna array operated from 4.5 to 8 GHz and had a maximum RF-to-DC conversion efficiency of 35% at 5.7 GHz for an incident CP power density of $7.78\,\mathrm{mW/cm^2}$. The spiral array operated from 8.5 to 15 GHz and used alternating RHCP and LHCP spirals to achieve a maximum RF-to-DC conversion efficiency of 45% at 10.7 GHz for $1.56\,\mathrm{mW/cm^2}$.

In 2018, the University of Waterloo and Prince Sattam University collaborated on ways to increase the absorption efficiency per unit

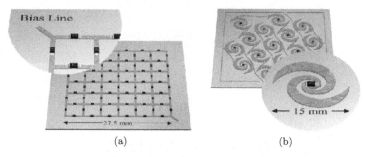

Figure B.1.   University of CO rectenna arrays: (a) grid and (b) spiral. Black rectangles are Schottky diodes.

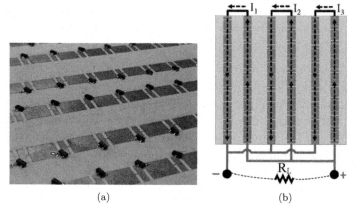

Figure B.2.   University of Waterloo/Prince Sattam University rectifying array.

area to near unity in order to channel more RF to the rectifying diodes. Using detailed Floquet analysis, they designed the 3.4 GHz energy harvesting rectenna array shown in Figure B.2. This array used asymmetric dipoles covered by a high-permittivity TMM-10i superstrate to increase the incident power delivered to the diodes. An overall array RF-to-DC conversion efficiency of 76% was obtained experimentally, which the authors claim is the highest ever recorded for an energy harvesting surface [3].

B.2   Retrodirective Rectennas

Retrodirectivity has been considered a standard for MPT systems since the late 1980s thanks to efforts by researchers at the

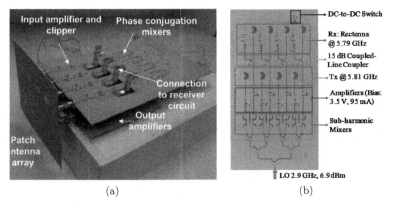

Figure labels:
- Input amplifier and clipper
- Phase conjugation mixers
- Connection to receiver circuit
- Output amplifiers
- Patch antenna array

(b):
- DC-to-DC Switch
- Rx: Rectenna @ 5.79 GHz
- 15 dB Coupled-Line Coupler
- Tx @ 5.81 GHz
- Amplifiers (Bias: 3.5 V, 95 mA)
- Sub-harmonic Mixers
- LO 2.9 GHz, 6.9 dBm

(a) (b)

Figure B.3. UCLA Retrodirective arrays designed: (a) Full-duplex retrodirective array and (b) adaptive-power-controllable retrodirective array.

Kobe University and the Kyoto University. Since then, a variety of novel rectenna arrays have been developed to extend retrodirectivity's usefulness. Figure B.3(a) array developed in 2004 by Itoh and his students at UCLA used a full duplex retrodirective array for high-speed beam tracking and pointing. This retrodirective array accomplished 10-Mb/s data reception and transmission as well as beam-steering functionality [4]. A year later in 2005, UCLA introduced (Figure B.3(b)) retrodirective antenna array which used four distinct rectennas, to awaken the overall array system from an idle mode [5]. Each rectenna used a circular sector antenna to provide the radiating capabilities as well as inherent low pass filter functionality for harmonic frequency suppression. The incoming RF power received by the antennas was split between the receiver and the rectenna. Most of this power was directed to the rectenna in order to rectify enough DC to activate a switch connected to a battery. Once the switch was activated, the retrodirective array could operate as intended.

B.3 Dual-band Rectennas

In 2002, an LP dual-frequency rectenna designed in a CPS layout and operating at both 2.45 and 5.8 GHz was developed by Suh and Chang at Texas A&M University [6]. To resonate at the two frequencies, slight modifications to the traditional dipole structure were made as shown in Figure B.4(a). The rectenna was placed 17 mm above a metal reflecting ground plane to focus the antennas'

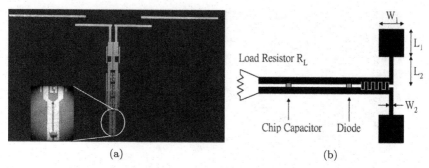

Load Resistor R$_L$

Chip Capacitor   Diode

(a)   (b)

Figure B.4.   Texas A&M LP rectenna designs: (a) Dual frequency rectenna operating at 2.45 and 5.8 GHz. (b) 5.8 GHz stepped-impedance dipole rectenna.

energy in one direction. The rectenna used a MACOM MA4E1317 Schottky diode and a chip capacitor to achieve 84.4% and 82.7% at 2.45 and 5.8 GHz, respectively. A combination of CPS low pass and band stop filters was used to suppress higher-order harmonic energy from re-radiating into free space. In 2007, Texas A&M University researchers Tu, Hsu, and Chang designed the compact 5.8 GHz rectenna seen in Figure B.4(b) to obtain an η_D of 76%. A stepped-impedance dipole antenna is used to facilitate a 23% reduction in dipole length [7].

At the National Central University in 2010, Jhongli, Taiwan, researchers designed the dual-band rectenna shown in Figure B.5(a). For an incident power density of 30 mW/cm^2, this rectenna achieves 53% and 37% conversion efficiency at 35 and 94 GHz, respectively. The linear tapered slot antenna has gains of 7.4 and 6.5 dBi at 35 and 94 GHz. The total rectenna size is 2.9 mm^2[8]. At the University of Hong Kong, C. Chin, Q. Xue, and C. Chan developed the 5.8 GHz rectenna shown in Figure B.5(b). This rectenna achieves 68.5% conversion efficiency [9].

B.3.1   *Dual Polarization Rectennas*

Japanese researchers at Kobe University and Kyoto University, in the 1980s and 1990s, introduced orthogonally fed dual-polarized circular patch radiators into etched rectenna arrays, most notably MILAX and ETHER. The circular patches were advantageous since their harmonic frequencies, determined from Bessel functions, were not integer multiples of the operating frequency. Since rectifying diodes

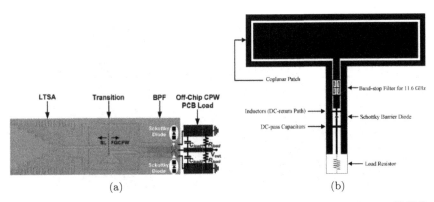

Figure B.5. (a) Dual-band 35 and 94 GHz rectenna in 0.13 μm CMOS. (b) University of Hong Kong design.

produce harmonics that are integer multiples, the circular patches operated as inherent harmonic rejection filters. Additionally, dual polarization allowed for the transmit array and rectenna array to be rotated with respect to each other while maintaining stable DC power at the output of the rectenna array. MILAX functioned at 2.411 GHz and ETHER's rectenna array operated at 2.45 GHz [10]. The ETHER rectenna array was 2.7 m × 3.4 m (1200 elements) and output 2.8 kW. Each of ETHER's rectenna subarray panels consisted of 20 elements, received an incident power density of 850 W/m², and produced maximum RF-to-DC efficiencies of 81% [11].

B.3.2 Flexible Rectennas

Powering aerial vehicles has been an application space for rectenna arrays since the 1960s. Being lightweight, conforming to an aircraft's surface, and having polarization diversity are advantages for rectenna arrays that are devoted to flyable systems. Researchers at Inha University in Korea, Norfolk State University in Norfolk, VA, and NASA Langley (LARC) have collaborated on several projects trying to develop such rectennas. In 2006, they introduced the rectenna arrays designed to operate from 9 to 12 GHz. Both were etched on 50-μm flexible polyimide film allowing lightweight conformality. One array was vertically polarized while the other layout allowed for polarization diversity. These arrays are placed directly onto the vehicles with no $\lambda_0/4$ reflecting plate [12].

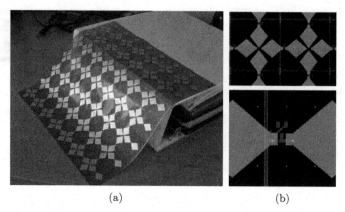

(a)   (b)

Figure B.6.   A&E Partnership flexible dual-polarized rectenna array: (a) proto-type and (b) layouts.

In 2018, A&E Partnership showed results from their flexible 7×7 element dual-polarized rectenna array pictured in Figure B.6(a). The 5.6 GHz prototype was printed on 5 mil Kapton and sized to 275 mm × 275 mm. An RF-to-DC conversion efficiency of 50% was achieved for an incident power density of approximately $2 \, \text{W/m}^2$ and input powers to the diodes of about 50 mW. This array can be directly applied to curved surfaces [13].

B.3.3   *Transparent Rectennas*

Transparency is another desirable characteristic for some rectenna applications. It enables the rectennas to be better integrated into urban environments. In 2014, French researchers developed Figure B.7 array printed on transparent plexiglass. This array used six loop antennas with each having a coplanar stripline RF-to-DC rectifier. DC voltages of 70 and 190 mV were measured for incident power densities of 1 and $5 \, \mu\text{W/cm}^2$, respectively [14].

B.3.4   *Optical Rectennas*

Optical rectennas are now being explored due to advances in nanoscale fabrication. To achieve a working optical rectenna, an antenna must be coupled to a diode that operates on the order of 1 PHz. To achieve these types of frequencies, diodes must

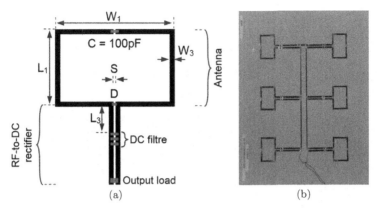

Figure B.7. Transparent rectenna array: (a) single element layout and (b) manufactured array.

have capacitance reduced to a few attofarads. This presents diode manufacturing challenges, as does the diode's coupling to the nanoscale antenna. Georgia Tech researchers overcame these challenges and published work on the carbon nanotube rectenna in Nature Nanotechnology in 2015. Visible and infrared light illuminated the carbon nanotubes resulting in measurable DC open-circuit voltages and short-circuit currents. Power rectification remained unchanged, even after numerous current-voltage scans between 5 and 77°C, indicating robustness and reliability [15].

B.3.5 *CMOS Rectennas*

CMOS technology has also found its way into rectenna design. CMOS becomes advantageous at shorter wavelengths where antennas can be integrated on-chip. This allows for the antenna and rectifier to be fully integrated. In 2017, researchers at Tel Aviv University introduced their 95 GHz energy harvesting rectenna array system seen in Figure B.8(a). It consisted of a 3 × 3 array of 65-nm CMOS rectennas on a PCB. A peripheral connector was used to access the DC, and a potentiometer was present to adjust the rectenna loading. Each rectenna had a ring antenna paired in CMOS to a rectifier as shown in Figure B.8(b). The 0.611 mm^2 3×3 rectenna array collected 2.5 mW, resulting in 6% RF-to-DC conversion efficiency at 95 GHz.

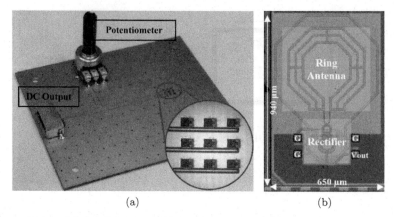

(a)   (b)

Figure B.8.   *Tel-Aviv University W-band rectenna: (a)* 3×3 *rectenna system and (b) CMOS antenna/rectifier.*

An individual rectifier produced 21.5% conversion at around 87 GHz for an incident power of 18.6 mW [16].

B.3.6   *Near Field Focused Rectennas*

Historically, rectennas have been designed to focus on the far field, especially for SBSP types of applications. In 2018, Chinese and British researchers designed a 5.8 GHz LHCP rectenna array system that functioned in the near field. An 8×8 truncated patch transmitting array, divided into 16 2×2 subarrays, was designed to focus on the near field. All 2×2 subarrays were connectorized to receive the necessary inputs for constructing a parabolic phase front across the transmitting array's surface. On the receiving end, a "focused" rectenna array used an 8×8 array of subwavelength radiating elements, as seen in Figure B.9, to provide harmonic filtering, decreased mutual coupling, and reduced backscatter. The rectenna array was able to achieve 57.74% RF-to-DC conversion efficiency in the near-field [17].

B.3.7   *High-Power Rectennas*

In 2018, Korean researchers once again put the focus on high power MPT and developed the 2.45 GHz 100 W rectenna array shown in Figure B.10(a). This array was made up of 24 rectenna array panels.

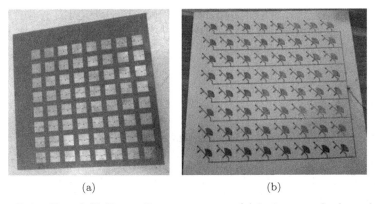

Figure B.9. Near-field "focused" rectenna array: (a) 8×8 array of subwavelength elements (front) and (b) rectifier circuits (back).

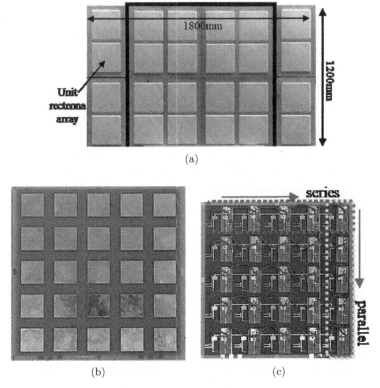

Figure B.10. Korean 100 W 2.45 GHz rectenna array: (a) full array, (b) 5 × 5 patch antennas (front), and (c) 5 × 5 array of rectifiers (back).

Each panel was a 5×5 rectenna array, like the one in Figures B.10(b) and B.10(c). The panels consisted of patch antennas with each patch having a designated rectifying circuit. The 2.16-m^2 paneled rectenna array, made of 600 rectenna elements, rectified enough DC power to turn on a 60-W LED bar at a 10-m distance [18].

B.3.8   *Broadband Rectennas*

Many of the past rectenna array designs have used narrowband radiators. In 2019, Shanghai University researchers developed a novel 35 GHz millimeter-wave rectenna array that was designed to have broadband performance. The array seen in Figure B.11 was composed of 16 subarrays. Each subarray had a 2×2 array of slot-coupled patch antennas backed by substrate-integrated waveguide (SIW) cavities and a single rectifier. The antennas worked well from 31 to 40 GHz. Each 2×2 rectenna subarray had an RF-to-DC conversion efficiency of 51% when 13 dBm was incident upon a 550 Ω load [19].

B.3.9   *Photovoltaic Rectennas*

The quest to increase energy harvesting capabilities has generated many publications, over the past decade, describing the merging of microwave frequency rectennas and photovoltaics. One such design, seen in Figure B.12, was published by Brazilian researchers

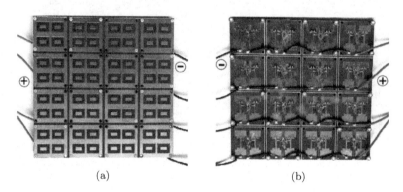

(a)   (b)

Figure B.11.   35 GHz millimeter-wave rectenna array.

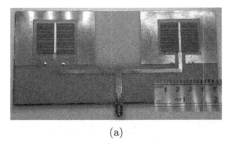 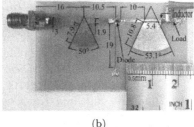

(a)   (b)

Figure B.12. Rectenna/photovoltaic array: (a) integrated rectenna with solar cells (dimensions in mm) and (b) rectifier.

in 2019. At 2.45 GHz, the rectifier accomplished RF-to-DC conversion efficiencies of 18% and 30% for incident power levels of −20 dBm and −10 dBm, respectively [20].

B.4 Conclusion

Rectennas have been developed in myriad varieties. Variants for different frequencies, polarizations, power densities, integration schemes, and operating circumstances appear in the literature. These versatile power receivers will continue to be investigated for a range of different applications.

References

[1] *The Task, by William Cowper.* 1785. [Online]. Available at: https://www.gutenberg.org/files/3698/3698-h/3698-h.htm (Accessed October 14, 2023).

[2] J. A. Hagerty and Z. Popovic, "An Experimental and Theoretical Characterization of a Broadband Arbitrarily-Polarized Rectenna Array," in *2001 IEEE MTT-S International Microwave Symposium Digest (Cat. No.01CH37157)*, Phoenix, AZ, USA: IEEE, 2001, pp. 1855–1858, doi: 10.1109/MWSYM.2001.967269.

[3] A. Z. Ashoor, T. S. Almoneef, and O. M. Ramahi, "A Planar Dipole Array Surface for Electromagnetic Energy Harvesting and Wireless Power Transfer," *IEEE Trans. Microw. Theory Tech.*, vol. 66, no. 3, pp. 1553–1560, March 2018, doi: 10.1109/TMTT.2017.2750163.

[4] K. M. K. H. Leong, Y. Wang, and T. Itoh, "A Full Duplex Capable Retrodirective Array System for High-Speed Beam Tracking and

Pointing Applications," *IEEE Trans. Microw. Theory Tech.*, vol. 52, no. 5, pp. 1479–1489, May 2004, doi: 10.1109/TMTT.2004.827025.

[5] S. Lim, K. M. K. H. Leong, and T. Itoh, "Adaptive Power Controllable Retrodirective Array System for Wireless Sensor Server Applications," *IEEE Trans. Microw. Theory Tech.*, vol. 53, no. 12, pp. 3735–3743, December 2005, doi: 10.1109/TMTT.2005.856086.

[6] Young-Ho Suh and Kai Chang, "A High-Efficiency Dual-Frequency Rectenna for 2.45- and 5.8-Ghz Wireless Power Transmission," *IEEE Trans. Microw. Theory Tech.*, vol. 50, no. 7, pp. 1784–1789, July 2002, doi: 10.1109/TMTT.2002.800430.

[7] W.-H. Tu, S.-H. Hsu, and K. Chang, "Compact 5.8-GHz Rectenna Using Stepped-Impedance Dipole Antenna," *IEEE Antennas Wirel. Propag. Lett.*, vol. 6, pp. 282–284, 2007, doi: 10.1109/LAWP.2007. 898555.

[8] H.-K. Chiou and I.-S. Chen, "High-Efficiency Dual-Band On-Chip Rectenna for 35- and 94-GHz Wireless Power Transmission in 0.13-μm CMOS Technology," *IEEE Trans. Microw. Theory Tech.*, vol. 58, no. 12, pp. 3598–3606, December 2010, doi: 10.1109/TMTT.2010. 2086350.

[9] C.-H. K. Chin, Quan Xue, and Chi Hou Chan, "Design of a 5.8-GHz Rectenna Incorporating a New Patch Antenna," *IEEE Antennas Wirel. Propag. Lett.*, vol. 4, pp. 175–178, 2005, doi: 10.1109/LAWP. 2005.846434.

[10] Y. Fujino *et al.*, "A Rectenna for MILAX," in *Proc. 1st Wireless Power Transmission Conference*, Texas, USA, February 1993, pp. 273–277.

[11] N. Kaya, S. Ida, Y. Fujino, and M. Fujita, "Transmitting Antenna System for Airship Demonstration (ETHER).," *Space Energy Transp.*, vol. 1, no. 4, pp. 237–245, 1996.

[12] J. Kim, S.-Y. Yang, K. D. Song, S. Jones, J. R. Elliott, and S. H. Choi, "Microwave Power Transmission Using a Flexible Rectenna for Microwave-Powered Aerial Vehicles," *Smart Mater. Struct.*, vol. 15, no. 5, pp. 1243–1248, October 2006, doi: 10.1088/0964-1726/15/5/012.

[13] A. Boryssenko and E. Boryssenko, "Rectenna Array on Flexible Substrate," in *2018 IEEE International Symposium on Antennas and Propagation & USNC/URSI National Radio Science Meeting*, Boston, MA: IEEE, July 2018, pp. 2525–2526, doi: 10.1109/APUS-NCURSINRSM.2018.8609135.

[14] H. Takhedmit, L. Cirio, F. Costa, and O. Picon, "Transparent Rectenna and Rectenna Array for RF Energy Harvesting at 2.45 GHz," in *The 8th European Conference on Antennas and Propagation (EuCAP 2014)*, The Hague, Netherlands: IEEE, April 2014, pp. 2970–2972, doi: 10.1109/EuCAP.2014.6902451.

[15] A. Sharma, V. Singh, T. L. Bougher, and B. A. Cola, "A Carbon Nanotube Optical Rectenna," *Nat. Nanotechnol.*, vol. 10, no. 12, pp. 1027–1032, December 2015, doi: 10.1038/nnano.2015.220.

[16] N. Weissman, S. Jameson, and E. Socher, "W-Band CMOS On-Chip Energy Harvester and Rectenna," in *2014 IEEE MTT-S International Microwave Symposium (IMS2014)*, Tampa, FL, USA: IEEE, June 2014, pp. 1–3, doi: 10.1109/MWSYM.2014.6848243.

[17] Y. Dong *et al.*, "Focused Microwave Power Transmission System with High-Efficiency Rectifying Surface," *IET Microw. Antennas Propag.*, vol. 12, no. 5, pp. 808–813, 2018, doi: 10.1049/iet-map.2017.0530.

[18] Y. Park, K. Kim, and D. Youn, "Rectenna Array Design for Receiving High Power in Beam Type Wireless Power Transmission," in *Proceedings of the 2018 Asia-Pacific Microwave Conference*, Kyoto, Japan, November 2018.

[19] W. Huang, J.-X. Du, G.-N. Tan, and X.-X. Yang, "A Novel 35-GHz Slot-Coupled Patch Rectenna Array Based on SIW Cavity for WPT," in *2019 International Conference on Microwave and Millimeter Wave Technology (ICMMT)*, Guangzhou, China: IEEE, May 2019, pp. 1–3, doi: 10.1109/ICMMT45702.2019.8992128.

[20] E. V. V. Cambero, H. P. Da Paz, V. S. Da Silva, H. X. De Araujo, I. R. S. Casella, and C. E. Capovilla, "A 2.4 GHz Rectenna Based on a Solar Cell Antenna Array," *IEEE Antennas Wirel. Propag. Lett.*, vol. 18, no. 12, pp. 2716–2720, December 2019, doi: 10.1109/LAWP.2019.2950178.

Further Exploration

In addition to the resources listed at the end of chapters of this book, the following conferences and websites offer avenues to follow recent developments concerning power beaming. Many journals feature articles concerning or relevant to power beaming, including those published by the Institute of Electrical and Electronics Engineers, the Directed Energy Professional Society, and others.

Conferences

IEEE Wireless Power Technology Conference and Expo (WPTCE)
https://ieee-wptce.org/

Optical Wireless and Fiber Power Transmission Conference (OWPT)
https://owpt.opicon.jp/

Directed Energy Science & Technology Symposium
https://www.deps.org/DEPSpages/events.html

The International Space Development Conference
https://space.nss.org/international-space-development-conference/

The International Astronautical Congress
https://www.iafastro.org/events/iac/

Websites

IEEE Wireless Power Transfer Project
https://wpt.ieee.org/

The Directed Energy Professional Society (DEPS)
https://www.deps.org/

PV Education
https://www.pveducation.org/

Antenna-Theory.com
http://www.antenna-theory.com/

Edmund Optics® Knowledge Center
https://www.edmundoptics.com/knowledge-center/

RP Photonics Encyclopedia
https://www.rp-photonics.com/encyclopedia.html

Microwave Encyclopedia
https://www.microwaves101.com/

National Space Society — Space Solar Power
https://space.nss.org/space-solar-power/

Glossary

Airy disk: The central region of the diffraction pattern that results when light passes through a uniformly illuminated circular aperture.

Active device: A type of circuit component that can electrically control electric charge flow [1].

Aperture: An opening or area from which electromagnetic waves are transmitted or received.

Attenuation: A lessening in the magnitude of a quantity.

Balun: A device for connecting balanced and unbalanced elements in electrical and RF systems [2, p. 538].

Bandwidth: The specified span of a continuous range of frequencies or wavelengths.

Beam aspect factor (BAF): A quotient or ratio defined by the distance traversed by a power beaming link divided by or compared to the sum of the largest dimensions of the transmit and receive structures.

Beam diameter or beamwidth: The diameter of a beam measured perpendicular to its propagation direction. The boundary of the beam may be defined in different ways.

Beam parameter product (BPP): The product of the beam radius at its waist and the far-field half-angle beam divergence [3].

Beam quality: A quantity that conveys the limit of how well a given beam can be focused as compared to an ideal Gaussian beam.

Beam waist: The location at which the beam is narrowest.

Beam collection efficiency (BCE): The quantity that results when using the Goubau relationship to approximate the theoretical efficiency of a power beaming link, and which depends only on the transmitter and receiver aperture sizes, the separation distance, and the operating wavelength.

Beam waveguide: A sequence of apertures that refocus a beam periodically.

Coherence length: A measure of temporal coherence, expressed as the propagation distance over which the coherence significantly decays [4].

Confocal: Having a shared focus.

Duty cycle: The ratio of time something is in an "on" state compared to the time is in an "off" state [5].

End-to-end efficiency: The efficiency found by dividing the power from the output of a power beaming system by the input power, to include all power needed by ancillary systems for thermal management, control, and other functions.

Energy: The quantity that relates the change in a system to an interaction with its surroundings [6, p. 216].

Étendue: A conserved quantity that describes both the angular and spatial propagation of flux through an optical system [7].

Far field: The region at some relatively great distance from a source of electromagnetic waves where the angular field distribution is essentially independent of the distance from a specified point near the source [8].

Field of regard: Also known as instantaneous access area, the area that an optic or antenna could potentially illuminate or see at a given instant [9, p. 164].

Field of view: Also known as footprint area, the area that an optic or antenna can illuminate or see at a given instant [9, p. 164]. The field of view is a subset of the field of regard.

Figure of merit: A numerical quantity based on one or more characteristics of a system or device that represents a measure of efficiency or effectiveness [10].

Flux density: See irradiance.

Fraunhofer region: The region in which the field of a source of electromagnetic waves is focused. For a system focused at infinity, this coincides with the far-field region [8, p. 16].

Fresnel region: The region or regions adjacent to the Fraunhofer region [8, p. 16].

Fried parameter: Also known as Fried's coherence length, a measure of the quality of optical transmission through the atmosphere due to random inhomogeneities in the atmosphere's refractive index [11].

Full width at half maximum (FWHM): A quantity given by the distance between two points on a curve representing a continuous amplitude distribution at which the amplitude reaches half its maximum value [12].

Gaussian beam: The beam that results from a monochromatic aperture irradiance distribution resembling that of a Gaussian function.

Grating lobes: Regions of radiated energy from an aperture that rival the main beam in intensity, and which are generally undesirable.

Illumination: The irradiance pattern on a surface or aperture, either emitted or received.

Intrabeam: In the beam.

Irradiance: The radiant flux received by a surface per unit area, also known as flux density [13], measured in W/m^2.

Key performance parameter: Also called Key Performance Indicator, an attribute of a system considered critical to the development of an effective capability or product [14].

Light curtain: Also sometimes called a safety light curtain, a sensing system employing light beams that is designed to protect people and objects from harm.

Maximum power point: The point on the current–voltage (I–V) curve of a photovoltaic cell or module under illumination where the product of current and voltage is maximum [15].

Metamaterial: An artificially structured material that exhibits electromagnetic properties not available or not easily obtainable in nature [16].

Metric: A standard of measurement or quantitative means of comparison.

Measurement: The size, length, or amount of something, as established by measuring [17].

Multimode: A medium or instance in which more than one mode of propagating electromagnetic energy can be sustained. Examples include some optical fibers and some microwave waveguides.

Noise floor: A quantity representing the sum of unwanted signals within a measurement system.

Optics: A science that is concerned with the study of light and associated phenomena.

Parameter: A numerical or other measurable factor forming one of a set that defines a system or sets conditions of its operation [18].

Paraxial: Close to and nearly parallel to an axis, generally satisfying $\sin \theta \approx \theta$.

Passive device: A component incapable of controlling electricity via another electrical signal [1].

Path loss: The reduction in power density of an electromagnetic wave as it propagates through space [19, p. 277].

Photon: A massless particle of quantized energy that travels at the speed of light.

Power: The amount of energy transferred or converted per unit of time.

Power beaming: Uncoupled wireless power transmission in which the link distance exceeds the sum of the largest dimensions of the transmit and receive structures, and which demonstrates at least one percent end-to-end energy transmission efficiency.

Propagation loss: Losses arising from propagation of energy through a medium [20].

Quantity: The amount or number of a material or immaterial thing not usually estimated by spatial measurement [21].

Radiance: the optical power per unit area and solid angle measured in $(\text{W cm}^{-2}\,\text{sr}^{-1})$ [22].

Radiating near field: The region between the far field and reactive near field of a source, typically an antenna. For an antenna focused at infinity, this region is sometimes also called the Fresnel region [8, p. 23].

Reactive near field: The region very near a source or antenna where the reactive field dominates, and in which energy can be thought of as stored rather than launched [8].

Safety light curtain: See light curtain.

Single mode: A medium or instance in which only one mode of propagating electromagnetic energy can be sustained. Examples include some optical fibers and some microwave waveguides.

Specular: Of, relating to, or having the qualities of a mirror [23].

Superdirectivity: A condition that occurs when the directivity of an antenna significantly exceeds that obtained from an aperture of the same dimension as that of the actual antenna but with uniform aperture illumination or, in the case of an array, from the array with elements uniformly excited [24].

Super-Gaussian: A distribution that exhibits greater uniformity than a Gaussian distribution, the degree of which is determined by the super-Gaussian number SG, where SG = 1 is a Gaussian and SG = ∞ is a top hat distribution.

Top hat distribution: Also tophat or top-hat, a uniform irradiance distribution across a circular aperture.

Wall-plug efficiency: The total electrical-to-optical power efficiency of a laser system [25].

Wireless power transmission (or transfer): The deliberate delivery of energy from a transmitter for capture and utilization at a receiver without using wires or the movement of matter.

References

[1] "Active Versus Passive Devices | Amplifiers and Active Devices | Electronics Textbook." [Online]. Available at: https://www.allabout circuits.com/textbook/semiconductors/chpt-1/active-versus-passive-devices/ (accessed November 5, 2022).

[2] C. A. Balanis, *Antenna Theory: Analysis and Design*, Third edition. Hoboken, NJ: John Wiley, 2005.

[3] D. R. Paschotta, "Beam Parameter Product." [Online]. Available at: https://www.rp-photonics.com/beam_parameter_product.html (accessed August 20, 2023).

[4] D. R. Paschotta, "Coherence Length." [Online]. Available at: https://www.rp-photonics.com/coherence_length.html (accessed August 20, 2023).

[5] "What is Duty Cycle?" [Online]. Available at: https://www.fluke.com/en-us/learn/blog/electrical/what-is-duty-cycle (accessed December 18, 2021).

[6] R. W. Chabay and B. A. Sherwood, *Matter & Interactions*, Fourth edition. Hoboken, NJ: John Wiley & Sons, 2015.

[7] R. John Koshel, "Étendue (Photon Snacks 11)." January 27, 2022. [Online]. Available at: https://wp.optics.arizona.edu/jkoshel/wp-cont ent/uploads/sites/78/2022/01/Photon-Snacks-11.pdf (accessed October 21, 2023).

[8] "IEEE Standard for Definitions of Terms for Antennas," *IEEE Std 145-2013 Revis. IEEE Std 145-1993*, pp. 1–50, March 2014, doi: 10.1109/IEEESTD.2014.6758443.

[9] J. R. Wertz and W. J. Larson, Eds., *Space Mission Analysis and Design*, Third edition, in Space technology library. El Segundo, Calif.: Dordrecht; Boston: Microcosm; Kluwer, 1999.

[10] "Definition of Figure of Merit." [Online]. Available at: https://www.merriam-webster.com/dictionary/figure+of+merit (accessed May 6, 2020).

[11] D. L. Fried, "Optical Resolution Through a Randomly Inhomogeneous Medium for Very Long and Very Short Exposures," *J. Opt. Soc. Am.*, vol. 56, no. 10, p. 1372, October 1966, doi: 10.1364/JOSA.56.001372.

[12] E. W. Weisstein, "Full Width at Half Maximum." [Online]. Available at: https://mathworld.wolfram.com/FullWidthatHalfMaximum.html (accessed May 14, 2020).

[13] D. R. Paschotta, "Irradiance." [Online]. Available at: https://www.rp-photonics.com/irradiance.html (accessed September 25, 2022).

[14] "Key Performance Parameters (KPP)," AcqNotes. [Online]. Available at: https://acqnotes.com/acqnote/acquisitions/key-perfrormance-parameter (accessed October 21, 2023).

[15] "Solar Energy Glossary," Energy.gov. [Online]. Available at: https://www.energy.gov/eere/solar/solar-energy-glossary (accessed May 19, 2020).

[16] "Metamaterial | Britannica." [Online]. Available at: https://www.britannica.com/topic/metamaterial (accessed November. 26, 2022).

[17] "Measurement | Definition of Measurement by Lexico," Lexico Dictionaries | English. [Online]. Available at: https://www.lexico.com/en/definition/measurement (accessed May 7, 2020).

[18] "Parameter | Definition of Parameter by Lexico," Lexico Dictionaries | English. [Online]. Available at: https://www.lexico.com/en/definition/parameter (accessed May 7, 2020).

[19] X. Yang, C. Hui, and L. F. Haizhon, *Handbook on Sensor Networks*. World Scientific, 2010.

[20] D. R. Paschotta, "Propagation Losses." [Online]. Available at: https://www.rp-photonics.com/propagation_losses.html (accessed October 21, 2023).

[21] "Quantity | Definition of Quantity by Lexico," Lexico Dictionaries | English. [Online]. Available at: https://www.lexico.com/en/definition/quantity (accessed May 7, 2020).

[22] D. R. Paschotta, "Radiance." [Online]. Available at: https://www.rp-photonics.com/radiance.html (accessed September 25, 2022).

[23] "Definition of SPECULAR." [Online]. Available: https://www.merriam-webster.com/dictionary/specular (accessed January 16, 2021).

[24] "IEC 60050 — International Electrotechnical Vocabulary — Details for IEV number 712-02-63: 'superdirectivity.'" [Online]. Available at: http://www.electropedia.org/iev/iev.nsf/display?openform&ievref=712-02-63 (accessed May 9, 2020).

[25] D. R. Paschotta, "Wall-plug Efficiency." [Online]. Available at: https://www.rp-photonics.com/wall_plug_efficiency.html (accessed May 16, 2020).

List of Acronyms

AC	Alternating Current
AEL	Accessible Emission Limit
AFRL	Air Force Research Laboratory
ANSI	American National Standards Institute
AO	Adaptive Optics
ARIEL	Atmospheric Remote-sensing Infrared Exoplanet Large-survey
ARL	Army Research Laboratory
A_{RX}	Area of the Receiver Aperture
A_{TX}	Area of the Transmitter Aperture
BAF	Beam Aspect Factor
BPP	Beam Parameter Product
CBC	Coherent Beam Combination
CDRH	Center for Devices and Radiological Health
CDW	Cycloidal Diffractive Waveplate
CFR	Code of Federal Regulations
CoP	Coefficient of Performance
CP	Circularly Polarized
CPS	Coplanar Strip Line
CW	Continuous Wave
DARPA	Defense Advanced Research Projects Agency
DC	Direct Current
DEWS	Directed Energy Weapons Systems
DOD	Department of Defense
DOE	Department of Energy
EM	Electromagnetic

ERL	Exposure Reference Level
ESA	European Space Agency
FAA	Federal Aviation Administration
FCC	Federal Communications Commission
FDA	Food and Drug Administration
FLPPS	Federal Laser Product Performance Standard
FO	Foreign Object
FOM	Figure of Merit
FOR	Field of Regard
FOV	Field of View
FSM	Fast Steering Mirror
FWHM	Full Width at Half Maximum
GaAs	Gallium Arsenide
GaN	Gallium Nitride
GEO	Geosynchronous Earth Orbit
ICNIRP	International Commission on Non-Ionizing Radiation Protection
IEC	International Electrotechnical Commission
IEEE	Institute of Electrical and Electronics Engineers
IoT	Internet of Things
IR	Infrared
ISC	Integrated Symmetrical Concentrator
ISM	Industrial, Scientific, and Medical
ISO	International Organization for Standardization
ITU	International Telecommunications Union
ITU-R	International Telecommunications Union Radiocommunications Sector
JAXA	Japan Aerospace Exploration Agency
JPL	Jet Propulsion Laboratory
JSpOC	Joint Space Operations Center
KERI	Korea Electrotechnology Research Institute
KPI	Key Performance Indicator
KPP	Key Performance Parameter
LC	Inductor–Capacitor
LCH	Laser Clearinghouse
LED	Light Emitting Diode
LEO	Low Earth Orbit
LeRC	Lewis Research Center
LIA	Laser Institute of America

LP	Linearly Polarized
LPB	Laser Power Beaming
LSO	Laser Safety Officer
MAPLE	Microwave Array for Power-Transfer Low-Orbit Experiment
MEO	Medium Earth Orbit
METI	Ministry of Economy, Trade, and Industry
MHI	Mitsubishi Heavy Industries
MIT	Massachusetts Institute of Technology
MPE	Maximum Permissible Exposure
MPP	Maximum Power Point
MPPT	Maximum Power Point Tracking
MPT	Microwave Power Transmission or Microwave Power Transfer
MSFC	Marshall Space Flight Center
MTT-S	Microwave Theory and Technology Society
NA	Numerical Aperture
NASA	National Aeronautics and Space Administration
NHZ	Nominal Hazard Zone
NIR	Near Infrared
NOHD	Nominal Ocular Hazard Distance
NRE	Non-Recurring Engineering
NRL	Naval Research Laboratory
NTIA	National Telecommunications and Information Administration
OAM	Orbital Angular Momentum
OD	Optical Density
OSC	Ontario Science Center
OSHA	Occupational Safety and Health Administration
OWPT	Optical Wireless Power Transmission
PAE	Power Added Efficiency
PB	Power Beaming
PCM	Phase-Controlled Magnetron
PLL	Phase-Locked Loop
PLT	Power Light Technologies
PMAD	Power Management and Distribution
PPE	Personal Protective Equipment
PPT	Power Point Tracking
PV	Photovoltaic

RC	Resistor–Capacitor
RDA	Retrodirective Antenna
RF	Radio Frequency
RFID	Radio Frequency Identification
RISH	Research Institute for Sustainable Humanosphere
RLC	Resistor–Inductor–Capacitor
RR	Radio Regulations
RXCV	Reception–Conversion
SABER	Semi-Autonomous BEam Rider
SBC	Spectral Beam Combination
SBSP	Space-based Solar Power
SERT	Space Solar Power Exploratory Research & Technology
SiC	Silicon Carbide
SiGe	Silicon Germanium
SG	Super-Gaussian
SPORTS	Space POwer Radio Transmission System
SPS	Solar Power Satellite
SPSS	Solar Power Satellite System
SSP	Space Solar Power
SSPIDR	Space Solar Power Incremental Development & Research
SSPS	Space Solar Power System
sUAS	small Uncrewed Air System
SWELL	Space Wireless Energy Laser Link
SWIPT	Simultaneous Wireless Information and Power Transfer
T/R	Transmit/Receive
TEC	Thermoelectric Cooler
TPC	Thermophotovoltaic
TWT	Traveling Wave Tube
UAH	University of Alabama in Huntsville
UAV	Uncrewed Air Vehicle
UHF	Ultrahigh Frequency
UMCP	University of Maryland College Park
UN	United Nations
USAF	United States Air Force
USEF	Unmanned Space Experiment Free-Flyer
USSF	United States Space Force

VCO	Voltage Controlled Oscillator
VHF	Very High Frequency
VSWR	Voltage Standing Wave Ratio
WBC	Wavelength Beam Combination
WPT	Wireless Power Transmission or Wireless Power Transfer
WRC	World Radiocommunication Conference

About the Authors

Paul Jaffe is an electronics engineer and researcher with 30 years of experience at the U.S. Naval Research Laboratory (NRL). He has led or held major roles on dozens of space missions and on breakthrough technology development projects for civilian, defense, and intelligence community sponsors, including SSULI, STEREO, TacSat-1, TacSat-4, ORS, MIS, PRAM, CARINA, RSGS, PTROL, S2FOBs, LEctenna, and SWELL. He was responsible for electrical system and spacecraft computer hardware development. He served as the coordinator and editor of two seminal solar power satellite study reports and was the principal investigator for ground-breaking power beaming and space solar research efforts. His current roles include program management and systems engineering of a portfolio of projects. He served as the Deputy Director of The Office of the Secretary of Defense's Operational Energy Innovation office from 2021–2022. Since 2017, he has taught the course "Introduction to Power Beaming and Space Solar" in the Aerospace Engineering department at the University of Maryland. He has over 60 journal, conference, and patent publications and is the recipient of numerous awards. Dr. Jaffe has made many international speaking and media appearances, including as a TEDx speaker, on MSNBC, CuriosityStream, and the Science Channel's "Through the Wormhole with Morgan Freeman." He is also active in educational and STEM outreach.

 **Tom Nugent**, an engineering physicist with over 30 years of experience in advanced technology development, currently serves as the Chief Technology Officer of PowerLight Technologies, a company he co-founded in 2007. His work focuses on designing, building, and testing laser power beaming systems, covering a broad range of technologies including lasers, high power optics, receiver tracking, beam steering, photovoltaic power converters, power management and distribution, innovative active laser safety systems, controls, and thermal management. Additionally, he leads PowerLight Technologies' intellectual property efforts, resulting in patent filings across more than 30 patent families in a range of technology domains related to laser power beaming.

Under Nugent's leadership, the PowerLight (formerly LaserMotive) team won NASA's Centennial Challenge in Power Beaming in 2009. Together, they have demonstrated advanced laser power beaming capabilities in free space (aka wireless) and in power over fiber, often in collaboration with the government, industry, and academia. He's helped secure and executed work with DOD sponsors including the U.S. Naval Research Laboratory (NRL), Defense Advanced Research Projects Agency (DARPA), and the Office of the Secretary of Defense's (OSD) Operational Energy Capability Improvement Fund (OECIF); research institutions including Johns Hopkins Applied Physics Laboratory and the National Renewable Energy Laboratory (NREL); and industrial partners, such as Lockheed Martin, Raytheon, Intel, Blue Origin, and Ericsson.

Prior to his role at PowerLight Technologies, Nugent's career spanned diverse technical areas. He began at the Jet Propulsion Laboratory (JPL), exploring an advanced fusion-based space propulsion concept. Later on, he worked on novel liquid fuel rocket engines at MIT and space elevator design at LiftPort. While at Intellectual Ventures Labs, he worked on medical device prototypes, as well as the Photonic Fence project aimed at eradicating malaria by employing laser technology to suppress mosquito populations.

Nugent holds a Bachelor of Science degree in Engineering Physics from the University of Illinois at Urbana-Champaign, and a Master of Science degree in Materials Science and Engineering from MIT. He also completed an accelerated program in Japanese through an intensive one-year joint program at the University of Pittsburgh and Carnegie Mellon University.

With over 70 issued U.S. patents and numerous international patents as well as more than 15 pending patent applications, Nugent is a recognized expert in his field and is frequently invited to speak at conferences and events. Beyond his professional pursuits, he is passionate about energy innovation and enjoys spending time with his family, reading science fiction, and listening to music.

 **Bernd Strassner II** received a B.S. degree in Electrical Engineering from the Rose-Hulman Institute of Technology, Terre Haute, IN, in 1995, and M.S. and Ph.D. degrees in Electrical Engineering from Texas A&M University, College Station, in 1997 and 2002, respectively. From 1996 to 1997, he was with Sandia National Laboratories, Albuquerque, NM, where he was involved with the study on how harmonic load-pull terminations improve power-amplifier performance. From 1998 to 2002, he was a Research Assistant at Texas A&M University's Electromagnetics Laboratory, where his work focused on passive-backscatter RFID tags for tracking pipe in oil drill strings, rectifying antenna arrays for microwave power reception, and reflecting antenna arrays for space platforms. In 2002, after completing his doctoral thesis work, Bernd returned to Sandia National Laboratories and joined their remote sensing group. For 20 years there, he designed a variety of wideband antenna arrays and unique radiating elements for synthetic-aperture radar and other novel communication systems. In 2023, Bernd joined the startup company Massive Light where he is involved in designing antennas for communication systems that have a low probability of being detected. He has authored numerous journal papers and book chapters and has a variety of patents covering microwave components and antennas. He currently resides in Albuquerque, NM, with his wife and four children.

 **Mitchel Szazynski** is the Chief Engineer of Virtus Solis Technologies. After running a successful neighborhood vending machine business from age 10 to 13, and spending years studying golf course architecture (including in-depth meetings with world-renowned architects), Mitchel completed his Bachelor's degree in Electrical Engineering from Purdue University (Indianapolis) at the age of 18 and his Master's degree at 19, writing his thesis on microwave power beaming.

He worked for five years in robotics and automation for Bastian Solutions, a Toyota Industries subsidiary, created a blockchain-based electronic tablet tracking device for ethically sourced gemstone mining at Trusted Inc., and developed an ultrasonic imaging system for Sonablate, a medical device company that uses high intensity focused ultrasound to treat prostate cancer. He holds two patents, on piezoelectric energy harvesting using magnetic tickling and voice control of autonomous mobile robots.

In his spare time, he spends time with his family, is a novelist (under a pen name), is an avid player of golf, football, and tennis, and remains a lifelong learner in many fields, including engineering, science, government, history, and linguistics.

Index

Printed in the United States
by Baker & Taylor Publisher Services